Annals of Mathematics Studies

Number 206

A Hierarchy of Turing Degrees

A Transfinite Hierarchy of Lowness Notions in the Computably Enumerable Degrees, Unifying Classes, and Natural Definability

Rod Downey

Noam Greenberg

PRINCETON UNIVERSITY PRESS

PRINCETON AND OXFORD

2020

Published by Princeton University Press
41 William Street, Princeton, New Jersey 08540
6 Oxford Street, Woodstock, Oxfordshire OX20 1TR

press.princeton.edu

ISBN 978-0-691-19965-8
ISBN (pbk.) 978-0-691-19966-5
ISBN (ebook) 978-0-691-20021-7

British Library Cataloging-in-Publication Data is available

Editorial: Susannah Shoemaker and Lauren Bucca
Production Editorial: Nathan Carr
Text Design: Leslie Flis
Jacket/Cover Design: Leslie Flis
Production: Brigid Ackerman
Publicity: Matthew Taylor and Katie Lewis

This book has been composed in LaTeX

Printed on acid-free paper. ∞

Printed in the United States of America

10 9 8 7 6 5 4 3 2 1

Dedicated to Gerald Sacks

Contents

Acknowledgments **ix**

1 Introduction **1**
1.1 Historical context 1
1.2 Background: unifying constructions and natural definability 3
1.3 Toward the hierarchy of totally α-c.a. degrees 8
1.4 The contents of this monograph 14
1.5 An application to admissible computability 16
1.6 Notation and general definitions 17

2 α-c.a. functions **23**
2.1 $\mathcal{R}$-c.a. functions 23
2.2 Canonical well-orderings and strong notations 29
2.3 Weak truth-table jumps and ω^α-c.a. sets and functions 37

3 The hierarchy of totally α-c.a. degrees **55**
3.1 Totally $\mathcal{R}$-c.a. degrees 55
3.2 The first hierarchy theorem: totally ω^α-c.a. degrees 58
3.3 A refinement of the hierarchy: uniformly totally ω^α-c.a. degrees 68
3.4 Another refinement of the hierarchy: totally $<\omega^\alpha$-c.a. degrees 74
3.5 Domination properties 80

4 Maximal totally α-c.a. degrees **84**
4.1 Existence of maximal totally ω^α-c.a. degrees 84
4.2 Limits on further maximality 94

5 Presentations of left-c.e. reals **106**
5.1 Background 106
5.2 Presentations of c.e. reals and non-total ω-c.a. permitting 110
5.3 Total ω-c.a. anti-permitting 123

6 m-topped degrees **134**
6.1 Totally ω-c.a. degrees are not m-topped 135
6.2 Totally ω^2-c.a. degrees are not m-topped 140
6.3 Totally $<\omega^\omega$-c.a. degrees are not m-topped 145

7 Embeddings of the 1-3-1 lattice **149**
7.1 Embedding the 1-3-1 lattice 150
7.2 Non-embedding critical triples 167
7.3 Defeating two gates 176
7.4 The general construction 184

8 Prompt permissions **188**
8.1 Prompt classes 188
8.2 Minimal pairs of separating classes 202
8.3 Prompt permission and other constructions 212

Bibliography **215**

Acknowledgments

We would like to thank the following people: our coauthors on papers related to this book, in particular, Rebecca Weber and Joseph Miller; former students who have worked on related material, Adam Day, Katherine Arthur, and Michael McInerney; the referees, for their close reading and helpful comments; our families, for their forbearance; and Andrew Marks, for providing the spur to finally finish the book.

Both authors were supported by grants from the Marsden Fund, administered by the Royal Society of New Zealand.

A Hierarchy of Turing Degrees

Chapter One

Introduction

WHAT DOES IT take to perform a certain construction? In computability theory, this question is the basis of a long-term programme which seeks to understand the relationship between dynamic properties of sets and their algorithmic complexity. Our main thesis in this monograph is that where the computably enumerable (c.e.) Turing degrees are concerned, *a degree can compute complicated objects if and only if some functions in the degree are difficult to approximate.* Computability-theoretic tools allow us to quantify precisely what we mean by "difficult to approximate." More specifically, we use a classification of Δ^0_2 functions defined by Ershov in [39–41]. While Ershov's hierarchy of complexity is orthogonal to complexity as measured by Turing reducibility, we show that combining these two notions of complexity yields a new, transfinite hierarchy inside the low_2 c.e. degrees, and that two levels of this hierarchy capture the dynamics of a number of seemingly unrelated constructions in different areas of computability. Further, some of these constructions show that these two levels are naturally definable in the c.e. degrees.

1.1 HISTORICAL CONTEXT

The roots of computability theory go back to the work of Borel [8], Dedekind [18], Hermann [50], Dehn [19], and others in the late nineteenth and early twentieth centuries. From a modern point of view, these authors were highly interested in algorithmic procedures in algebra. Around the same time, Hilbert famously posed the *Entscheidungsproblem*, which asked whether there was an algorithmic procedure to decide the validity of statements in first-order logic. To show that the answer is yes, we would need to give such an algorithm, as we do with truth tables in propositional logic. However, to demonstrate that there is *no such algorithm*, we would first need to mathematically specify what an algorithm is. Culminating in the work of Turing [97], several authors gave proofs that first-order logic is undecidable; there is no such algorithm. Turing's work built on Gödel's First Incompleteness Theorem, and gave a beautiful conceptual analysis which convincingly laid the foundations of computability theory. Turing machines gave a universal model of computation.

Following these early results, many problems, such as Hilbert's 10th problem, the word problem for groups, or DNA self-assembly, have been shown to be undecidable. These proofs mostly followed a familiar pattern. They used Turing's notion of a *reduction* [98], and typically showed that the halting problem is reducible to the algorithmic decision problem at hand by some effective coding process.

A major impetus for the development of computability theory was Post's [78] which gave an analysis of the fine structure of reductions, and set a research agenda in the "structure theory" of computation. This paper was also famous as it "stripped away the formalism associated with the development of recursive functions in the 1930's and revealed in a clear informal style the essential properties of recursively enumerable sets and their role in Gödel's incompleteness theorem" (Soare [91]). Following Post's paper, there were three major developments:

- The Kleene-Post development of the finite extension method. This and related techniques demonstrated the richness of the structure of the Turing degrees, and were arguably a precursor to Cohen's method of forcing.
- The Friedberg-Muchnik theorem showing that there were intermediate computably enumerable Turing degrees. This result introduced the priority method to computability theory and is a hallmark of the area to this day.
- Sacks's work [81, 82] which culminated in his book [80] which proved a number of penetrating results on the structure of degrees, and developed the infinite-injury priority method, first introduced by Shoenfield [85]. Sacks's book famously proposed a research agenda with a number of difficult questions still open.

There were subsequent books by, for example, Rogers [79], Lerman [66], and Soare [91] exploring the universe of the degrees of unsolvability. But conceptual clarification provided by this early work has seen a flowering of applications of computability theory to many areas of mathematics. These include computable analysis [102] (a subject going back to Turing's [97]), computable algebra and model theory (see, for example, [38]), algorithmic randomness [27, 69, 74], algorithmic learning theory ([45]), and reverse mathematics [89], to name but a few. (See [22] for a general historical discussion of this development, mainly focussing on randomness.) Each of these areas has its own subareas, and hence the area of computability has become remarkably diverse.

This monograph has several goals. Some are in the spirit of Sacks's book. That is, we wish to introduce new techniques and classification tools for understanding the complexity of computation. These include some new nonuniform methods and certain symmetric games in the sense of Lachlan [60], in which obstacles in constructions turn out to reflect the boundary between what is and what is not possible. These games allow us to prove new definability results in the computably enumerable degrees. Another goal is in the spirit of Soare's book; we carefully guide the graduate student through complex techniques involving modern arguments. Our final goal is to formalize the persistent intuition that many of the constructions in the diverse areas of computability theory seem to have common combinatorics. How should we explain that? We will draw several areas back together by showing that the hierarchy we introduce can be used to explain, classify, and *unify* combinatorics in these areas.

1.2 BACKGROUND: UNIFYING CONSTRUCTIONS AND NATURAL DEFINABILITY

1.2.1 Unifying constructions and levels of permitting

Computability theory has a small number of classes of degrees which capture the underlying dynamics of a number of apparently similar constructions. A good example is the class of high degrees, the degrees $\mathbf{d}$ satisfying $\mathbf{d}' \geqslant \mathbf{0}''$. Martin [70] showed that a c.e. degree is high if and only if

(1) it contains a function dominating all computable functions;
(2) it contains a maximal set;
(3) it contains a hyperhypersimple set.

Another example would be the class of the promptly simple degrees (Ambos-Spies, Jockusch, Shore, and Soare [2]), which coincide with the low-cuppable degrees and the non-cappable degrees. A more recent example of current interest is the class of K-trivial degrees (see, for example, [28, 72, 73]), which have several characterisations arising from lowness constructions.

The example most relevant to this monograph is the class of array computable degrees, defined by Downey, Jockusch, and Stob [30, 31]. Recall that by Shoenfield's Limit Lemma [84], a function $g\colon \omega \to \omega$ is Δ^0_2 if and only if it has a *computable approximation*: a uniformly computable sequence $\langle g_s \rangle$ of functions which converge to g in the discrete topology, that is, for which for all n, $g_s(n) = g(n)$ for all but finitely many s. We think of each g_s as a stage s approximation for g. Associated to every computable approximation $\langle g_s \rangle$ is its *mind-change function*, which maps each n to the number of stages s such that $g_{s+1}(n) \neq g_s(n)$.

A c.e. Turing degree $\mathbf{a}$ is array computable if every function $g \in \mathbf{a}$ has a computable approximation $\langle g_s \rangle$ such that for all n there are at most n many stages s such that $g_{s+1}(n) \neq g_s(n)$, that is, whose mind-change function is bounded by the identity function. The array computable degrees capture the combinatorics of a wide class of constructions. To wit, we observe that a c.e. degree is array noncomputable if and only if

(1) it is the degree of a perfect thin Π^0_1 class (Cholak, Coles, Downey, and Herrmann [12]);
(2) it bounds a disjoint pair of c.e. sets which have no separator computing $\varnothing'$ (Downey, Jockusch, and Stob [30]);
(3) it contains a c.e. set with maximal Kolmogorov complexity (Kummer [57]);

(4) it does not have a strong minimal cover in the Turing degrees (Ishmukhametov [51]);
(5) it has effective packing dimension 1 (Downey and Greenberg [24]);
(6) it contains two left-c.e. reals with no common upper bound in the cl-degrees of left-c.e. reals (Barmpalias, Downey, and Greenberg [7]);
(7) it contains a set which is not reducible to the halting problem with tiny use (Franklin, Greenberg, Stephan, and Wu [43]).

The dynamics captured by classes of degrees are often phrased in terms of *permitting*. We perform some computable construction, often using the priority method. To make the construction succeed, we need to satisfy infinitely many requirements, and to meet each requirement, we need to enumerate some numbers into a c.e. set A that we are building. The question is whether we can perform the construction "below" a given c.e. degree $\mathbf{d}$, which means, can we make $A \leqslant_T \mathbf{d}$? In the standard framework, we choose a c.e. set $D \in \mathbf{d}$, and along with the construction we define a Turing reduction Φ of A to D. Then, when we want to enumerate a number n into A, we seek *permission* from D to do so, which means that we want to see some number enter D below the use $\varphi(n)$ that we declared for computing $A(n)$ from D using Φ. Naturally, we will not always receive such permission, and so we need to make several attempts at meeting the requirement, using different potential numbers n to enumerate into A.

The "amount" of permitting that is required to carry out the construction (that is, to meet every requirement) corresponds to the class of degrees $\mathbf{d}$ below which we can perform the construction. The most common notion is *simple* permitting, which is given by any nonzero c.e. degree $\mathbf{d}$. Here it suffices for at least one of the attempts made by a given requirement to receive permission. This argument then shows, for example, that every c.e. degree bounds two incomparable c.e. degrees (the Friedberg-Muchnik construction can be performed using simple permitting), or that every c.e. degree bounds a 1-generic sequence.

Prompt permission, given by any promptly simple degree, also needs just one attempt to receive permission, but this permission must be given quickly: the required change in D needs to happen within some computable bound given the stage number. At the other extreme from simple permitting is high permitting, in which every requirement makes infinitely many attempts, and to meet the requirement, all but finitely many of these attempts need to be permitted.

Array noncomputable permitting, originally called "multiple permitting," is an intermediate version, in which for each attempt at meeting a requirement, a number of required permissions is stated in advance. The connection with the complexity of approximations of functions in the degree is direct: mind-changes essentially correspond to instances of permission; the computable bound on the number of mind-changes is the same bound on the number of permissions required to meet a requirement. The remarkable fact is that in many cases it is shown that the level of permitting is not only sufficient but also necessary for the construction to succeed.

As we shall see, in this monograph we introduce a transfinite hierarchy of classes, each of which has its own level of permitting; these classes generalise the array noncomputable degrees.

1.2.2 Natural definability and lattice embeddings

Ever since Lachlan and Yates's [59, 105] construction of a minimal pair refuted Shoenfield's conjecture [86] that the c.e. degrees are homogeneous, research in the c.e. degrees tended toward showing that they are as complicated as can be. For example, their theory (as a partial ordering) is computationally equivalent to full first-order arithmetic (see [49, 75]). This paradigm leads us to study definability in the partial ordering of the c.e. degrees, with the expectation that full bi-interpretability with arithmetic would hold. That would entail that a relation in the c.e. degrees is definable if and only if it is induced by a degree-invariant, arithmetic relation on indices of c.e. sets. Currently, this has almost been achieved, up to double jump classes:

Theorem 1.1 (Nies, Shore, Slaman [75]). *Any relation on the c.e. degrees which is invariant under the double jump is definable in the c.e. degrees if and only if it is definable in first-order arithmetic.*

The proof of Theorem 1.1 involves interpreting the standard model of arithmetic in the structure of the c.e. degrees without parameters, and obtaining a definable map from degrees to indices (in the model) which preserves the double jump. The result gives a definition of a large collection of classes of degrees (for example, all jump classes high_n and low_n, the latter for $n \geqslant 2$).

Theorem 1.1 has two shortcomings. One is the reliance on the invariance of the relation under the double jump. It follows that no collection of c.e. degrees that contains some, but not all, low_2 degrees can be defined using the theorem; these are the kinds of collections that we investigate in this monograph.

Another issue is that the definitions provided by the theorem are not *natural*, as discussed by Shore [88]. The definitions given by Theorem 1.1 are not structural; they do not give insights into the role of the relations being defined in the structure of the c.e. degrees. To date, there are not many examples of natural definitions in the c.e. degrees. Among them are:

- the promptly simple degrees are defined as the non-cappable ones (Ambos-Spies, Jockusch, Shore, and Soare [2]);
- the contiguous degrees are defined as the locally distributive ones (Downey and Lempp [33]) and also as the ones which are not the top of a copy of the pentagon lattice (the nonmodular, 5-element lattice N_5) in the c.e. degrees (Ambos-Spies and Fejer [1]);
- a third example takes place in the truth-table c.e. degrees rather than the Turing c.e. degrees: a c.e. truth-table degree is low_2 if and only if it has no minimal cover in the c.e. truth-table degrees (Downey and Shore [34]).

The example of the contiguous degrees (Turing c.e. degrees all of whose c.e. elements are weak truth-table equivalent) shows that natural definability results can be found when considering lattice embeddings into the c.e. degrees (see, for example, [63, 64, 67]). The question of which finite lattices can be embedded into the c.e. degrees (preserving join and meet) is also closely related to the problem of determining how much of the theory of the c.e. degrees is decidable. For example, Kleene and Post [55] showed that every finite partial ordering is embeddable into the c.e. degrees, and so that the 1-quantifier theory of the c.e. degrees is decidable. Deciding 2-quantifier questions involves lattice embeddings and extensions of embeddings.

All distributive finite lattices are embeddable into the c.e. degrees (Thomason [96], and independently Lerman, unpublished). All nondistributive lattices contain copies of one of the two following lattices:

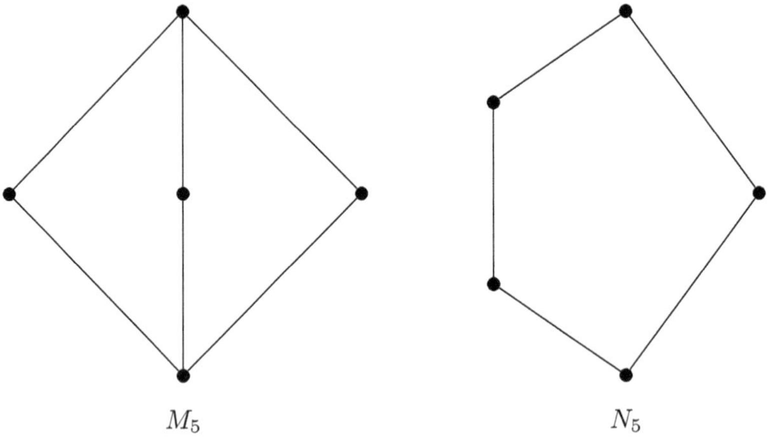

Figure 1.1. The two basic nondistributive lattices.

As mentioned, the lattice N_5 is nonmodular (the relation $a \vee (x \wedge b) = (a \vee x) \wedge b$ fails for some $a \leqslant b$), and every nonmodular lattice contains a copy of N_5. The lattice M_5, also known as the *1-3-1 lattice*, is modular, and every nondistributive, modular lattice contains a copy of the 1-3-1 lattice. Both lattices are embeddable into the c.e. degrees (Lachlan [61]).

The general question of which finite lattices are embeddable into the c.e. degrees remains open. The 1-3-1 is a significant obstacle, in that a slightly more complicated formation, known as the lattice S_8 (fig. 1.2), is not embeddable into the c.e. degrees (Lachlan and Soare [62]).

Thus, the 1-3-1 lattice is "just barely embeddable" in the c.e. degrees. Recalling our discussion above about permitting, the next natural question is how much computational power is required to embed this lattice. The point is that the embedding of the 1-3-1 lattice is quite complicated. Such an embedding, which is often done preserving the bottom element, involves the enumeration of

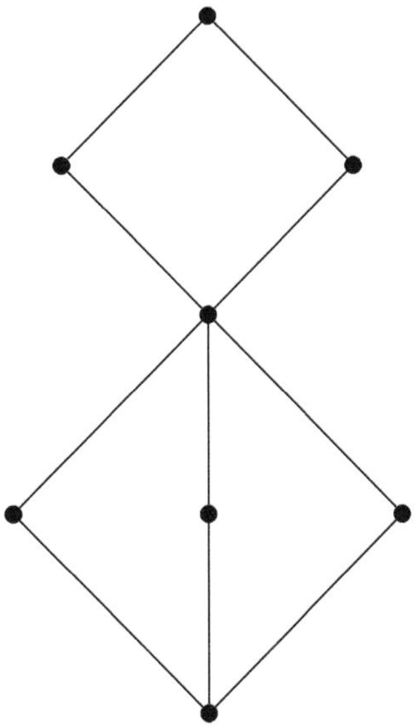

Figure 1.2. The lattice S_8.

three c.e. sets, A_0, A_1 and A_2, which pairwise form a minimal pair, and pairwise join above the third. The join and meet requirements interact very badly, and to overcome the difficulties, Lachlan used what became known as "continuous tracing." These difficulties were exploited by Downey [21], who showed that not every c.e. degree bounds a copy of the 1-3-1 lattice. In that paper, Downey noted that the embedding of the 1-3-1 lattice seemed to be tied up with multiple permitting in a way that was similar to non-low$_2$-ness. This intuition was verified by Downey and Shore [35], who showed that every non-low$_2$ c.e. degree bounds a copy of the 1-3-1 lattice in the c.e. degrees.

In attempting to synthesize the exact lattice structure which creates the embedding problems, Downey [21] and Weinstein [103] isolated the notion of a *critical triple.* A critical triple in a lattice consists of elements a_0, a_1 and b such that $a_0 \vee b = a_1 \vee b$ but $a_0 \wedge a_1 \leqslant b$ (fig. 1.3)

More generally, in an upper semilattice (which may fail to be a lattice), the meet requirement is replaced by $c \leqslant a_0, a_1 \rightarrow c \leqslant b$. Weinstein also introduced the notion of a *weak critical triple* (which we will not use in this manuscript); there the meet requirement is replaced by $c \leqslant a_0, a_1 \rightarrow a_0 \nleqslant b \vee c$. In the 1-3-1 lattice, the middle three elements, in any order, form a critical triple. Downey actually constructed an initial segment of the c.e. degrees in which there are no critical triples, and Weinstein did the same for weak critical triples.

The notion of non-low$_2$-ness seemed too strong to capture the class of degrees which bound a copy of the 1-3-1 lattice, but it was felt that something like that should suffice. On the other hand, Walk [101] constructed an array noncomputable c.e. degree bounding no weak critical triples, and hence it was already known that array noncomputability was not enough for such embeddings. In any

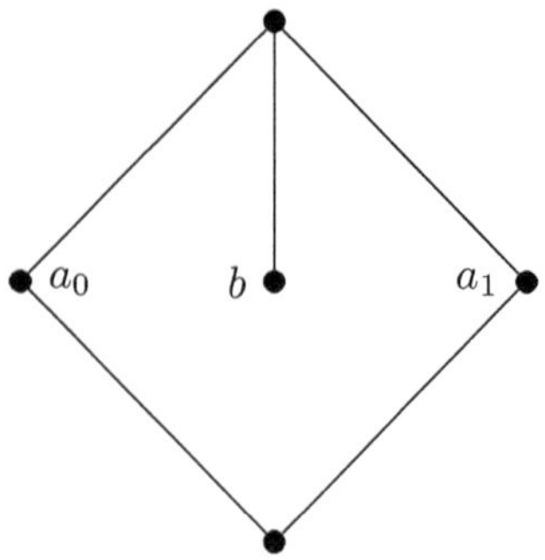

Figure 1.3. A critical triple.

case it was presumed that bounding the 1-3-1 lattice is equivalent to bounding a critical triple (or a weak critical triple). Our main result in this monograph implies that this presumption is false, and completely characterises the amount of permitting required to embed the 1-3-1 lattice.

1.3 TOWARD THE HIERARCHY OF TOTALLY α-C.A. DEGREES

We now turn to discussing two levels of the new hierarchy that we introduce. Some preliminary ideas and results appear in the companion papers [23, 25], and some related results appeared after our work was discussed with colleagues. Now we will discuss these ideas and results together in a mathematically, rather than historically, coherent way. Later we will discuss in detail the content of this monograph.

1.3.1 Totally ω-c.a. degrees

In 2005, J. Miller (unpublished) defined a non-uniform version of the class of array computable degrees. We call a function *ω-computably approximable* (ω-c.a.) if it has a computable approximation whose mind-change function is bounded by some computable function. This is equivalent to the function being weak truth-table reducible to $\varnothing'$. The notion is widely used in computability, with applications in algorithmic randomness as well (for example, in [42, 44, 47, 48]).[1]

This first step toward our new hierarchy is inspired by the above characterisation of array computable c.e. degrees as those which only contain functions with computable approximations with mind-change functions bounded by

[1]The terminology "ω-c.a." is new. In the literature one usually finds "ω-c.e.," although "ω-computable" is also used. In Chapter 2 below we justify the new terminology.

the identity. This is in some sense a "forced marriage" between two notions of complexity: complexity in terms of Turing degree; and complexity in terms of simplicity of approximations. ω-c.a. functions are in some sense relatively simple, in that we can guess them with few mistakes; on the other hand, they can be Turing equivalent to $\mathbf{0}'$, making them as complicated as possible among all Δ_2^0 functions when we consider Turing reducibility. When we consider approximations of *all* functions in a Turing degree, we get a new, useful concept. Thus Miller defined:

Definition 1.2. A c.e. degree is *totally ω-c.a.* if every function in it is ω-c.a.

Array computability is a uniform version of this notion: it requires the same bound on the mind-change function for all functions in the degree.

As discussed, this notion naturally aligns itself with a level of permitting. Recall that in array noncomputable permitting (previously named "multiple permitting"), each requirement plans infinitely many attempts at meeting it. Roughly, for the nth attempt to succeed, it needs n many permissions on numbers associated with this attempt. This corresponds to the identity bound on the number of mind changes. In *non-totally ω-c.a. permitting*, we again set up infinitely many attempts, but we are allowed to wait to declare how many permissions each attempt requires. Thus, for example, if the nth attempt is set up at stage s, then we could require s many permissions; and s could be much larger than n. For each requirement, the function mapping n to the number of permissions required to meet the nth attempt is computable, but different requirements will define different computable functions, likely with no uniform computable bound on these functions when all requirements are considered.

Using this notion of permitting, the class of totally ω-c.a. degrees captures the dynamics of a number of constructions. The first result appeared in [25], in which the authors, together with R. Weber, proved:

Theorem 1.3. *The following are equivalent for a c.e. degree* $\mathbf{d}$*:*

(a) $\mathbf{d}$ *bounds a critical triple in the c.e. degrees;*
(b) $\mathbf{d}$ *bounds a weak critical triple in the c.e. degrees;*
(c) $\mathbf{d}$ *is not totally ω-c.a.*

Note that this theorem shows that the totally ω-c.a. degrees are naturally definable in the c.e. degrees.

In this book we show another equivalence, characterising the dynamics of an existing construction. It considers presentations of left-c.e. reals in the unit interval $[0,1]$. A real is left-c.e. if the left cut it defines in the rationals is c.e. These reals are the measures of effectively open subsets of Cantor space; equivalently, each such real equals the sum $\sum_{\sigma\in A} 2^{-|\sigma|}$ for some prefix-free c.e. set

$A \subset 2^{<\omega}$. Such a set A is called a *presentation* of the sum, and is always computable from the sum. However, presentations can be simpler than the sum; in fact, every left-c.e. real has a computable presentation, even though the left-c.e. real itself may be noncomputable. The question is whether we can always code the complexity of a left-c.e. real into one of its presentations. In [32], Downey and LaForte answered this question negatively in a strong way: they constructed a noncomputable left-c.e. real, all of whose presentations are computable. The dynamics of coding complexity into presentations are captured by the totally ω-c.a. degrees:

Theorem 1.4.

(1) If a c.e. degree $\mathbf{d}$ *is not totally* ω*-c.a. then there is a left-c.e. real* $\varrho \leqslant_{\mathrm{T}} \mathbf{d}$ *and a c.e. set* $B <_{\mathrm{T}} \varrho$ *such that every presentation of* ϱ *is* B*-computable.*
(2) If a left-c.e. real ϱ *has totally* ω*-c.a. degree then there is a presentation of* ϱ *which is Turing equivalent to* ϱ*.*

For more background and details see Chapter 5, where we prove Theorem 1.4.

After our results were announced, Barmpalias and the authors [7] obtained yet another construction whose dynamics are captured by this class. Their results concern the interaction of Turing and weak truth-table reducibility. They showed that a c.e. degree is totally ω-c.a. if and only if every set in that degree is weak truth-table reducible to a ranked set (equivalently, to a hyperimmune set, or to a proper initial segment of a computable, scattered linear ordering). In further work, Brodhead, Downey, and Ng [9] showed that the totally ω-c.a. degrees capture a finite form of randomness.

Also, Adam Day [17] proved that a c.e. degree bounding a generic set which could compute an indifferent subset for itself cannot be totally ω-c.a. In his Ph.D. thesis, McInerney [71] has established similar results relating "multiple genericity" and "integer valued martingales" to being totally ω-c.a.

In the same way that array computability has become a central area of computability theory and its applications, we are confident that once researchers become sensitized to the combinatorics involving the notion of total ω-c.a.-ness, many further applications will be found.

1.3.2 Totally $<\omega^\omega$-c.a. degrees

As mentioned, contrary to expectation, we show in this monograph that in the c.e. degrees, bounding critical triples is not equivalent to bounding the 1-3-1 lattice. Very roughly speaking, the "continuous tracing" used in the embedding of the 1-3-1 lattice requires layers over layers of permitting. We now describe the dynamics of the construction, without connecting them to the requirements; more details will be given in Chapter 7.

The basic cycle in the construction of a critical triple goes as follows. A requirement starts defining a sequence $x_0, x_1, x_2, \ldots$ of numbers which it may want to enumerate into a c.e. set that we are building. At each stage s, we choose another number x_s and add it to the list. Then, possibly, at some stage t, a primary Σ_1 event happens (the realisation of a follower), and we want to enumerate these numbers into the sets, starting with x_t and working backwards. For each such number we need to wait for a secondary Σ_1 event (a new length of agreement of a minimal pair requirement). The requirement is met when the first number x_0 is enumerated. In a permitting argument, each such enumeration needs permission, so to meet the requirement we need t many permissions. This kind of permitting is precisely the kind given by non-totally ω-c.a. degrees.

In the 1-3-1 embedding, though, the number of minimal pair requirements stronger than the one we are looking at makes the process more complicated. If there is just one such requirement to contend with, the behaviour is just like the critical triple embedding. If there are two, though, the process is as follows:

(a) Define a sequence $x_0, x_1, x_2, \ldots$, adding a new number at each stage.
(b) When the primary Σ_1 event happens, start with x_t, and repeat the following t times:
  (i) If we are currently dealing with x_j (for $j \leqslant t$), start appointing a sequence $y_0^j, y_1^j, y_2^j, \ldots$, adding a new number at each stage.
  (ii) When a secondary Σ_1 event happens at some stage $s = s^j$, say, we start enumerating the numbers $y_s^j, y_{s-1}^j, \ldots$, each time waiting for some tertiary Σ_1 event.
  (iii) When all numbers y_i^j for $i < s^j$ have been enumerated, we also enumerate x_j, and repeat the cycle with x_{j-1}. If $j = 0$, the requirement is met.

When dealing with three minimal pair requirements, we add a layer:

(a) Define a sequence $x_1, x_2, \ldots$, adding a new number at each stage.
(b) When the primary Σ_1 event happens, start with x_t, and repeat the following t times:
  (i) If we are currently dealing with x_j, start appointing a sequence y_0^j, y_1^j, y_2^j, $\ldots$, adding a new number at each stage.
  (ii) When a secondary Σ_1 event happens at some stage $s = s^j$, start with y_s^j, and repeat the following s times:
    (1) If we are currently dealing with y_i^j (for $i < s^j$), we appoint a sequence $z_0^{j,i}, z_1^{j,i}, z_2^{j,i}, \ldots$, adding a new number at each stage.
    (2) When a tertiary Σ_1 event happens, at some stage $r = r^{j,i}$, we start enumerating the numbers $z_r^{j,i}, z_{r-1}^{j,i}, \ldots$, each time waiting for a quaternary Σ_1 event.

(3) When all numbers $z_k^{j,i}$ have been enumerated, we also enumerate y_i^j, and repeat the cycle with y_{i-1}^j. If $i=0$ then we exit this cycle.

(iii) We enumerate x_j; we repeat the outer cycle with x_{j-1}. If $j=0$, the requirement is met.

How many permissions are needed to meet the requirement? With two minimal pair requirements constraining us, we need $t+s^0+s^1+\cdots+s^t$ many permissions; with three, we need

$$\begin{aligned}
&t+\\
&s^0+r^{0,0}+r^{0,1}+r^{0,2}+\cdots+r^{0,s^0}+\\
&s^1+r^{1,0}+r^{1,1}+r^{1,2}+\cdots+r^{0,s^1}+\\
&\vdots\\
&s^t+r^{t,0}+r^{t,1}+r^{t,2}+\cdots+r^{t,s^t}.
\end{aligned}$$

We come now to the key insight. The real question is not how many permissions are required, but what is the reason that the process of meeting a requirement requires only finitely many steps. And the answer to the latter question is that we can attach a transfinite ordinal number to the process, and count down the ordinal along with the steps. With two minimal pair requirements, we start with the ordinal ω^2. When stage t is discovered, we go down to $\omega(t+1)$. When s^t is discovered, we descend to $\omega t+s^t$, and then decrease by 1 each time we enumerate another y_i^t. When y_0^t is enumerated, we are at ωt; when s^{t-1} is discovered, we go down to $\omega(t-1)+s^{t-1}$, and repeat. When three minimal pair requirements are present, we need to start at ω^3; then we go down to $\omega^2(t+1)$, then $\omega^2 t+\omega(s^t+1)$, then $\omega^2 t+\omega s^t+r^{t,s^t}$, then decrease by 1 each time some z_k^{t,s^t} is enumerated, and so on. Each time an inner cycle is finished (we enumerate some y_i^j) we go past some multiple of ω; each time an outer cycle is finished (we enumerate some x_j) we go past some multiple of ω^2.

In terms of permitting, the corresponding notion comes from Ershov's hierarchy of Δ_2^0 functions. We give exact details in Chapter 2, but informally, for a computable ordinal α, an *α-computable approximation* is a computable approximation $\langle g_s\rangle$ of a Δ_2^0 function g equipped with a counting down α which witnesses the fact that $g_s(n)$ changes only finitely many times: it is a uniformly computable sequence $\langle o_s\rangle$ of functions from $\mathbb{N}$ to α such that for all n, $o_s(n)<\alpha$, $o_{s+1}(n)\leqslant o_s(n)$, and if $g_{s+1}(n)\neq g_s(n)$ then $o_{s+1}(n)<o_s(n)$. The function g is called *α-computably approximable*, or α-c.a. Note that for $\alpha=\omega$ the notion coincides with the definition above. We thus see that in the 1-3-1 embedding, very roughly, to meet a requirement which has to contend with n stronger minimal pair requirements, we need permission from a function which is not ω^n-c.a. Thus we define:

Definition 1.5. A c.e. degree is *totally* $<\omega^\omega$*-c.a.* if every function in it is ω^n-c.a. for some n.

And the main theorem in this monograph, which realises the intuitive description above, is:

Theorem 1.6. *A c.e. degree bounds a copy of the 1-3-1 lattice if and only if it is not totally* $<\omega^\omega$*-c.a.*

Note that, as above, Theorem 1.6 shows that the class of totally $<\omega^\omega$-c.a. degrees is naturally definable in the c.e. degrees.

Nonuniform anti-permitting arguments

When we show that a class of degrees captures the dynamics of a construction (such as we do in Theorems 1.3 and 1.6) the argument has two parts: a permitting argument, which shows that the construction can be performed below a degree which permits accordingly; and an *anti-permitting* argument, which shows the converse. The latter is not a priority argument; we usually have different attempts at constructing objects which give that direction of the theorem, but these have very little interaction with each other. On the other hand, there is a certain nonuniformity to the construction, in that one of the attempts will succeed, but we cannot computably tell which. In the case of totally $<\omega^\omega$-c.a. degrees, we have ω levels of nonuniformity, which means that even though no injury occurs, only the oracle $\varnothing^{(\omega)}$ can tell which of the constructions we performed actually succeeds. This kind of argument, which we hinted at in [23], is presented in this monograph (in Chapters 6 and 7) in full for the first time. We believe that it will have wider applications.

1.3.3 The hierarchy of totally α-c.a. degrees

We have characterised the degrees which bound critical triples and degrees which bound a copy of the 1-3-1 lattice; but we have not yet argued that these classes are distinct, that is, that there is a degree which bounds a critical triple but not a copy of the 1-3-1. This will come out of a general investigation into a hierarchy of classes of degrees. The two classes under discussion are two levels of this hierarchy.

Armed with the definition of α-c.a. functions (which, as discussed, will require clarification, which we give in Chapter 2), we can extend the definitions above and define a degree to be *totally* α*-c.a.* if every function in it is α-c.a.; and more generally, to be *totally* $<\alpha$*-c.a.* if every function in it is β-c.a. for some $\beta<\alpha$. All such degrees are low_2. In the first part of this monograph, we give a detailed investigation of these classes, and in particular we find which are the proper levels of the hierarchy. For example, we show:

- there is a totally α-c.a. degree which is not totally β-c.a. for any $\beta < \alpha$ if and only if α is a power of ω;
- there is a totally $<\alpha$-c.a. degree which is not totally $<\beta$-c.a. for any $\beta < \alpha$ if and only if α is a limit of powers of ω.

This, in particular, shows that there are ω many distinct levels between the totally ω-c.a. degrees and the totally $<\omega^\omega$-c.a. degrees.

1.4 THE CONTENTS OF THIS MONOGRAPH

In the first part of the monograph, we introduce and investigate our new hierarchy.

In Chapter 2, we give a rigorous treatment of the notion of α-c.a. functions. The main issue is to properly define what we mean by a computable function o from $\mathbb{N}$ to α, which is required for the definition of α-computable approximations. Naturally, to deal with an ordinal α computably, we need a notation for this ordinal, or more generally, a computable well-ordering of order-type α. To form the basis of a solid hierarchy, the notion of α-c.a. should not depend on which well-ordering we take, rather it should only depend on its order-type. Thus we cannot consider all computable copies of α. Rather, we restrict ourselves to a class of particularly well-behaved well-orderings, in a way that ensures that they are all computably isomorphic. For example, when considering copies of ω^2, we must compute not only the collection of limit points and the successor function, but we also need to know which copy of ω inside ω^2 is which. In general, we need the Cantor normal form to be computable. This turns out to be sufficient for small enough ordinals; we develop the theory for ordinals $\alpha \leqslant \varepsilon_0$. The theory can be pushed further, but not all the way up to ω_1^{CK}; we do not pursue such extensions here.

Having defined α-c.a. functions, we also (in Section 2.3) relate these functions to iterations of the bounded jump (the jump inside the weak truth-table degrees). This extends and solidifies work by Coles, Downey, and LaForte [15], and independently Anderson and Csima [3]. Extending the familiar result for ω, we show (Theorem 2.40) that a function is ω^α-c.a. if and only if it is weak truth-table reducible to the αth iteration of the bounded *function* jump; an analogous result holds for sets.

In Chapter 3 we investigate the hierarchy of totally α-c.a. degrees. As mentioned above, we show precisely when this hierarchy collapses (Theorem 3.6), and refine this hierarchy when we consider totally $<\alpha$-c.a. degrees (Theorem 3.25). We further consider *uniform* versions of our classes. Recall that the array computable degrees were a uniform version of the totally ω-c.a. degrees, in that we took a single computable bound on the mind-change function of approximations of functions in the degree. We find the right formulation that generalises this to define *uniformly totally α-c.a. degrees*, and show (Theorem 3.20) how they fit in our hierarchy. For a general picture, see Figure 3.3.

1.4.1 Maximality

It is not common to find maximal elements of classes in the c.e. degrees; usually, density prevails. However, in Chapter 4 we show that at every level of our main hierarchy there are maximal degrees (Theorem 4.1). Thus, for example, there are maximal degrees with respect to not bounding a critical triple, namely, maximal totally ω-c.a. degrees. Since the totally ω-c.a. degrees are naturally definable, we obtain a naturally definable antichain in the c.e. degrees; the only previously known such antichain consisted of the maximal contiguous degrees (Cholak, Downey, and Walk [14]).

On the other hand, we show (Theorem 4.12) that maximality cannot go too far, that is, to the next level. For example, no totally ω-c.a. degree can be maximal totally ω^2-c.a. A corollary of the argument shows that there are no maximal totally $<\omega^\omega$-c.a. degrees, that is, no degrees maximal with respect to not bounding a 1-3-1.

We remark that in further work with Katherine Arthur [5] we investigate bounding by maximal degrees. For example, there are totally ω-c.a. degrees bounded by no such maximal degrees. The general picture is interesting. We suspect that, in general, the following holds:

- Let $\alpha \leqslant \beta \leqslant \varepsilon_0$ be powers of ω. Then every totally α-c.a. degree is bounded by a maximal totally β-c.a. degree if and only if $\beta \geqslant \alpha^\omega$.

Further questions consider collapse of our hierarchy in upper cones. Theorem 4.12 implies that every totally ω-c.a. degree is bounded by a strictly greater degree which is totally ω^2-c.a. However we do not know if we can always make that degree not totally ω-c.a. The best result so far, which appears in [5], implies that every totally ω-c.a. degree is bounded by a totally ω^4-c.a. degree which is not totally ω-c.a. Is it ω^2, or ω^3? We cannot yet tell.

1.4.2 Calibrating the dynamics of constructions

The second part of the monograph consists of Chapters 5, 6, and 7, in which we discuss and calibrate the dynamics of three different constructions. In Chapter 5 we prove Theorem 1.4 about presentations of left-c.e. reals. In Chapter 7 we prove our main Theorem 1.6. In Chapter 6 we consider *m-topped degrees*, continuing [23]. The notion of m-topped degrees comes from a general study of the interaction between Turing reducibility and stronger reducibilities among c.e. sets. For example, this study includes the contiguous degrees. A c.e. Turing degree $\mathbf{d}$ is m-topped if it contains a greatest degree among the many-one degrees of c.e. sets in $\mathbf{d}$. Such degrees (other than $\mathbf{0}'$) were constructed in Downey and Jockusch [29]. They are all low_2. In [23] we showed that there are totally ω^ω-c.a. m-topped degrees. Here we show that this is the best possible: no m-topped degree is totally $<\omega^\omega$-c.a. (Theorem 6.1). We remark though that in this case we cannot hope to get full equivalence: we cannot prove that every degree which is not totally $<\omega^\omega$-c.a. bounds an m-topped degree. This is because m-topped degrees cannot be low, whereas every level of our hierarchy contains both low degrees and degrees which are low_2 but not low.

1.4.3 Promptness

One can ask, regarding the embedding of the 1-3-1 lattice, what it would take to get an embedding preserving the bottom, that is, an embedding whose bottom degree is $\mathbf{0}$ (as is obtained in Lachlan's original construction). We discuss this in Chapter 8, where we introduce prompt versions of all levels in our hierarchy. This generalises the already familiar notion of prompt permitting, which is the prompt version of simple permitting. Prompt array noncomputable permission, for example, allows us to construct a pair of separating classes whose elements form minimal pairs (Theorem 8.22); whereas traditional (non-prompt) array noncomputable permission only gives Turing incomparability [30]. Similarly, a degree which is promptly not totally $<\omega^\omega$-c.a. bounds a copy of the 1-3-1 lattice with bottom $\mathbf{0}$.

This however cannot be reversed: every high degree bounds a copy of the 1-3-1 lattice with bottom $\mathbf{0}$, and there are high degrees which are not promptly simple (let alone promptly non-totally $<\omega^\omega$-c.a.). Informally what this says is that there are at least two ways to get such an embedding: either by quickly getting the precise number of permissions required; or by getting many permissions (cofinitely many), in which case we can wait for the permissions and don't need them promptly.

It would be interesting to find a common generalisation.

1.5 AN APPLICATION TO ADMISSIBLE COMPUTABILITY

Combined with results of the second author, our work has an application to admissible computability. This is a generalisation of traditional computability to ordinals beyond ω. In [46] it is shown that for any admissible ordinal α, the α-c.e. degrees are not elementarily equivalent to the c.e. degrees. This was done in cases, depending on the proximity of α to ω. In one case the separation between the theories is not natural but relies on coding models of arithmetic. However one result is:

Theorem 1.7 ([46]). *Let $\alpha > \omega$ be an admissible ordinal, and let* $\mathbf{a}$ *be an incomplete α-c.e. degree. The following are equivalent:*

(1) $\mathbf{a}$ *computes a cofinal ω-sequence in α.*
(2) $\mathbf{a}$ *bounds a copy of the 1-3-1 lattice.*
(3) $\mathbf{a}$ *bounds a critical triple.*

Again, it is the analysis of continuous tracing that underlies this result. The basic idea is the following. Consider again the dynamic aspect of the embedding of a critical triple which we discussed above. We start by appointing elements $x_0, x_1, x_2, \ldots$, adding one at each stage. When the primary Σ_1 event happens (the follower is realised), it is important (because of use considerations) that we attempt to enumerate the elements x_j *starting with the last number* x_t and working backwards.

Trying to do this when time goes beyond ω presents a completely new problem: after ω many stages, we will have elements x_j for all $j < \omega$, that is, we will not have a last element. We cannot then peel it back, each step removing only the last element. It turns out that this blockage is fundamental. The only case it might be possible for a degree **a** to bound a critical triple is if it itself can see that α is far from being a regular cardinal—if it can essentially re-order time and space to order-type ω, so that the construction can be (at least after the fact) seen to have taken ω steps, avoiding infinite sequences of numbers. In one direction, effectively closed and unbounded sets are used to show that this is necessary. In the other direction, a fine-structural result of Shore's [87] says that an incomplete degree of computable cofinality ω must be high, and can compute a bijection between α and ω. Working below such a degree, we can translate back to ω-computability, and use non-low$_2$ permitting to embed the 1-3-1 lattice (for a technical reason, we cannot quite use high permitting).

To sum, what this says is that once we go beyond ω, the fine distinctions between totally ω-c.a. degrees and totally $<\omega^\omega$-c.a. degrees completely disappear. Combined with the current work, this gives us a single, natural sentence which separates the elementary theory of the c.e. degrees from the theory of the α-c.a. degrees for any admissible $\alpha > \omega$.

Theorem 1.8. *Let $\alpha \geqslant \omega$ be admissible. The following are equivalent:*

(1) There is an incomplete α-c.e. degree which bounds a critical triple but not the 1-3-1 lattice.
(2) $\alpha = \omega$.

1.6 NOTATION AND GENERAL DEFINITIONS

We recap some notions that we discussed above, and introduce terminology and conventions that will be used throughout the monograph. First, though, we comment on the expected mathematical background a reader will need. We assume that the reader has mastered the basics of computability theory, up to and including basic finite-injury priority arguments, in particular the Friedberg-Muchnik theorem, and basic infinite-injury priority constructions, mainly the construction of a minimal pair of c.e. degrees, as performed on a priority tree. For years, the standard reference in this area has been Soare's [91]. Other possible sources are the second chapter of [27], the first chapter of [74], Cooper's [16], Odifreddi's [77], or Steffen Lempp's unpublished notes on priority arguments in computability theory, available on his website. We also assume some basic information on ordinals and ordinal arithmetic; any standard set theory text would be more than sufficient.

1.6.1 Computable approximations and enumerations

A *computable approximation* for a function $f\colon \omega \to \omega$ is a uniformly computable sequence $\langle f_s \rangle_{s<\omega}$ of functions such that for all x, for almost all s, $f_s(x) = f(x)$. In other words, $f = \lim_s f_s$ when we equip ω with the discrete topology. Shoenfield's limit lemma [84] states that a function f is Δ^0_2-definable if and only if it is computable from the halting set $\varnothing'$ if and only if it has a computable approximation. If A is a *set* (a subset of ω, identified with an element of Cantor space) then a computable approximation of A is a sequence of sets.

A *computable enumeration* of a c.e. set A is a computable, $\subseteq$-increasing sequence of finite sets $\langle A_s \rangle$ such that $A = \bigcup_s A_s$. We can also think of a computable enumeration as a computable approximation of A, again by taking characteristic functions. We say that a number x is *enumerated into* A_s if $x \in A_s \setminus A_{s-1}$.

1.6.2 Turing functionals

A (Turing) *functional* is a c.e. set of triples $\langle \sigma, x, y \rangle$ consisting of a finite sequence σ of natural numbers and a pair of natural numbers x and y. We consider such triples as *axioms*, and sometimes write them as $\sigma \mapsto (x, y)$. If $f\colon \omega \to \omega$ and Φ is a functional, then we define the multi-valued function (i.e., relation) $\Phi(f) \subseteq \omega \times \omega$ by letting $\Phi(f, x) = y$ if there is some finite $\sigma \prec f$ such that the axiom $\sigma \mapsto (x, y)$ is in Φ. We write $\Phi(f, x)\downarrow$ for $x \in \operatorname{dom} \Phi(f)$ and $\Phi(f, x)\uparrow$ for $x \notin \operatorname{dom} \Phi(f)$.

In general we allow functionals, especially the ones that we build, to be *inconsistent*. That is, we allow them to contain contradictory axioms: a pair of axioms $\sigma \mapsto (x, y)$ and $\tau \mapsto (z, w)$ such that σ and τ are comparable (that means that $\sigma \preccurlyeq \tau$ or $\tau \preccurlyeq \sigma$), $x = z$ but $y \neq w$. A functional Φ is called *consistent relative to an oracle* f if $\Phi(f)$ is a partial function, i.e., is not multi-valued. A functional is consistent if and only if it is consistent relative to every oracle.

The following are equivalent for $f, g\colon \omega \to \omega$:

(1) there is a consistent functional Φ such that $\Phi(f) = g$;
(2) there is a functional Φ, consistent relative to f, such that $\Phi(f) = g$;
(3) $g \leqslant_{\mathrm{T}} f$.

If $\langle \Phi_s \rangle$ is a computable enumeration of a functional Φ, then each Φ_s is also a functional. If $\langle f_s \rangle$ is a computable approximation of a function $f\colon \omega \to \omega$, then the finite multi-valued function $\Phi_s(f_s)$ can be effectively obtained from s. If for all s, Φ_s is consistent relative to f_s, then Φ is consistent relative to f. Note that if, further, $\Phi(f)$ is a total function, then we can extend $\langle \Phi_s(f_s) \rangle$ to a computable approximation of $\Phi(f)$, since $\langle \operatorname{dom} \Phi_s(f_s) \rangle$ is uniformly computable. When the notation $\Phi_s(f_s)$ becomes unwieldy, we sometimes write $\Phi(f)[s]$, and in general may use Lachlan's square bracket notation.

Suppose that Φ is a functional which is consistent relative to an oracle f. If $x \in \operatorname{dom}\Phi(f)$, we also refer to $\Phi(f,x)=y$ as a "computation." Let σ be the shortest initial segment of f for which $\sigma \mapsto (x,y)$ is an axiom in Φ. Often in fact there will be a unique such initial segment. The string σ determines the *use* of the computation, denoted by $\varphi(f,x)$ (and when f is clear from the context, by $\varphi(x)$). We will use two conflicting notions:

- If either f or Φ are given, then the use is the length of σ.
- If both f and Φ are built by us then we let the use be $|\sigma|-1$, the "greatest number queried during the computation." In this case f is usually a c.e. set A. The idea is that we may want to void the computation by enumerating the use $\varphi(x)$ into A.

If $\langle \Phi_s \rangle$ is a computable enumeration of a Turing functional Φ, and $\langle f_s \rangle$ is a computable approximation of a function f (and again we assume that for all s, Φ_s is consistent relative to f_s), $s<\omega$ and $x \in \operatorname{dom}\Phi_s(f_s)$, then we say that the computation $\Phi_s(f_s,x)$ is *destroyed* (or *injured*) at stage $s+1$ if $\sigma \not\prec f_{s+1}$, where σ as above is the shortest axiom applying to f giving the computation at stage s. That is, if $f_{s+1}\restriction_u \neq f_s\restriction_u$ where $u=\varphi_s(f_s,x)$ is the use of the computation, in the case in which either f or Φ are given; if both are built by us, then the computation is destroyed if $f_s\restriction_{u+1} \neq f_{s+1}\restriction_{u+1}$, and as described above, this will often happen because we enumerate u into f_{s+1}.

In contrast, we say that a computation $\Phi_s(f_s,x)=y$ is *f-correct* if $\sigma \prec f$. The fundamental fact about Turing computations, used without mention throughout computability theory, is that $x \in \operatorname{dom}\Phi(f)$ if and only if there is a stage s (equivalently, for almost all stages s) such that $x \in \operatorname{dom}\Phi_s(f_s)$ by an f-correct computation. When working with c.e. sets we often use the fact that correct computations never go away: if $\langle A_s \rangle$ is a computable enumeration of a c.e. set A, and $\Phi_s(A_s,x)$ is an A-correct computation, then for all $t \geqslant s$, $x \in \operatorname{dom}\Phi_t(A_t)$ by the same computation.

The following lemma is used when we build functionals which apply to c.e. sets that we enumerate.

Lemma 1.9. *Let $\langle \Phi_s \rangle$ be a computable enumeration of a functional Φ, and let $\langle A_s \rangle$ be a computable enumeration of a c.e. set A. Suppose that for all s,*

(1) if an axiom $\sigma \mapsto (x,y)$ is enumerated into Φ_s, then $\sigma \prec A_s$;
(2) for each x, at most one axiom $\sigma \mapsto (x,y)$ is enumerated into Φ_s.

Let $s<\omega$, and suppose that Φ_s is consistent for A_s. Suppose that for all $x<\omega$,

(3) if an axiom $\sigma \mapsto (x,y)$ is enumerated into Φ_{s+1}, and $x \in \operatorname{dom}\Phi_s(A_s)$, then some number $u \leqslant \varphi_s(A_s,x)$ is enumerated into A_{s+1}.

Then Φ_{s+1} is consistent for A_{s+1}.

Hence if conditions (1)–(3) hold at every stage s, then Φ is consistent for A. Note that usually Φ will not be consistent for all oracles: we could void a computation $\Phi_s(A_s, x)$ by enumerating $u = \varphi_s(A_s, x)$ into A_{s+1}, and then define a new computation $\Phi_{s+1}(A_{s+1}, x)$ with smaller use, so Φ_{s+1} may be inconsistent for A_s.

Convention 1.10. We often assume that for a given consistent functional Φ, for any oracle f, $\operatorname{dom}\Phi(f)$ is an initial segment of ω. That is, we require that if $\sigma \mapsto (x, y)$ is in Φ, then for all $x' < x$ there is some $\sigma' \preccurlyeq \sigma$ and some y' such that $\sigma' \mapsto (x', y')$ is also in Φ. We simply prevent $\sigma \mapsto (x, y)$ from entering Φ until we see the other necessary axioms.

In this situation we also assume that if $\langle \Phi_s \rangle$ is a computable enumeration of a Turing functional Φ, then for all s and f, $\operatorname{dom}\Phi_s(f)$ is an initial segment of ω.

The point is that if we are only interested in *total* functions computable from an oracle f, then we can restrict ourselves to functionals of the type described.

We let $\langle \Phi_e \rangle$ be some enumeration of all *consistent* functionals; associated with which we are given uniformly computable enumerations $\langle \Phi_{e,s} \rangle$ of Φ_e.

Convention 1.11. We sometimes identify natural numbers with the von Neumann ordinals isomorphic to them; that is, we identify the natural number n with the set $\{0, 1, 2, \ldots, n-1\}$. In particular, if for some functional Φ and oracle f, $\operatorname{dom}\Phi(f)$ is an initial segment of ω (per Convention 1.10), then we write $x < \operatorname{dom}\Phi(f)$ for $x \in \operatorname{dom}\Phi(f)$, and $x \leqslant \operatorname{dom}\Phi(f)$ for $\{0, 1, \ldots, x-1\} \subseteq \operatorname{dom}\Phi(f)$.

Functionals which take more than one oracle are treated in a similar fashion. For example, when taking two oracles, axioms will be of the form $(\sigma, \tau) \mapsto (x, y)$. Usually, for a pair of oracles f, g in which we are interested, for each x there will be at most one pair of strings $\sigma \prec f$ and $\tau \prec g$ such that $(\sigma, \tau) \mapsto (x, y)$ is in the functional Φ we are building or examining. These determine the *f-use* and the *g-use* of the computation $\Phi(f, g, x)$, according to the notational convention discussed above. When Φ is not built by us we often assume that the f-use and the g-use are the same, and that common value is referred to simply as the use $\varphi(f, g, x)$ of the computation.

1.6.3 Priority arguments and tree constructions

In our constructions we keep the *convention of small numbers.*

Convention 1.12. At stage s of a construction, all numbers played by the "opponent" are bounded by s. These are the values of functions that are not defined by us during the construction.

On the other hand, the constructions would often call on us to define new values for functions that are *large.* This means that the new values are picked to be numbers that are larger than any other number previously used or observed in the construction, including the stage number.

Most terminology we will use in priority constructions is common. We will attempt to meet *requirements*. *Positive requirements* are those which can be met by enumerating numbers into c.e. sets we are enumerating. *Negative requirements* are met by imposing *restraint* on other actors. The numbers enumerated into the c.e. sets are sometimes called *followers*. In the standard Friedberg-Muchnik construction, for example, a requirement attempting to ensure that $\Phi(A) \neq B$ will *appoint* a follower x, which means choose some number x (that will not be used by any other requirement), wait until we see that $\Phi(A, x)\downarrow = 0$, and then enumerate it into B. The prototypical negative requirements, on the other hand, are met in the Lachlan-Yates minimal pair construction. In most of our constructions, restraint will be imposed by *initialising* other requirements. Typically, initialising a positive requirement means that any follower x it appointed is *cancelled*: this means that the number x will not be involved in the construction any longer. Any new follower will be chosen to be large.

Tree constructions, namely priority constructions done with the aid of a *tree of strategies*, are now standard; a reference is Chapter XIV of [91]. Elements of the tree are called *strategies*, or *nodes*; these are finite sequences of symbols. To describe the tree of strategies, we give two pieces of information:

(a) An association of requirements for nodes; we say that a node *works for* the requirement associated with it. Often, but not always, all nodes of a given level of the tree work for the same requirement.
(b) For nodes working for some requirement, the list of *outcomes* of these nodes.

The tree is then defined recursively. The empty node is always on the tree of strategies; if a node σ has already been determined to lie on the tree of strategies, and a requirement R has been associated with it, then the immediate successors of σ on the tree are the nodes of the form $\sigma\hat{\ }o$, where o is a possible outcome for nodes working for R.

The collection of possible outcomes of any node will be linearly ordered; we say that an outcome o is *stronger* than an outcome o' if $o < o'$. This ordering induces a linear ordering of the tree of strategies, by taking a lexicographic amalgamation of the orderings of outcomes: $\sigma < \tau$ if $\sigma \prec \tau$, or if there are η, o and o' such that $\sigma \succcurlyeq \eta\hat{\ }o$, $\tau \succcurlyeq \eta\hat{\ }o'$, and $o < o'$. We say that a node σ is *stronger* than a node τ if $\sigma < \tau$, and that a node σ *lies to the left* of a node τ if $\sigma < \tau$ but $\sigma \not\prec \tau$. We sometimes write $\sigma <_L \tau$; this has nothing to do with the constructible universe.

At any stage s, the construction describes the (finite) collection δ_s of nodes that are *accessible* at stage s. In our constructions this will always be an initial segment of the tree of strategies, linearly ordered by extension of nodes. We will not use constructions with links. Usually, the empty node $\langle\rangle$ is accessible at every stage.

We then say that a node σ lies on the *true path* δ_ω if there are infinitely many stages s of the construction such that $\sigma \in \delta_s$ (that is, such that σ is accessible at stage s), but the same is not true for any node τ that lies to the left of σ. The true path δ_ω will be a linearly ordered initial segment of the tree of strategies. We will need to prove that the true path is infinite.

As with simpler constructions, tree constructions will involve initialisations, this time of nodes rather than of requirements. Again, when a node is initialised, all parameters associated with the node (such as followers) are removed (or *cancelled*), and new ones will have to be defined, either immediately, or more often, at the next stage at which the node is accessible. When a stage ends, every node which lies to the right of an accessible node (a node in δ_s) is initialised. Often, but not always, nodes extending the longest node in δ_s are also initialised at the end of stage s. We ensure that whenever a node σ is initialised, and τ is a node weaker than σ, then τ is also initialised at the same time.

We say that the construction is *fair* to a node σ if σ is initialised only finitely many times (i.e., at only finitely many stages of the construction). The main *fairness lemma* for each construction will state that the construction is fair to every node on the true path δ_ω. If σ is a node on the true path and the construction is not fair to σ then there will be some node $\tau \prec \sigma$ on the true path which initialises σ at infinitely many stages. This is because initialisation has to respect the priority ordering; no node weaker than σ can initialise σ.

Other standard conventions of priority constructions are employed without mention. For example, we use "stickiness" or "persistence" of parameters: if, for example, a requirement R or strategy σ has a follower at some stage s, and the requirement or node is not tampered with (e.g., initialised) at stage $s+1$, say, then that follower is still considered to be a follower for the requirement or strategy at stage $s+1$.

Chapter Two

α-c.a. functions

ERSHOV ([40], SEE ALSO [4]) extended the hierarchy of differences of c.e. sets into the transfinite, based on Kleene's notations for computable ordinals. Unfortunately, the levels of this hierarchy depend heavily on the choice of notation. To get around this problem, based on ideas from [15], we focus on lower levels of the hierarchy, using canonical well-orderings. We then, extending [3], relate these lower, canonical levels, to iterations of a jump in the weak truth-table degrees.

We remark that Kleene's notations suffice for the purposes they were designed for. For example, Spector's theorem states that the iteration of the Turing jump along a computable ordinal does not depend on the choice of notation for that ordinal. But as soon as we have finer distinctions such as those of the present monograph, we need more sensitive notions of notations. The fact that ours robustly and invariantly capture the combinatorics of many constructions shows that they seem to be good choices.

2.1 $\mathcal{R}$-C.A. FUNCTIONS

Let $\mathcal{R} = (R, <_{\mathcal{R}})$ be a computable well-ordering of a computable set R. An $\mathcal{R}$-*computable approximation* of a function f is a computable approximation $\langle f_s \rangle$ of f, equipped with a uniformly computable sequence $\langle o_s \rangle_{s<\omega}$ of functions from ω to R such that for all x and s:

- $o_{s+1}(x) \leqslant_{\mathcal{R}} o_s(x)$; and
- if $f_{s+1}(x) \neq f_s(x)$, then $o_{s+1}(x) <_{\mathcal{R}} o_s(x)$.

The sequence $\langle o_s \rangle_{s<\omega}$, together with the well-foundedness of $\mathcal{R}$, witnesses the fact that the approximation $\langle f_s \rangle$ indeed reaches a limit.

Definition 2.1. A function $f\colon \omega \to \omega$ is $\mathcal{R}$-*computably approximable* (or $\mathcal{R}$-*c.a.*) if it has an $\mathcal{R}$-computable approximation.

The following equivalent formulation is sometimes taken as a definition:

Proposition 2.2. *A function $f\colon \omega \to \omega$ is $\mathcal{R}$-c.a. if and only if there is a partial computable function ψ such that for all x, $f(x) = \psi(x, z)$ for the $\mathcal{R}$-least z such that $(x, z) \in \operatorname{dom} \psi$.*

(In particular, the totality of f implies that for all $x<\omega$ there is some $z\in R$ such that $(x,z)\in\operatorname{dom}\psi$.)

Proof. Let $\langle f_s, o_s\rangle$ be an $\mathcal{R}$-computable approximation of f. For $x<\omega$ and $z\in R$, let $\psi(x,z)=f_s(x)$ for any $s<\omega$ such that $o_s(x)=z$; if there is no such s, we let $\psi(x,z)\uparrow$. The fact that $f_s(x)$-changes have to be accompanied by an $o_s(x)$-change implies that ψ is well-defined. Then ψ witnesses that f is $\mathcal{R}$-c.a.

Suppose that ψ is a partial computable function as in the proposition. Define a uniformly computable sequence $\langle o_s\rangle$ as follows. Let $A=\operatorname{dom}\psi$. Since A is c.e., let $\langle A_s\rangle$ be some effective enumeration of A. Since f is total, for all $x<\omega$ there is some $t_x<\omega$ such that $(x,z)\in A_{t_x}$ for some $z\in R$. For any $x<\omega$ and $s<\omega$ we let $o_s(x)$ be the $\mathcal{R}$-least z such that $(x,z)\in A_{\max\{s,t_x\}}$.

Since $A_t\subseteq A_s$ whenever $t\leqslant s$, we have $o_{s+1}(x)\leqslant o_s(x)$ for all x and s. Let $f_s(x)=\psi(x,o_s(x))$. Then $\langle f_s,o_s\rangle$ is an $\mathcal{R}$-computable approximation of f. □

2.1.1 $\mathcal{R}$-c.e. sets

For sets, Ershov refined the hierarchy of $\mathcal{R}$-c.a. functions to levels resembling the arithmetic hierarchy. For $z\in R$, let $R\restriction_z=\{w\in R : w<_{\mathcal{R}} z\}$, which is a computable $\mathcal{R}$-initial segment of R; and let $\mathcal{R}\restriction_z$ be the restriction of $<_{\mathcal{R}}$ to $R\restriction_z$. Recall that an ordinal is *even* if it is of the form $\alpha+2n$ for some limit ordinal α (or $\alpha=0$), where $n<\omega$; and *odd* otherwise. We say that $\mathcal{R}$ is *even* if the order-type $\operatorname{otp}(\mathcal{R})$ is even, and *odd* otherwise; and we say that an element $z\in R$ is $\mathcal{R}$-*even* if $\mathcal{R}\restriction_z$ is even, and $\mathcal{R}$-*odd* otherwise. If $\mathcal{R}$ is even, we write $\texttt{parity}(\mathcal{R})=0$; otherwise we write $\texttt{parity}(\mathcal{R})=1$. Similarly, we write $\texttt{parity}_{\mathcal{R}}(z)=\texttt{parity}(\mathcal{R}\restriction_z)$.

Definition 2.3. Suppose that the collection of $\mathcal{R}$-even elements of R is computable. A set $A\subseteq\omega$ is *$\mathcal{R}$-c.e.* if there is a uniformly c.e. sequence $\langle A_z\rangle_{z\in R}$ such that:

- if $z<_{\mathcal{R}} w$ then $A_z\subseteq A_w$; and
- for all $x<\omega$, $x\in A$ if and only if $x\in\bigcup_{z\in R}A_z$, and for the $\mathcal{R}$-least z such that $x\in A_z$ we have $\texttt{parity}_{\mathcal{R}}(z)\neq\texttt{parity}(\mathcal{R})$.

We let $\Sigma^{-1}_{\mathcal{R}}$ denote the collection of all $\mathcal{R}$-c.e. sets.

The definition should be understood dynamically. Indexed by some *late* element z of $\mathcal{R}$ we see a number x enter the "playground" $\bigcup_w A_w$. We then move backwards in $\mathcal{R}$, so to speak, and at each step we change our mind about whether x is in the target set or not. Thus, this notion extends the finite difference hierarchy. For $n\geqslant 1$, let n also denote a computable linear ordering which has exactly n elements. Then a set is 1-c.e. if it is c.e., is 2-c.e. if it is the (set theoretic) difference of two c.e. sets (also known as d.c.e.), and, in general, is $(n+1)$-c.e. if it is of the form $A\setminus B$, where A is c.e. and B is n-c.e.

Ershov lets $\Pi^{-1}_{\mathcal{R}}$ be the collection of complements of $\mathcal{R}$-c.e. sets, and lets $\Delta^{-1}_{\mathcal{R}} = \Sigma^{-1}_{\mathcal{R}} \cap \Pi^{-1}_{\mathcal{R}}$ be the collection of sets which are both $\mathcal{R}$-c.e. and co-$\mathcal{R}$-c.e.

Proposition 2.4. *Suppose again that the parity function* $\mathtt{parity}_{\mathcal{R}}$ *is computable. Then every set in* $\Delta^{-1}_{\mathcal{R}}$ *is* $\mathcal{R}$*-c.a. If further the order-type of* $\mathcal{R}$ *is a limit ordinal, then* $\Delta^{-1}_{\mathcal{R}}$ *coincides with the collection of* $\mathcal{R}$*-c.a. sets.*

Proof. Suppose that $A \in \Delta^{-1}_{\mathcal{R}}$. Suppose, for simplicity of notation, that $\mathcal{R}$ is even; the odd case is identical. Let $\langle A_z \rangle_{z\in R}$ witness that $A \in \Sigma^{-1}_{\mathcal{R}}$, and $\langle B_z \rangle_{z\in R}$ witness that $A \in \Pi^{-1}_{\mathcal{R}}$. Define a partial computable function ψ as follows. Let $x < \omega$ and $z \in R$. If $x \notin A_z \cup B_z$, we let $\psi(x,z)\uparrow$. Otherwise, x shows up first in either A_z or B_z.

- If x shows up first in A_z, then we let $\psi(x,z) = \mathtt{parity}_{\mathcal{R}}(z)$.
- If x shows up first in B_z, then we let $\psi(x,z) = 1 - \mathtt{parity}_{\mathcal{R}}(z)$.

Fix $x < \omega$. Then $x \in \bigcup_{z\in R}(A_z \cup B_z)$ because $A \subseteq \bigcup_z A_z$ and $\omega \setminus A \subseteq \bigcup_z B_z$. Hence there is some $z \in R$ such that $(x,z) \in \operatorname{dom}\psi$. Let z be the $\mathcal{R}$-least element of R such that $(x,z) \in \operatorname{dom}\psi$. If $x \in A_z$, then z is the $\mathcal{R}$-least such that $x \in A_z$; so $x \in A$ if and only if $\mathtt{parity}_{\mathcal{R}}(z) \neq \mathtt{parity}(\mathcal{R}) = 0$. So if x shows up first in A_z, then we let $\psi(x,z) = 1$ if and only if $\mathtt{parity}_{\mathcal{R}}(z) = 1$ if and only if $A(x) = 1$. If $x \in B_z$, then z is $\mathcal{R}$-least such that $x \in B_z$, and so $x \notin A$ if and only if $\mathtt{parity}_{\mathcal{R}}(z) = 1$; so if x shows up first in B_z, then we let $\psi(x,z) = 0$ if and only if $\mathtt{parity}_{\mathcal{R}}(z) = 1$ if and only if $A(x) = 0$. Overall, we see that for all x, $A(x) = \psi(x,z)$ for the $\mathcal{R}$-least z such that $(x,z) \in \operatorname{dom}\psi$. By Proposition 2.2, A is $\mathcal{R}$-c.a.

For the other direction, it is sufficient to show that every $\mathcal{R}$-c.a. set is in $\Sigma^{-1}_{\mathcal{R}}$; the result would follow from the fact that the complement of an $\mathcal{R}$-c.a. set is also $\mathcal{R}$-c.a. Let A be an $\mathcal{R}$-c.a. set; by Proposition 2.2, let ψ be a partial computable function such that for all x, $A(x) = \psi(x,z)$ for the $\mathcal{R}$-least z such that $(x,z) \in \operatorname{dom}\psi$. We assume now that $\mathcal{R}$ has no greatest element. In particular, $\mathcal{R}$ is even.

We define the sequence $\langle A_z \rangle_{z\in R}$ which will show that $A \in \Sigma^{-1}_{\mathcal{R}}$. Let $(x,z) \in \operatorname{dom}\psi$.

- if $\psi(x,z) = \mathtt{parity}_{\mathcal{R}}(z)$ then we let $x \in A_w$ for all $w \geqslant_{\mathcal{R}} z$.
- if $\psi(x,z) \neq \mathtt{parity}_{\mathcal{R}}(z)$ then we let $x \in A_w$ for all $w >_{\mathcal{R}} z$.

It is clear that if $z <_{\mathcal{R}} w$ then $A_z \subseteq A_w$. Let $x < \omega$. We know that there is some $z \in R$ such that $(x,z) \in \operatorname{dom}\psi$. Since $\mathcal{R}$ has no greatest element, no matter what the parity of z is, we enumerate x into some A_w; so $x \in \bigcup_w A_w$. Let w be the $\mathcal{R}$-least element of R such that $x \in A_w$. We want to show that $x \in A$ if and only if w is odd in $\mathcal{R}$, in other words, that $A(x) = \mathtt{parity}_{\mathcal{R}}(w)$.

Let z be the $\mathcal{R}$-least element of R such that $(x,z) \in \operatorname{dom}\psi$. Either $\psi(x,z) = \mathtt{parity}_{\mathcal{R}}(z)$, in which case $z = w$; or $\psi(x,z) \neq \mathtt{parity}_{\mathcal{R}}(z)$, in which case w is the $\mathcal{R}$-successor of z. In the first case,

$$A(x) = \psi(x,z) = \mathtt{parity}_{\mathcal{R}}(z) = \mathtt{parity}_{\mathcal{R}}(w)$$

as required. In the second case,

$$A(x) = \psi(x,z) = 1 - \mathtt{parity}_{\mathcal{R}}(z) = \mathtt{parity}_{\mathcal{R}}(w),$$

again as required. □

Ash and Knight [6] refer to the sets in $\Delta^{-1}_{\mathcal{R}}$ as "$\mathcal{R}$-computable." However, in common yet misleading terminology, many authors refer to $\mathcal{R}$-c.a. sets as "$\mathcal{R}$-c.e." We prefer to be careful and not confuse the two notions.[1]

2.1.2 Listing $\mathcal{R}$-c.a. functions

For any computable well-ordering $\mathcal{R}$, we can effectively list all $\mathcal{R}$-c.a. functions. To do this we need to consider a nice class of $(\mathcal{R}+1)$-computable approximations. We of course let $\mathcal{R}+1$ denote a computable well-ordering extending $\mathcal{R}$ by one element at the end.

Definition 2.5. Let $\mathcal{R}$ be a computable well-ordering. An $(\mathcal{R}+1)$-computable approximation $\langle f_s, o_s\rangle$ is *tidy* if:

- for all n, $f_0(n) = 0$; and
- for all n and s, if $o_s(n+1) \in R$ then $o_s(n) \in R$.

The idea is that we have a "partial" $\mathcal{R}$-computable approximation, in that $\langle f_s\rangle$ is total but we may wait a while to declare the elements of R that we use; while we wait we let $o_s(n)$ be the new element beyond $\mathcal{R}$. And further, at every stage we will have declared our "true ordinals" (elements of R) for an initial segment of inputs.

Lemma 2.6. *If f has a tidy $(\mathcal{R}+1)$-computable approximation then f is $\mathcal{R}$-c.a.*

Proof. Let $\langle g_s, m_s\rangle_{s<\omega}$ be a tidy $(\mathcal{R}+1)$-computable approximation of f. There are two cases. In the first, for all x there is some s such that $m_s(x) \in R$. We say that the approximation is *eventually* $\mathcal{R}$-computable. We then modify the approximation $\langle g_s, m_s\rangle$ by waiting until we see this happen. Formally, for each x we let $t(x)$ be the least t such that $m_t(x) \in R$; we then let, for all x and s,

[1] On a fundamental level, we believe that "c.e." denotes the Σ-side of a hierarchy, in this case $\Sigma^{-1}_{\mathcal{R}}$, rather than the "ambiguous" class $\Delta^{-1}_{\mathcal{R}}$. Note that for $n < \omega$, the standard terminology "n-c.e." is correct, and indeed refers to the class Σ^{-1}_n rather than the class Δ^{-1}_n. It is therefore regrettable that many (but not all!) authors use "ω-c.e." to denote Δ^{-1}_ω rather than Σ^{-1}_ω.

$o_s(x) = m_{\max\{s,t(x)\}}(x)$ and $f_s(x) = g_{\max\{s,t(x)\}}(x)$; $\langle f_s, o_s\rangle$ is an $\mathcal{R}$-computable approximation of f.

In the second case, for all but finitely many x, $o_s(x)$ is constant and equals the extra element of $\mathcal{R}+1$. In that case $f(x)=0$ for all such x, so f is computable. □

It is clear from the proof of Lemma 2.6 that passing from a tidy $(\mathcal{R}+1)$-computable approximation for a function f to an $\mathcal{R}$-computable approximation for f cannot be done uniformly. Indeed a diagonalisation argument shows that there cannot be an effective list of $\mathcal{R}$-computable approximations listing all $\mathcal{R}$-c.a. functions. However we can make a list of tidy $(\mathcal{R}+1)$-computable approximations that yields all $\mathcal{R}$-c.a. functions.

Proposition 2.7. *There is a computable list $\langle\langle f_s^e, o_s^e\rangle_{s<\omega}\rangle_{e<\omega}$ of tidy $(\mathcal{R}+1)$-computable approximations such that letting $f^e = \lim_s f_s^e$, the sequence $\langle f^e\rangle_{e<\omega}$ lists the $\mathcal{R}$-c.a. functions.*

Proof. There is an effective list of all pairs $\langle h_s, m_s\rangle$ of uniformly computable sequences of partial functions. We show how to convert any such pair, uniformly, to a tidy $(\mathcal{R}+1)$-computable approximation $\langle f_s, o_s\rangle$, such that if $\langle h_s, m_s\rangle$ is an $\mathcal{R}$-computable approximation, then $\lim h_s = \lim f_s$.

Fix such $\langle h_s\rangle$ and $\langle m_s\rangle$. The idea is to define $\langle f_s\rangle$ by copying $\langle h_s\rangle$ with delays, until we see evidence that a change is allowed. Let ∞ denote the extra element of $\mathcal{R}+1$. Let $x<\omega$. We start with $f_0(x)=0$ and $o_0(x)=\infty$. Let $s>0$. To define $f_s(x)$ and $o_s(x)$, we enumerate the graphs of $\langle h_s\rangle$ and $\langle m_s\rangle$ for s many steps. We let $t_s(x)$ be the greatest $t \leqslant s$ such that for all $r \leqslant t$ and all $y \leqslant x$,

- at stage s we see that $h_r(y)\downarrow$ and $m_r(y)\downarrow$;
- $m_r(y) \in R$, and if $r>0$, $m_r(y) \leqslant_{\mathcal{R}} m_{r-1}(y)$;
- if $r>0$ and $h_r(y) \neq h_{r-1}(y)$ then $m_r(y) <_{\mathcal{R}} m_{r-1}(y)$.

If there is no such t, then we leave $t_s(x)$ undefined. If $t_s(x)$ is defined then we let $f_s(x) = h_{t_s(x)}(x)$ and $o_s(x) = m_{t_s(x)}(x)$. If $t_s(x)$ is not defined then we let $f_s(x)=0$ and $o_s(x)=\infty$. □

Note that restricting our approximations to sets, we also get a listing of all $\mathcal{R}$-c.a. sets.

Corollary 2.8. *The collection of $\mathcal{R}$-c.a. functions is uniformly computable from $\mathbf{0}'$. That is, there is a uniformly $\mathbf{0}'$-computable sequence $\langle f^e\rangle_{e<\omega}$ of all $\mathcal{R}$-c.a. functions.*

Remark 2.9. The reader may wonder why, in the case that $\operatorname{otp}(\mathcal{R})$ is a successor ordinal, we cannot list all $\mathcal{R}$-c.a. functions, each with an $\mathcal{R}$-computable approximation. After all, now we do not need to guess which ordinal to start with, we always start with $\max \mathcal{R}$. However we still need to guess what the initial value of our approximation is; we allowed f_0 to be any computable function. If we require that f_0 is the constant function 0 then we know the initial value but when attempting to diagonalise are restricted to keep our initial value 0 as well, and so may never be allowed to diagonalise.

2.1.3 Effective embeddings and isomorphisms

Proposition 2.10. *Let $\mathcal{R}$ and $\mathcal{S}$ be computable well-orderings. If there is a computable embedding of $\mathcal{R}$ into $\mathcal{S}$, then every $\mathcal{R}$-c.a. function is $\mathcal{S}$-c.a.*

Proof. Let $j\colon R \to S$ be an embedding of $\mathcal{R}$ into $\mathcal{S}$. Let $\langle f_s, o_s \rangle$ be an $\mathcal{R}$-computable approximation. Then $\langle f_s, j \circ o_s \rangle$ is an $\mathcal{S}$-computable approximation. □

Corollary 2.11. *Let $\mathcal{R}$ and $\mathcal{S}$ be computable well-orderings. If there is a computable isomorphism between $\mathcal{R}$ and $\mathcal{S}$, then a function is $\mathcal{R}$-c.a. if and only if it is $\mathcal{S}$-c.a.*

2.1.4 Bounds on mind-change functions

Let $\langle f_s \rangle_{s<\omega}$ be a computable approximation of a function f. The associated *mind-change function* is

$$m_{\langle f_s \rangle}(x) = \# \{s : f_{s+1}(x) \neq f_s(x)\}.$$

For any function $g\colon \omega \to \omega$, we say that the approximation $\langle f_s \rangle$ is a *g-bounded* approximation if for all x, $m_{\langle f_s \rangle}(x) \leqslant g(x)$, that is, if g majorises $m_{\langle f_s \rangle}$.

Recall that if $\mathcal{A} = (A, <_{\mathcal{A}})$ and $\mathcal{B} = (B, <_{\mathcal{B}})$ are linear orderings, then the product linear ordering $\mathcal{A} \cdot \mathcal{B}$ is the right-lexicographic ordering on $A \times B$. Its order-type is obtained by replacing every point in $\mathcal{B}$ by a copy of $\mathcal{A}$.

Proposition 2.12. *Let $\mathcal{R}$ be a computable well-ordering. A function is $\omega \cdot \mathcal{R}$-c.a. if and only if it has a computable approximation which is g-bounded for some $\mathcal{R}$-c.a. function g.*

Proof. Let $\langle f_s \rangle$ be a computable approximation of a function f.

Suppose that $\langle f_s, o_s \rangle$ is an $\omega \cdot \mathcal{R}$-computable approximation. For any x and s, let $o_s(x) = (n_s(x), l_s(x)) \in \omega \times R$. For any x and s, we let $t_s(x)$ be the least stage $t \leqslant s$ such that $l_s(x) = l_t(x)$. We then let

$$g_s(x) = n_{t_s(x)}(x) + \# \{r < t_s(x) : f_{r+1}(x) \neq f_r(x)\}.$$

Then $\langle g_s, l_s \rangle$ is an $\mathcal{R}$-computable approximation of a bound on $m_{\langle f_s \rangle}$.

Suppose that we are given an $\mathcal{R}$-computable approximation $\langle g_s, l_s\rangle$ for a bound g on $m_{\langle f_s\rangle}$. We may assume that for all x and s,

$$g_s(x) \geqslant \#\{r<s : f_{r+1}(x) \neq f_r(x)\},$$

since otherwise we can just wait until $g_t(x)$ changes at some $t>s$. We can therefore let

$$n_s(x) = g_s(x) - \#\{r<s : f_{r+1}(x) \neq f_r(x)\},$$

and $o_s(x) = (n_s(x), l_s(x))$. If $t<s$ and $l_s(x)=l_t(x)$ then $g_s(x)=g_t(x)$ which shows that if $f_{s+1}(x) \neq f_s(x)$ then $o_{s+1}(x) <_{\omega\cdot\mathcal{R}} o_s(x)$, so $\langle f_s, o_s\rangle$ is an $\omega\cdot\mathcal{R}$-computable approximation. □

Since the computable functions are characterised as those functions which are $\mathcal{R}$-c.a. for $\mathcal{R}$ of order-type 1, and since for any such $\mathcal{R}$, $\omega\cdot\mathcal{R}$ is computably isomorphic to ω, we see that Proposition 2.12 generalises the well-known fact that a function is ω-c.a. if and only if it has a computable approximation whose mind-change function is bounded by a computable function.

2.2 CANONICAL WELL-ORDERINGS AND STRONG NOTATIONS

Ershov proved the following:

Theorem 2.13. *Every Δ^0_2 function is $\mathcal{R}$-c.a. for some computable well-ordering $\mathcal{R}$ of order-type ω.*

Proof. Let f be a Δ^0_2 function. By Shoenfield's limit lemma, let $\langle f_s\rangle$ be a computable approximation for f. Let

$$R = \{(x,s) \in \omega\times\omega : s=0 \text{ or } f_s(x) \neq f_{s-1}(x)\}.$$

For (x,s) and $(y,t) \in R$, let $(x,s) <_{\mathcal{R}} (y,t)$ if $x<y$ or if $x=y$ and $s>t$. For any $x<\omega$ let R_x be the collection of pairs (x,s) in R; so R is the disjoint union of the R_x's, each R_x is finite (as $\langle f_s(x)\rangle$ reaches a limit), and the ordering $\mathcal{R} = (R, <_{\mathcal{R}})$ orders $R_0 < R_1 < R_2 < \cdots$; so $\operatorname{otp}(\mathcal{R}) = \omega$.

For $x, s<\omega$, let $t(x,s)$ be the least $t \leqslant s$ such that for all $u \in [t,s]$, $f_u(x) = f_s(x)$. For all x and s, $(x, t(x,s)) \in R_x$, and so we can let $o_s(x) = (x, t(x,s))$. It is clear that $\langle f_s, o_s\rangle$ is an $\mathcal{R}$-computable approximation for f. □

Ershov's theorem is displeasing as we try to define a hierarchy of complexity inside the Δ^0_2 functions. Its meaning is that calibrating the complexity of a function f by the length of a computable well-ordering $\mathcal{R}$ such that f is $\mathcal{R}$-c.a. is not very informative: the hierarchy collapses at level ω. The reason for this collapse is not that all Δ^0_2 functions have simple approximations, but that the complexity of these approximations can be coded into the isomorphism between $\mathcal{R}$ and ω. In other words, if $\mathcal{R}$ is complicated then $\mathcal{R}$-c.a. functions may be complicated as well, even if $\mathcal{R}$ is short. In terms of the algebraic complexity of $\mathcal{R}$ itself, we notice that key functions associated with $\mathcal{R}$, such as the predecessor and successor function, may be far from computable.

One possible solution is to restrict the computable well-orderings to those given by notations on some Π^1_1 path through Kleene's $\mathcal{O}$. This is less than satisfying on two accounts. The first is that even though the path may be cofinal in $\mathcal{O}$ (so have notations for every computable ordinal), this does not exhaust all Δ^0_2 functions [41]. The other is that there is no canonical way to choose a path through Kleene's $\mathcal{O}$, and so any such choice is arbitrary, and different choices give different hierarchies of functions.

Another way forward is to give up any claim to exhausting all Δ^0_2 functions, but restrict our attention to a particularly well-behaved class of computable well-orderings. We will require that all orderings in the class that have the same length are computably isomorphic, so Corollary 2.11 will ensure that we will have a good notion of α-c.a. functions for some class of computable ordinals α. The criterion for canonicity of these orderings is the computability of all reasonable associated functions, such as the predecessor, successor, and so on. It turns out that up to ε_0, the function which encapsulates all the required information is Cantor's normal form.

2.2.1 Cantor normal form

Recall that every ordinal α has a unique expression as the sum

$$\omega^{\alpha_1} n_1 + \omega^{\alpha_2} n_2 + \cdots + \omega^{\alpha_k} n_k$$

where $n_i < \omega$ are nonzero and $\alpha_1 > \alpha_2 > \cdots > \alpha_k$ are ordinals. Recall also that

$$\varepsilon_0 = \sup\left\{\omega, \omega^\omega, \omega^{\omega^\omega}, \omega^{\omega^{\omega^\omega}}, \ldots\right\}$$

is the least ordinal γ such that $\omega^\gamma = \gamma$, so for all $\alpha < \varepsilon_0$, every ordinal appearing as an exponent in the Cantor normal form of α is strictly smaller than α.

Let $\mathcal{R} = (R, <_\mathcal{R})$ be a computable well-ordering of order-type $\leqslant \varepsilon_0$, and let $|\cdot|\colon R \to \mathrm{otp}(\mathcal{R})$ be the unique isomorphism between $\mathcal{R}$ and its order-type. The pullback to $\mathcal{R}$ of the Cantor normal form function is the function $\mathtt{nf}_\mathcal{R}$ whose domain is R and is defined by letting

$$\mathtt{nf}_\mathcal{R}(z) = \langle (z_1, n_1), (z_2, n_2), \ldots, (z_k, n_k) \rangle$$

where $n_i < \omega$ are nonzero, $z_i \in R$, $z_1 >_{\mathcal{R}} z_2 >_{\mathcal{R}} \cdots >_{\mathcal{R}} z_k$, and

$$|z| = \omega^{|z_1|} n_1 + \omega^{|z_2|} n_2 + \cdots + \omega^{|z_k|} n_k.$$

Definition 2.14. A computable well-ordering $\mathcal{R}$ is *canonical* if its associated Cantor normal form function $\mathtt{nf}_{\mathcal{R}}$ is also computable.

Remark 2.15. Suppose that $\operatorname{otp}(\mathcal{R}) \leqslant \varepsilon_0$. Then $\mathcal{R}$ is canonical if and only if the relations

$$\left\{(x, y) \in R^2 : |x| = \omega^{|y|}\right\}$$

and

$$\{(x, y, z) \in R^3 : |x| = |y| + |z|\}$$

are computable. That is, if the possibly partial operations of ordinal addition and exponentiation with base ω are partial computable. In the second direction, it is clear that if addition and exponentiation with base ω are partial computable, then identifying $\mathtt{nf}_{\mathcal{R}}(x)$ is computable. In the first direction, $|x| = \omega^{|y|}$ if and only if $\mathtt{nf}_{\mathcal{R}}(x) = \langle (y, 1) \rangle$. For addition, we note that we can compute addition from the Cantor normal form, using the fact that ordinal addition is associative, and the fact that if $\beta < \gamma$ then for all m, $\omega^\beta m + \omega^\gamma = \omega^\gamma$.

Proposition 2.16. *Let $\mathcal{R}$ and $\mathcal{S}$ be canonical computable well-orderings, with $\operatorname{otp}(\mathcal{R}) \leqslant \operatorname{otp}(\mathcal{S}) \leqslant \varepsilon_0$. Then the unique embedding of $\mathcal{R}$ as an initial segment of $\mathcal{S}$ is computable.*

Proof. For every ordinal $\alpha \leqslant \varepsilon_0$ let $J(\alpha)$ be the $\subseteq$-least set of ordinals J such that $\alpha \in J$ and for every $\beta \in J$, every exponent appearing in β's Cantor normal form is also in J. The set $J(\alpha)$ is finite. To see this, consider the tree T_α of finite sequences of ordinals defined recursively as follows. We declare that $\langle \alpha \rangle \in T_\alpha$. Then, if $\sigma\hat{\ }\beta \in T_\alpha$, $\beta > 0$ and $\beta = \omega^{\gamma_1} n_1 + \ldots \omega^{\gamma_k} n_k$ is β's Cantor normal form, then for all $i \leqslant k$, $\sigma\hat{\ }\beta\hat{\ }\gamma_i \in T_\alpha$. The tree is finitely branching and each sequence on the tree is a decreasing sequence of ordinals, and so T_α does not have a path. By König's lemma, T_α is finite; and $J(\alpha)$ is the set of ordinals appearing in a sequence in T_α. For a computable well-ordering $\mathcal{R}$ with $\operatorname{otp}(\mathcal{R}) \leqslant \varepsilon_0$, for each $z \in R$, let $J_{\mathcal{R}}(z) = \{w \in R : |w| \in J(|z|)\}$ be the pullback of $J(|z|)$ to elements of $\mathcal{R}$. If $\mathcal{R}$ is canonical then the function $J_{\mathcal{R}}$ is computable. This means that from z we effectively obtain a "strong index" for $J_{\mathcal{R}}(z)$.

Let $j\colon R \to S$ be the embedding of $\mathcal{R}$ into $\mathcal{S}$ as an initial segment. Then $j(z) = w$ if and only if there is a bijection $i\colon J_{\mathcal{R}}(z) \to J_{\mathcal{S}}(w)$ such that $i(z) = w$ and which preserves Cantor normal form: for all $x \in J_{\mathcal{R}}(z)$, if

$$\mathtt{nf}_{\mathcal{R}}(x) = \langle (y_1, n_1), (y_2, n_2), \ldots, (y_k, n_k) \rangle$$

(where necessarily $y_1, y_2, \ldots, y_k \in J_{\mathcal{R}}(z)$) then

$$\mathtt{nf}_{\mathcal{S}}(i(x)) = \langle (i(y_1), n_1), (i(y_2), n_2), \ldots, (i(y_k), n_k) \rangle.$$

This shows that j is computable. □

Beyond ε_0, we need to strengthen canonicity to obtain an extension of Proposition 2.16. We do not develop this further here, as ε_0 is well beyond the ordinals that come up in the constructions we examine.

2.2.2 Existence of canonical well-orderings

For a computable well-ordering $\mathcal{R}$, the computable well-ordering $\omega^{\mathcal{R}}$, whose order-type is $\omega^{\mathrm{otp}(\mathcal{R})}$, is defined using Cantor normal form. The field of $\omega^{\mathcal{R}}$ is the collection of all sequences of pairs $\langle (z_1, n_1), (z_2, n_2), \ldots, (z_k, n_k) \rangle$ from $R \times (\omega \setminus \{0\})$ such that $z_1 >_{\mathcal{R}} z_2 >_{\mathcal{R}} \cdots >_{\mathcal{R}} z_k$. We let

$$\langle (z_1, n_1), (z_2, n_2), \ldots, (z_k, n_k) \rangle <_{\omega^{\mathcal{R}}} \langle (w_1, m_1), (w_2, m_2), \ldots, (w_l, m_l) \rangle$$

if $k < l$ and for all $i \leqslant k$, $(z_i, n_i) = (w_i, m_i)$; or if for the least $i \leqslant k$ such that $(z_i, n_i) \neq (w_i, m_i)$ we have $w_i <_{\mathcal{R}} z_i$ or $w_i = z_i$ and $n_i < m_i$ (that is, if $(n_i, z_i) <_{\omega \cdot \mathcal{R}} (m_i, w_i)$).

Lemma 2.17. *Let $\mathcal{R}$ be a canonical computable well-ordering. Then the embedding of $\mathcal{R}$ into $\omega^{\mathcal{R}}$ as an initial segment is computable.*

Proof. In fact, this embedding is exactly $\mathtt{nf}_{\mathcal{R}}$. □

Lemma 2.18. *If $\mathcal{R}$ is a canonical computable well-ordering, then so is $\omega^{\mathcal{R}}$.*

Indeed, a computable index for $\mathtt{nf}_{\omega^{\mathcal{R}}}$ can be effectively obtained from a computable index for $\mathtt{nf}_{\mathcal{R}}$.

Proof. Let $j = \mathtt{nf}_{\mathcal{R}}$ be the canonical embedding of $\mathcal{R}$ into $\omega^{\mathcal{R}}$. For any $\langle (z_1, n_1), (z_2, n_2), \ldots, (z_k, n_k) \rangle$ in the field of $\omega^{\mathcal{R}}$, we have

$$\mathtt{nf}_{\omega^{\mathcal{R}}} (\langle (z_1, n_1), (z_2, n_2), \ldots, (z_k, n_k) \rangle) = \langle (j(z_1), n_1), (j(z_2), n_2), \ldots, (j(z_k), n_k) \rangle.$$

□

Lemma 2.19. *Let $\langle \mathcal{R}_n \rangle$ be a sequence of uniformly computable, uniformly canonical well-orderings (that is, the functions $\mathtt{nf}_{\mathcal{R}_n}$ are uniformly computable). Suppose that for all n, $\mathrm{otp}(\mathcal{R}_n) \leqslant \mathrm{otp}(\mathcal{R}_{n+1})$; let $i_n \colon R_n \to R_{n+1}$ be the embedding of $\mathcal{R}_n$ into $\mathcal{R}_{n+1}$ as an initial segment, and suppose that the sequence $\langle i_n \rangle$ is uniformly computable.*

Then the direct limit of the system $\langle \mathcal{R}_n, i_n \rangle_{n<\omega}$ has a canonical copy.

Proof. For $m \leqslant n$, let $i^n_m = i_{n-1} \circ i_{n-2} \circ \cdots \circ i_m$ be the initial segment embedding of $\mathcal{R}_m$ into $\mathcal{R}_n$ (and $i^n_n = \mathrm{id}_{R_n}$).

Let

$$\Gamma = \bigcup_n R_n \times \{n\}.$$

For $(w,m),(z,n) \in \Gamma$ where $m \leqslant n$, we let $(w,m) \sim (z,n)$ if $i^n_m(w) = z$. Then $\sim$ is an equivalence relation on Γ, and the universe of the direct limit of $\langle \mathcal{R}_n, i_n \rangle$ is $\Gamma/\sim$, the collection of $\sim$-equivalence classes. To get a computable copy, we pick out representatives to be the ones that appear earliest in an effective enumeration $\langle \Gamma_s \rangle$ of Γ, using the fact that $\langle \sim\restriction_{\Gamma_s} \rangle$ is uniformly computable. We let R be this computable set of representatives. The ordering $<_{\mathcal{R}}$ is defined by letting, for $(w,m),(z,n) \in R$ such that $m \leqslant n$, $(w,m) <_{\mathcal{R}} (z,n)$ if $i^n_m(w) <_{\mathcal{R}_n} z$. Certainly $\mathcal{R} = (R, <_{\mathcal{R}})$ is computable, and isomorphic to the direct limit of the system $\langle \mathcal{R}_n, i_n \rangle$. Note also that the representation function $c\colon \Gamma \to R$ defined by requiring that $c(z,n) \sim (z,n)$ is also computable.

Let $(z,n) \in R$, and let $\mathtt{nf}_{\mathcal{R}_n}(z) = \langle (z_1,m_1), \ldots, (z_k,m_k) \rangle$. Then

$$\mathtt{nf}_{\mathcal{R}}(z,n) = \langle (c(z_1,n),m_1), (c(z_2,n),m_2), \ldots, (c(z_k,n),m_k) \rangle$$

and so $\mathtt{nf}_{\mathcal{R}}$ is computable. □

Corollary 2.20. *There is a canonical computable well-ordering of order-type* ε_0.

Proof. Let $\mathcal{R}_0 = (\omega, <)$ and $\mathcal{R}_{n+1} = \omega^{\mathcal{R}_n}$, and apply Lemmas 2.17, 2.18, and 2.19. □

If $\mathcal{R}$ is a canonical computable well-ordering, then for all $z \in R$, the restriction of $\mathcal{R}$ to the initial segment of $\mathcal{R}$ defined by z is also a canonical computable well-ordering. Hence the collection of ordinals α for which there is a canonical computable well-ordering of length α forms an initial segment of the ordinals. Corollary 2.20 implies the following:

Proposition 2.21. *For every* $\alpha \leqslant \varepsilon_0$, *there is a canonical computable well-ordering of order-type* α.

In view of Propositions 2.16 and 2.21, we identify ordinals $\alpha \leqslant \varepsilon_0$ with canonical well-orderings of order-type α.

Definition 2.22. Let $\alpha \leqslant \varepsilon_0$. A function f is *α-c.a.* if it is $\mathcal{R}$-c.a. for some (all) canonical well-ordering $\mathcal{R}$ of order-type α.

This notion is well-defined by Corollary 2.11 and Propositions 2.16 and 2.21. By Propositions 2.10 and 2.16, if $\alpha < \beta \leqslant \varepsilon_0$, every α-c.a. function is β-c.a.

We go further and fix a canonical well-ordering $\mathcal{R}_{\varepsilon_0}$ of order-type ε_0. We identify $\alpha < \varepsilon_0$ with the element $z \in R_{\varepsilon_0}$ such that $|z|_{\mathcal{R}_{\varepsilon_0}} = \alpha$. As from z we can effectively obtain the initial segment $\mathcal{R}_{\varepsilon_0}\restriction_z$ of $\mathcal{R}_{\varepsilon_0}$ determined by z, we say that effectively from $\alpha < \varepsilon_0$ we can get a canonical well-ordering $\mathcal{R}_\alpha$ of order-type α. The identification of α with both $\mathcal{R}_\alpha$ and with $\mathcal{R}_\alpha$'s least upper bound in $\mathcal{R}_{\varepsilon_0}$ is true to von Neumann's definition of ordinals: an ordinal here is identified with the collection of its predecessors.

Note that the listing of tidy $(\mathcal{R}+1)$-computable approximations provided by the proof of Proposition 2.7 is uniform in an index for $\mathcal{R}$. Hence, uniformly in $\alpha < \varepsilon_0$, we can fix an effective list $\langle f^{e,\alpha}_s, o^{e,\alpha}_s \rangle$ of tidy $(\alpha+1)$-computable approximations, where, letting $f^{e,\alpha} = \lim_s f^{e,\alpha}_s$, the sequence $\langle f^{e,\alpha} \rangle_{e<\omega}$ is a listing of all α-c.a. functions.

Proposition 2.12 allows us to define some levels of the hierarchy of α-c.a. functions:

Proposition 2.23.

(1) Let $n < \omega$. A function is ω^{n+1}-c.a. if and only if it has a computable approximation which is bounded by an ω^n-c.a. function.

(2) Let $\alpha \geqslant \omega$, $\alpha \leqslant \varepsilon_0$. A function is ω^α-c.a. if and only if it has a computable approximation which is bounded by an ω^α-c.a. function.

2.2.3 On ordinal notations

One of the main uses of Kleene's system of ordinal notations [54] is to define effective transfinite iterations of the Turing jump, giving rise to the hyperarithmetic hierarchy. Roughly speaking, a notation for an ordinal corresponds to a computable well-ordering on which the successor function is computable, and which associates with every limit element a computable cofinal sequence. We briefly recall the definition. The set $\mathcal{O}$ and the partial ordering $<_\mathcal{O}$ are defined by (very much non-effective) recursion. We start with $1 \in \mathcal{O}$. If $a \in \mathcal{O}$ then $2^a \in \mathcal{O}$ and $w <_\mathcal{O} 2^a \leftrightarrow w \leqslant_\mathcal{O} a$. If φ_e is total and for all n, $\varphi_e(n) <_\mathcal{O} \varphi_e(n+1)$ then $3 \cdot 5^e \in \mathcal{O}$ and $w <_\mathcal{O} 3 \cdot 5^e \leftrightarrow (\exists n)\, w <_\mathcal{O} \varphi_e(n)$.[2] The relation $<_\mathcal{O}$ is in fact a transfinite tree, and each $a \in \mathcal{O}$ is considered a *notation* for the order-type of the well-ordered initial segment $I(a) = \{b \in \mathcal{O} : b <_\mathcal{O} a\}$. The ordinal $|a|_\mathcal{O}$ is computable, and every computable ordinal has a notation. Note that $|2^a|_\mathcal{O} = |a|_\mathcal{O} + 1$ and $|3 \cdot 5^e|_\mathcal{O} = \sup_n |\varphi_e(n)|_\mathcal{O}$.

For a notation $o \in \mathcal{O}$, the set of predecessors $I(o)$ of o according to $<_\mathcal{O}$ is c.e., uniformly in o, but not necessarily computable; and the restriction of $<_\mathcal{O}$ to $I(o)$ is also c.e., again uniformly in o. The uniformity allows us to pull back $<_\mathcal{O}\restriction_{I(o)}$ by an effective enumeration of $I(o)$ to give a computable well-ordering $\mathcal{R}_o$ (with computable domain) isomorphic to $<_\mathcal{O}\restriction_{I(o)}$.

---

[2]Why $3 \cdot 5^e$ and not 3^e? Because if $e = 0$ then $3^e = 1$, and 1 was already used.

Spector's theorem [93] is in some sense a version of Proposition 2.16: if $a,b\in\mathcal{O}$ and $|a|_{\mathcal{O}}\leqslant|b|_{\mathcal{O}}$ then H_a, the iteration of the jump along $\mathcal{R}_a$, is Turing reducible to H_b, the iteration of the jump along $\mathcal{R}_b$. This suffices to give a precise definition of an increasing sequence of degrees $\mathbf{0}^{(\alpha)}$ for all computable ordinals α.

For the purposes of defining α-c.a. functions and, later, totally α-c.a. degrees, general notations are not sufficient, as the well-orderings $\mathcal{R}_o$ are not necessarily canonical. For example, Ershov [40], and, later, Epstein, Haas, and Kramer [37], define a function to be α-c.a. if it is $\mathcal{R}_o$-c.a. for some notation $o\in\mathcal{O}$ for α. Under this definition, every Δ^0_2 function is ω^2-c.a., and as we shall see below, every Δ^0_2, low$_2$ degree is totally ω^2-c.a. For the small ordinals we are interested in, there is a natural choice for a system of notations: we say that a notation $o\in\mathcal{O}$ is a *strong notation* if $\mathcal{R}_o$ is canonical. This method was also chosen by Coles, Downey, and LaForte [15] in unpublished work looking at hierarchies based on truth-table reductions below $\mathbf{0}'$, and by Diamondstone, Hirschfeldt, and Nies (unpublished) for variations on Demuth randomness. Note that every notation for an ordinal below ω^2 is strong, but as we shall see, there are notations for ω^2 which are not strong.

Let us say that a computable well-ordering $\mathcal{R}$ of successor order-type is *notation-like* if:

- the successor function on $\mathcal{R}$ is computable; and
- the collection $L(\mathcal{R})$ of limit points of $\mathcal{R}$ is computable.

Lemma 2.24. *Let $\mathcal{R}$ be notation-like. Then there is an effective map giving, for every $z\in L(\mathcal{R})$, an index for a computable $<_{\mathcal{R}}$-increasing sequence (of order-type ω) cofinal in $\mathcal{R}\restriction_z$.*

Proof. For each n consider in turn the $\mathcal{R}$-greatest element of $R\restriction_z\cap\{0,\dots,n\}$. □

The reason that we only consider successor order-types is that if otp$(\mathcal{R})$ is a limit then we would need to add the requirement that there is a computable increasing sequence cofinal in $\mathcal{R}$.

Lemma 2.25. *A computable well-ordering of successor order-type is computably isomorphic to $\mathcal{R}_o$ for some $o\in\mathcal{O}$ if and only if $\mathcal{R}$ is notation-like.*

Proof. Of course, $\mathcal{R}$ is computably isomorphic to $\mathcal{R}_o$ if and only if the order-preserving bijection between $\mathcal{R}$ and $(I(o),<_{\mathcal{O}}\restriction_{I(o)})$ is computable.

If $j\colon R\to I(o)$ is order-preserving, then for all $z\in R$ except for the top element of $\mathcal{R}$, the successor of z in $\mathcal{R}$ is w where $j(w)=2^{j(z)}$; the collection $L(\mathcal{R})$ of limit points of $\mathcal{R}$ is the collection of $z\in R$ such that $j(z)=3\cdot 5^e$ for some e. This shows that if $\mathcal{R}$ is isomorphic to $\mathcal{R}_o$ for some $o\in\mathcal{O}$ then $\mathcal{R}$ is notation-like.

Suppose now that $\mathcal{R}$ is notation-like. By Lemma 2.24, let f be a computable function such that for $z \in L(\mathcal{R})$, $\varphi_{f(z)}$ is an $<_{\mathcal{R}}$-increasing and cofinal sequence in $\mathcal{R}\restriction_z$.

By effective transfinite recursion (as in [83]) we define a computable injection $j\colon R \to \mathcal{O}$ by letting:

(1) $j(z) = 1$, where z is the $\mathcal{R}$-least element of R;
(2) if z is the successor of w in $\mathcal{R}$, then we let $j(z) = 2^{j(w)}$;
(3) if $z \in L(\mathcal{R})$ then $j(z) = 3 \cdot 5^e$ where $\varphi_e = j \circ \varphi_{f(z)}$.

Specifically, we define a partial computable function $F\colon \omega \times R \to \omega$ as follows:

- If z is the $\mathcal{R}$-least element of R, then for all e we let $F(e, z) = 1$.
- If z is the $\mathcal{R}$-successor of w, then we let $F(e, z) = 2^{\varphi_e(w)}$.
- Let g be a computable function such that for all a and b, $\varphi_{g(a,b)} = \varphi_a \circ \varphi_b$. If $z \in L(\mathcal{R})$, then we let $F(e, z) = 3 \cdot 5^{g(e, f(z))}$.

By the recursion theorem, there is an index e such that for all $z \in R$, $F(e, z) = \varphi_e(z)$. Then $j = \varphi_e\restriction_R$ satisfies the conditions (1)–(3) above. The main point is that $R \subseteq \operatorname{dom} \varphi_e$: otherwise, since $\mathcal{R}$ is well-founded, there is an $\mathcal{R}$-least $z \in R$ for which $\varphi_e(z)\uparrow$, which, by definition of F, must be an $\mathcal{R}$-successor element of some $w \in R$; but then $\varphi_e(w)\downarrow$ implies that $F(e, z)\downarrow$ for a contradiction.

Now the fact that $R \subseteq \operatorname{dom} \varphi_e$ implies that for all $z \in L(\mathcal{R})$, $\varphi_e(z) = 3 \cdot 5^d$ where φ_d is indeed an increasing and cofinal sequence in $I(\varphi_e(z))$, so by transfinite induction on the elements of $\mathcal{R}$ we can show that j is an order-preserving bijection between $\mathcal{R}$ and $I(o)$, where $o = 2^{j(z)}$ for z being the $\mathcal{R}$-maximal element of R. □

Lemma 2.26. *Every canonical well-ordering of successor order-type $\beta < \varepsilon_0$ is notation-like.*

Proof. Let $\alpha \leqslant \varepsilon_0$, and let $\alpha = \omega^{\alpha_1} n_1 + \cdots + \omega^{\alpha_k} n_k$ be the Cantor normal form of α. Then α is a limit ordinal if and only if $\alpha_k \neq 0$. If α is a limit, then the successor of α is the ordinal β whose Cantor normal form is $\omega^{\alpha_1} n_1 + \cdots + \omega^{\alpha_k} n_k + \omega^0 1$; otherwise, it is the ordinal β whose Cantor normal form is $\omega^{\alpha_1} n_1 + \cdots + \omega^{\alpha_k}(n_k + 1)$. □

Corollary 2.27 (Coles, Downey, LaForte). *For every $\alpha < \varepsilon_0$, there is a strong notation $o \in \mathcal{O}$ for α.*

Proof. Suppose that $\alpha < \varepsilon_0$ is a successor ordinal. By Theorem 2.21, let $\mathcal{R}$ be a canonical computable well-ordering of order-type α. By Lemmas 2.26 and 2.25, there is some $o \in \mathcal{O}$ with $\mathcal{R}_o$ computably isomorphic to $\mathcal{R}$. Then $\mathcal{R}_o$ is also canonical, whence o is a strong notation for α.

If α is a limit ordinal, let o be a strong notation for $\alpha + 1$; then $\log_2 o$ is a strong notation for α. □

We show that some notations are not strong.

Lemma 2.28. *Let $\mathcal{R}$ be a computable well-ordering of order-type ω. Then $\omega\cdot\mathcal{R}+1$ is notation-like.*

Proof. The successor of $(n,z)\in\omega\times R$ in $\omega\cdot\mathcal{R}$ is $(n+1,z)$. Let z_0 be the $\mathcal{R}$-least element of R. Then the collection of limit points of $\omega\cdot\mathcal{R}$ is $(\omega\setminus\{0\})\times\{z_0\}$. □

Let $\mathcal{R}$ be a computable well-ordering of order-type ω. Certainly $z\mapsto(0,z)$ is a computable embedding of $\mathcal{R}$ into $\omega\cdot\mathcal{R}$. By Proposition 2.10, every $\mathcal{R}$-c.a. function is $\omega\cdot\mathcal{R}$-c.a. By Lemmas 2.25 and 2.28, every $\mathcal{R}$-c.a. function is $\mathcal{R}_o$-c.a. for some notation $o\in\mathcal{O}$ for ω^2. Ershov's Theorem 2.13 now implies:

Corollary 2.29 (Ershov). *For every Δ^0_2 function f there is a notation $o\in\mathcal{O}$ for ω^2 such that f is $\mathcal{R}_o$-c.a.*

Most Δ^0_2 functions are not ω^2-c.a., and so there are many notations for ω^2 which are not strong.

2.3 WEAK TRUTH-TABLE JUMPS AND ω^α-C.A. SETS AND FUNCTIONS

Coles, Downey, and LaForte [15], and independently Anderson and Csima [3], examined the analogue of the Turing jump in the weak truth-table degrees. Anderson and Csima went on to tie levels of sets in the Ershov hierarchy to finite iterations of this bounded jump, generalising the well-known fact that a set is ω-c.a. if and only if it is weak truth-table reducible to $\varnothing'$. If $\varnothing^{\langle n\rangle}$ is the result of iterating the bounded jump operation n times, starting with $\varnothing$ (we give a precise definition below), then Anderson and Csima showed that a set $A\in 2^\omega$ is ω^n-c.a. if and only if it is weak truth-table reducible to $\varnothing^{\langle n\rangle}$.

Coles, Downey, and LaForte defined strong notations in order to define an analogue of the H-sets in the Δ^0_2 weak truth-table degrees, namely to find a way to define transfinite iterations of the bounded jump operator which are invariant in the weak truth-table degrees. We carry out their programme for ordinals below ε_0, and extend Anderson and Csima's result to all such ordinals (Theorem 2.40(1)). We further discuss what happens when we pass from sets to functions (Theorem 2.40(2)).

2.3.1 Bounded g-c.e. sets and the bounded jump

Recall that a function f is *weak truth-table* reducible to a function g if there is a Turing functional Φ such that $\Phi(g)=f$, and the use function of this reduction is bounded by a computable function. We can extend this to partial functions: for any function $g\colon\omega\to\omega$, we say that a partial function $\psi\colon\omega\to\omega$ is *bounded*

g-computable if there is a Turing functional Φ and a partial computable function φ such that for all x and y, $\psi(x)=y$ if and only if $\varphi(x)\downarrow$ and $\Phi(g\restriction_{\varphi(x)}, x)=y$; and $x\notin \operatorname{dom}\psi$ if $\varphi(x)\uparrow$ or if there is no such y. A total function f is bounded g-computable if and only if $f\leqslant_{\mathrm{wtt}} g$. Note that in this section, we abandon the convention that for a Turing functional Φ and an oracle X, $\operatorname{dom}\Phi(X)$ is an initial segment of ω.

A *weak truth-table functional* is a pair (Φ,φ) consisting of a Turing functional and a partial computable function. If (Φ,φ) is a weak truth-table functional, $x<\omega$ and $g\colon \omega\to\omega$, then we write $\hat{\Phi}(g,x)=y$ if $\varphi(x)\downarrow$ and $\Phi(g\restriction_{\varphi(x)}, x)=y$. We write $\hat{\Phi}(g,x)\downarrow$ if $\hat{\Phi}(g,x)=y$ for some y. The notation $\hat{\Phi}$ assumes that the partial function φ is clear from context.

We say that a set $A\in 2^\omega$ is *bounded g-c.e.* if it is the domain of a partial bounded g-computable function.

We can enumerate all partial bounded X-computable functions, and all bounded X-c.e. sets, by giving an effective enumeration $\langle \Phi_e,\varphi_e\rangle_{e<\omega}$ of all weak truth-table functionals. We fix such an enumeration which is moreover *acceptable*: if $\langle \Psi_e,\psi_e\rangle_{e<\omega}$ is any effective list of weak truth-table functionals, then there is an (injective) computable function g such that for all e, $(\Phi_{g(e)},\varphi_{g(e)})=(\Psi_e,\psi_e)$. For all $g\colon\omega\to\omega$, $\left\langle \hat{\Phi}_e(g)\right\rangle_{e<\omega}$ is a g-effective list of all partial bounded g-computable functions, and letting $\hat{W}^g_e=\operatorname{dom}\hat{\Phi}_e(g)$, $\left\langle \hat{W}^g_e\right\rangle_{e<\omega}$ is a list of all bounded g-c.e. sets.

Some of the basic properties of partial computable functions and c.e. sets do not carry over to the bounded realm. The following proposition is meant as a cautionary tale.

Proposition 2.30. *Let $g\colon\omega\to\omega$.*

(1) Every nonempty bounded g-c.e. set is the range of some function $f\leqslant_{\mathrm{wtt}} g$; but there is a function $f\leqslant_{\mathrm{wtt}}\varnothing'$ whose range is not bounded $\varnothing'$-c.e.
(2) The graph of any partial bounded g-computable function is bounded g-c.e.; but there is a (total) function f which is not bounded $\varnothing'$-computable, but whose graph is bounded $\varnothing'$-c.e.
(3) If $A\leqslant_{\mathrm{wtt}} g$, then A is bounded g-c.e. (and so is its complement). However, there is a c.e. set C and a set A such that both A and its complement $\omega\setminus A$ are bounded C-c.e., but $A\nleqslant_{\mathrm{wtt}} C$. For the set C we cannot *choose $\varnothing'$: if both A and $\omega\setminus A$ are bounded $\varnothing'$-c.e., then $A\leqslant_{\mathrm{wtt}}\varnothing'$.*

Note, however, that with a computable oracle the distinctions disappear: a partial function is bounded $\varnothing$-computable if and only if it is partial computable, and a set is c.e. if and only if it is bounded $\varnothing$-c.e.

Sketch of proof. For (1), we note that for any g, if A is g-c.e. and nonempty, then there is some $f\leqslant_{\mathrm{wtt}} g$ such that $A=\operatorname{range} f$. In fact, the use function for reducing f to g can grow as slowly as we like; we simply wait with enumerating some $x \in A$ into the range of f until the input of f is large enough for A to see that x is in A. Hence, every nonempty Σ^0_2 set is the range of some ω-c.a. function. On the other hand, below we see that every bounded $\varnothing'$-c.e. set is Δ^0_2 (in fact, the Anderson-Csima result implies that a set is bounded $\varnothing'$-c.e. if and only if it is ω^2-c.a.). The result follows from the fact that there are Σ^0_2 sets that are not Δ^0_2.

For (2), note that if f is a Δ^0_2 function which has an increasing approximation, that is, a computable approximation $\langle f_s \rangle$ such that for all x and s, $f_s(x) \leqslant f_{s+1}(x)$, then the graph of f is d.c.e., and so ω-c.a., and so weak truth-table reducible to $\varnothing'$, and so certainly bounded $\varnothing'$-c.e. For any $\alpha \leqslant \varepsilon_0$ it is easy to define an increasing approximation for a function f which is not α-c.a. by diagonalising against all partial α-computable approximations (Proposition 2.7), always *increasing* the value of f if we want to change it. If we choose $\alpha = \omega$, then we get a function which is not ω-c.a., and so not weak truth-table reducible to $\varnothing'$, and so, since it is total, not bounded $\varnothing'$-computable.

We sketch the proofs of (3). First, we enumerate a c.e. set C and define a set A such that both A and $\omega \setminus A$ are bounded C-c.e. For $e < \omega$, the requirement R_e seeks a witness x such that $A(x) \neq \hat{\Phi}_e(C, x)$ if the latter converges. After picking a new witness x, we state that $x \notin A$ with fresh C-use $\psi_{\texttt{no}}(x)$, and freeze $C\restriction_{\psi_{\texttt{no}}(x)}$. If later $\hat{\Phi}_e(C_s, x)\downarrow = 0$ (i.e., "$\texttt{no}$"), then we enumerate $\psi_{\texttt{no}}(x) - 1$ into C and declare that $x \in A$ with A use $\psi_{\texttt{yes}}(x) > \varphi_e(x), \psi_{\texttt{no}}(x)$. Of course this enumeration into C may free the opponent to change their mind and later still let $\hat{\Phi}_e(C_s, x) = 1$ (i.e., "$\texttt{yes}$"). In that case we enumerate $\psi_{\texttt{yes}}(x) - 1$ into C but freeze C below that number, and declare that $x \notin A$ with the old use $\psi_{\texttt{no}}(x)$. The point is that $\psi_{\texttt{yes}}(x) > \varphi_e(x)$, so our freezing C means that the opponent cannot change their mind again and is stuck with declaring that $x \in A$, whereas we leave $x \notin A$ forever after that. Each time a requirement acts, all weaker requirements are initialised and are forced to pick new witnesses; so this is a finite-injury construction.

The difference between C and $\varnothing'$ is that unlike an arbitrary c.e. set C, the opponent in the previous construction, that is us in the current construction, controls a portion of $\varnothing'$. That is, we enumerate an auxiliary c.e. set E, and by the recursion theorem we know an index e such that $E = W_e$ which is the eth column of $\varnothing'$. Suppose that we are given that $A = \operatorname{dom} \hat{\Phi}_1(\varnothing')$ and $\omega \setminus A = \operatorname{dom} \hat{\Phi}_0(\varnothing')$. To reduce A to $\varnothing'$, given x, we wait for some i and s such that $\hat{\Phi}_i(\varnothing'_s, x)\downarrow$, i.e., $\varphi_i(x)\downarrow$ at stage s and $\Phi_i(\varnothing'_s, x)$ converges with use below $\varphi_i(x)$. We then set $\psi(x)$ to be some number large enough so that the agent which is responsible for computing $A(x)$ can control $\varphi_i(x)$ many elements of $\varnothing'$ (via E) with no interference from other agents which have already staked their claims for portions of E. We show that this control is sufficient to compute $A(x)$ from $\varnothing'\restriction_{\psi(x)}$. As long as $\hat{\Phi}_i(\varnothing'_s, x)\downarrow$, we keep stating that $A(x) = i$, with use $\varnothing'_s\restriction_{\psi(x)}$. If $\varnothing'$ changes below $\varphi_i(x)$, and we then see that $\hat{\Phi}_{1-i}(\varnothing'_s, x)\downarrow$, then we declare that $A(x) = 1 - i$ with use the new version of $\varnothing'\restriction_{\psi(x)}$ (as $\psi(x) > \varphi_i(x)$). If the computation $\hat{\Phi}_{1-i}(\varnothing', x)$ fizzles, we wait to see if we next get a new computation

$\hat{\Phi}_i(\varnothing'_s, x)\downarrow$. If not, then we will later get a new $\hat{\Phi}_{1-i}(\varnothing', x)\downarrow$ computation, and we didn't need to do anything. Otherwise, we enumerate one of our agitators into E so that we can redefine $A(x)=i$ with the new version of $\varnothing'\upharpoonright_{\psi(x)}$. The point is that no matter how large $\varphi_{1-i}(x)$ (it may be much larger than $\psi(x)$), every enumeration into E on behalf of computing $A(x)$ is tied to a failed $\hat{\Phi}_i(\varnothing'_s, x)$ computation, and so to some historic version of $\varnothing'\upharpoonright_{\varphi_i(x)}$. Thus we never run out of agitators and we can keep up with the changes in A and record them into $\varnothing'\upharpoonright_{\psi(x)}$ correctly. □

With the perils of bounded oracle computations in mind, we turn to define the bounded jump and a universal "jump function." For an oracle $g\colon \omega \to \omega$, we let

$$g^\dagger = \bigoplus_{e<\omega} \hat{W}_e^g = \left\{(e,x) : x \in \hat{W}_e^g\right\}.$$

In analogy with the jump function J, we define a function I^g as follows:

$$I^g(e,x) = \begin{cases} 0, & \text{if } x \notin \hat{W}_e^g; \\ \hat{\Phi}_e(g,x)+1, & \text{otherwise.} \end{cases}$$

Elementary properties of these jump operations are analogous to those of the Turing jump.

Lemma 2.31. *Let $g\colon \omega \to \omega$.*

(1) $g^\dagger$ is 1-complete for the class of bounded g-c.e. sets.

(2) $g^\dagger$ is computably isomorphic to the set $\left\{e : e \in \hat{W}_e^g\right\}$.

Proof. (1)—the fact that $g^\dagger$ is bounded g-c.e.—follows from the fact that the enumeration $\langle \Phi_e, \varphi_e\rangle_{e<\omega}$ is effective: $\langle \Phi_e\rangle$ is uniformly c.e., and $\langle \varphi_e\rangle$ are uniformly partial computable.

Let $g^* = \left\{e : e \in \hat{W}_e^g\right\}$. Since g^* is bounded g-c.e., to show (2) it is sufficient to show that g^* is also 1-complete for the class of bounded g-c.e. sets. Let (Φ, φ) be a weak truth-table functional. To reduce $\hat{\Phi}(g)$ to g^*, given any $x<\omega$ we define a partial computable function ψ_x such that for all w, $\psi_x(w)\downarrow$ if and only if $\varphi(x)\downarrow$, in which case, $\psi_x(w)=\varphi(x)$ for all w; and also define a Turing functional Ψ_x such that if $\varphi(x)\uparrow$, then $\Psi_x(h,w)\uparrow$ for all $w<\omega$ and all oracles h, and if $\varphi(x)\downarrow$, then $\Psi_x(h,w)=\Phi(h,w)$ for all oracles h and all $w<\omega$, with the same use. Since the numbering $\langle \Phi_e, \varphi_e\rangle$ is acceptable, there is an injective computable function f such that for all $x<\omega$, $(\Psi_x, \psi_x) = (\Phi_{f(x)}, \varphi_{f(x)})$. Then f witnesses that $\operatorname{dom}\hat{\Phi}(g) \leqslant_1 g^*$. □

For functions $f, g\colon \omega \to \omega$, we say that $f \leqslant_{\mathrm{m}} g$ if there is a computable function h such that $f = g\circ h$. Note that this definition extends the familiar one for sets. If $f \leqslant_{\mathrm{m}} g$ then $f \leqslant_{\mathrm{wtt}} g$.

Lemma 2.32. *Let $g\colon \omega \to \omega$. A set A is bounded g-c.e. if and only if $A \leqslant_{\mathrm{m}} g^{\dagger}$.*

Proof. Let $A \leqslant_{\mathrm{m}} g^{\dagger}$; so there is a computable function h such that $A = h^{-1}g^{\dagger}$. Let $x < \omega$; let $(e, y) = h(x)$. Then we let $\psi(x) = \varphi_e(y)$ and $\Psi(h, x) = \Phi_e(h, y)$ for every oracle h, with the same use. Then $A = \operatorname{dom} \hat{\Psi}(g)$.

In the other direction, let (Φ, φ) be a weak truth-table functional, and let $A = \operatorname{dom} \hat{\Phi}(g)$. There is some e such that $\hat{\Phi} = \hat{\Phi}_e$, and so for all x, $x \in A$ if and only if $(e, x) \in g^{\dagger}$, so the map $x \mapsto (e, x)$ shows that $A \leqslant_{\mathrm{m}} g^{\dagger}$. □

Lemma 2.33. *For all $g\colon \omega \to \omega$,*

(1) $g^{\dagger} \leqslant_{\mathrm{wtt}} I^g$;
(2) I^g is many-one equivalent to the "diagonal function" $e \mapsto I^g(e, e)$.

Proof. For (1), we have $(e, x) \in g^{\dagger}$ if and only if $I^g(e, x) \neq 0$. For (2), the reduction f of the proof of Lemma 2.31 satisfies $I^g(e, x) = I^g(f(e, x), f(e, x))$ for all e and x. □

Lemma 2.34. *Let $g\colon \omega \to \omega$.*

(1) $g <_{\mathrm{wtt}} I^g$.
(2) For any set *A, $A <_{\mathrm{wtt}} A^{\dagger}$.*

Proof. Let $\psi = \mathrm{id}$ be the identity function, and let Ψ be a Turing functional which maps any sequence σ to itself. So for all $g\colon \omega \to \omega$, $\hat{\Psi}(g) = g$. Hence there is some e such that $g(x) = I^g(e, x)$, so $g \leqslant_{\mathrm{wtt}} I^g$.

Every set A is bounded A-c.e., and so by Lemma 2.32, $A \leqslant_{\mathrm{m}} A^{\dagger}$. It follows that $A \leqslant_{\mathrm{wtt}} A^{\dagger}$.

The proof of (1) and (2) will be complete with the aid of Lemma 2.33(1), once we show that for any function g, $g^{\dagger} \nleqslant_{\mathrm{wtt}} g$. This is Cantor's argument, as the set

$$\left\{ e : e \notin \hat{W}^g_e \right\}$$

is weak truth-table reducible to $g^{\dagger}$, and is not bounded g-c.e., so cannot be weak truth-table reducible to g (Proposition 2.30(3)). □

Lemma 2.35. *Let $f, g\colon \omega \to \omega$.*

(1) $f \leqslant_{\mathrm{wtt}} g$ if and only if $I^f \leqslant_{\mathrm{m}} I^g$.
(2) If $f \leqslant_{\mathrm{wtt}} g$ then $f^{\dagger} \leqslant_{\mathrm{m}} g^{\dagger}$. The converse fails, even when restricting to sets rather than functions.

If $f, g\colon \omega \to \omega$ and $f \leqslant_{\mathrm{wtt}} g$, then from an index e such that $\hat{\Phi}_e(g) = f$ we can effectively obtain indices c and d such that $\hat{\Phi}_c(I^g) = I^f$ and $\hat{\Phi}_d(g^{\dagger}) = f^{\dagger}$.

It follows that the operations $g \mapsto I^g$ and $g \mapsto g^\dagger$ induce well-defined, strictly increasing functions on the partial ordering of the weak truth-table degrees.

Proof. It is easy to show that if $f \leqslant_{\mathrm{wtt}} g$, then $I^f \leqslant_{\mathrm{m}} I^g$ and $f^\dagger \leqslant_{\mathrm{m}} g^\dagger$. One simply composes the reduction of f to g with any weak truth-table functional; this composition is uniform in an index for a reduction of f to g.

Let $f, g\colon \omega \to \omega$, and suppose that h is computable and $I^f = I^g \circ h$. Fix e such that $f = \hat{\Phi}_e(f)$. Let $x < \omega$, and let $(d, y) = h(e, x)$. Since

$$I^g(d, y) = I^f(e, x) = f(x) + 1 > 0,$$

we have $\hat{\Phi}_d(g, y)\downarrow = f(x)$, which shows that $f \leqslant_{\mathrm{wtt}} g$.

The failure of the converse to (2) is exhibited by an argument similar to the one proving the first part of Proposition 2.30(3), and so we only sketch it. We enumerate a c.e. set B and approximate a d.c.e. set B such that $A \not\leqslant_{\mathrm{wtt}} B$ but $A^\dagger \leqslant_{\mathrm{m}} B^\dagger$. Instances $R_{e,x}$ of a global requirement for coding $A^\dagger$ into $B^\dagger$ define the value at (e, x) of a partial computable function ψ and enumerate axioms with use $B_s\restriction_{\psi(e,x)}$ into a functional Ψ; we then can find a computable function h such that for all e and x, $(h(e, x), h(e, x)) \in B^\dagger$ if and only if $\hat{\Psi}(B, e, x)\downarrow$; we need to ensure that this happens if and only if $(e, x) \in A^\dagger$. Requirements P_i diagonalise $A(z)$ against $\hat{\Phi}_i(B, z)$ for some appointed follower z. The priorities of the $R_{e,x}$ requirements are interspersed between the P_i requirements. In a typical scenario, $R_{e,x}$ observes that $\hat{\Phi}_e(A_s, x)\downarrow$ for the first time; it sets $\psi(e, x)$ to be some large number, and lets $\hat{\Psi}(B_s, e, x)\downarrow$. The size of $\psi(e, x)$ allows the requirement $R_{e,x}$ to enumerate a number into B once for each follower $z < \varphi_e(x)$ for a requirement P_i stronger than $R_{e,x}$. Of course followers for weaker requirements are cancelled and new ones are chosen to be larger than $\varphi_e(x)$. Whenever a strong P_i enumerates its follower z into A, the opponent may change whether $\hat{\Psi}_e(A_s, x)$ converges or not. If the change is from convergence to divergence, then we need to enumerate some number below $\psi(e, x)$ into B. This, in turn, may cause $\hat{\Phi}_i(B, z)$ to change, making P_i want to extract z from A. It does so, this time freezing $B\restriction_{\varphi_i(z)}$. The fact that $\varphi_i(z)$ may be larger than $\psi(e, x)$ does not disturb us: the extraction of z from A gives our opponent an opportunity to make $\hat{\Phi}_e(A, x)$ converge again, but we can then make $\hat{\Psi}(B, e, x)$ converge without changing $B\restriction_{\psi(e,x)}$, simply by enumerating a new axiom into Ψ. If later an even stronger requirement P_j acts, the process repeats, injuring P_i, but any new P_i follower will be greater than $\varphi_e(x)$, and so never disturb $R_{e,x}$ again. Hence we can fix $\psi(e, x)$ based on the priority of $R_{e,x}$ whenever we see $\varphi_e(x)$ converge. $\square$

Lemma 2.36. *For all Δ^0_2 functions g, I^g is also Δ^0_2.*

And so $g^\dagger$ is also Δ^0_2.

Proof. Let $\langle g_s \rangle$ be a computable approximation for g. For $e, x, s < \omega$, let $h_s(e,x) = 0$ if $\varphi_{e,s}(x)\uparrow$, or if $\Phi_{e,s}(g_s\restriction_{\varphi_e(x)})\uparrow$. Otherwise let $h_s(e,x) = \Phi_{e,s}(g_s\restriction_{\varphi_e(x)})$. Then $\langle h_s \rangle$ is a computable approximation of I^g. The point, of course, is that if $\varphi_e(x)\downarrow$, then $g_s\restriction_{\varphi_e(x)}$ eventually stabilizes. □

Since bounded $\varnothing$-c.e. sets are simply c.e. sets, $\varnothing^\dagger$ and $\varnothing'$ are computably isomorphic. Both sets are weak truth-table equivalent to $I^\varnothing$, since if we know that $\hat{\Phi}_e(\varnothing, x)\downarrow$, then finding the value $\hat{\Phi}_e(\varnothing, x)$ can be done effectively. Hence, for any Δ^0_2 function g we have $g^\dagger \equiv_{\mathrm{T}} I^g \equiv_{\mathrm{T}} \varnothing'$.

2.3.2 Transfinite iterations of the bounded jump

Let $g\colon \omega \to \omega$. We define, for a computable well-ordering $\mathcal{R} = (R, <_{\mathcal{R}})$, the iteration of the bounded jump set and function along $\mathcal{R}$, by induction on the order-type of $\mathcal{R}$.

- If $\mathcal{R}$ is empty, then we let $g^{\langle \mathcal{R} \rangle} = I^g_{\mathcal{R}} = g$.

Suppose that $\mathcal{R}$ is nonempty, and that by recursion, for all $z \in R$, both $g^{\langle \mathcal{R}\restriction_z \rangle}$ and $I^g_{\mathcal{R}\restriction_z}$ have already been defined.

- If the order-type of $\mathcal{R}$ is a successor ordinal, let z be the $\mathcal{R}$-greatest element of R; we then let $g^{\langle \mathcal{R} \rangle} = \left(g^{\langle \mathcal{R}\restriction_z \rangle}\right)^\dagger$ and $I^g_{\mathcal{R}} = I^{I^g_{\mathcal{R}\restriction_z}}$.
- If the order-type of $\mathcal{R}$ is a limit ordinal, we let $g^{\langle \mathcal{R} \rangle} = \bigoplus_{z \in R} g^{\langle \mathcal{R}\restriction_z \rangle}$ and $I^g_{\mathcal{R}} = \bigoplus_{z \in R} I^g_{\mathcal{R}\restriction_z}$. By this we mean that for all z and x, $(z,x) \in g^{\langle \mathcal{R} \rangle}$ if and only if $z \in R$ and $x \in g^{\langle \mathcal{R}\restriction_z \rangle}$ (so we consider $g^{\langle \mathcal{R} \rangle}$ as an element of 2^ω); and if $z \in R$, then $I^g_{\mathcal{R}}(z,x) = I^g_{\mathcal{R}\restriction_z}(x)$, whereas if $z \notin R$ then $I^g_{\mathcal{R}}(z,x) = 0$; so we can consider $I^g_{\mathcal{R}}$ as a function from ω to ω.

Proposition 2.37. *Let $\mathcal{R}$ and $\mathcal{S}$ be computable well-orderings. Suppose that* $\mathrm{otp}(\mathcal{R}) \leqslant \mathrm{otp}(\mathcal{S})$. *Also suppose that the embedding of $\mathcal{R}$ as an initial segment of $\mathcal{S}$ is computable. Suppose further that $\mathcal{R}$ is notation-like. Then for all $g\colon \omega \to \omega$, $g^{\langle \mathcal{R} \rangle} \leqslant_{\mathrm{wtt}} g^{\langle \mathcal{S} \rangle}$ and $I^g_{\mathcal{R}} \leqslant_{\mathrm{wtt}} I^g_{\mathcal{S}}$.*

Proof. Let $j\colon R \to S$ be the initial segment embedding of $\mathcal{R}$ into $\mathcal{S}$. We show that there are computable functions f and h such that for all $z \in R$, for all $g\colon \omega \to \omega$

$$g^{\langle \mathcal{R}\restriction_z \rangle} = \hat{\Phi}_{f(z)}(g^{\langle \mathcal{S}\restriction_{j(z)} \rangle})$$

and

$$I^g_{\mathcal{R}\restriction_z} = \hat{\Phi}_{h(z)}(I^g_{\mathcal{S}\restriction_{j(z)}}).$$

Replacing $\mathcal{R}$ and $\mathcal{S}$ by one element extensions $\mathcal{R}+1$ and $\mathcal{S}+1$ then yields the desired conclusion.

The definitions of f and h are done by effective transfinite recursion along $\mathcal{R}$. Directly, we define as follows:

(1) If z is the $\mathcal{R}$-least element of R, then we let $f(z)=h(z)=e$ where $\hat{\Phi}_e(g)=g$ for all $g\colon \omega\to\omega$.
(2) If z is the $\mathcal{R}$-successor of w, then by Lemma 2.35, from $f(w)$ we can effectively find a number $f(z)$ such that for all g,

$$\hat{\Phi}_{f(z)}\left(g^{\langle \mathcal{S}\restriction_{j(z)}\rangle}\right)=\hat{\Phi}_{f(z)}\left(\left(g^{\langle \mathcal{S}\restriction_{j(w)}\rangle}\right)^{\dagger}\right)=\left(g^{\langle \mathcal{R}\restriction_w\rangle}\right)^{\dagger}=g^{\langle \mathcal{R}\restriction_z\rangle},$$

and from $h(w)$ we can effectively find a number $h(z)$ such that for all g,

$$\hat{\Phi}_{h(z)}\left(I^g_{\mathcal{S}\restriction_{j(z)}}\right)=\hat{\Phi}_{h(z)}\left(I^{I^g_{\mathcal{S}\restriction_{j(w)}}}\right)=I^{I^g_{\mathcal{R}\restriction_w}}=I^g_{\mathcal{R}\restriction_z}.$$

(3) If z is a limit point of $\mathcal{R}$, then from $g^{\langle \mathcal{S}\restriction_{j(z)}\rangle}$ and $I^g_{\mathcal{S}\restriction_{j(z)}}$ we can obtain, uniformly in g and in $w<_{\mathcal{R}} z$, $g^{\langle \mathcal{S}\restriction_{j(w)}\rangle}$ and $I^g_{\mathcal{S}\restriction_{j(w)}}$, respectively, in a weak truth-table fashion. Thus from $f\restriction_{R\restriction_z}$ and $h\restriction_{R\restriction_z}$ we can compute indices $f(z)$ and $h(z)$ such that for all $w<_{\mathcal{R}} z$, for all $x<\omega$, for all g,

$$\begin{aligned}\hat{\Phi}_{f(z)}\left(g^{\langle \mathcal{S}\restriction_{j(z)}\rangle},(w,x)\right)&=\hat{\Phi}_{f(w)}\left(g^{\langle \mathcal{S}\restriction_{j(w)}\rangle},x\right)\\&=g^{\langle \mathcal{R}\restriction_w\rangle}(x)=g^{\langle \mathcal{R}\restriction_z\rangle}(w,x)\end{aligned}$$

and

$$\hat{\Phi}_{h(z)}\left(I^g_{\mathcal{S}\restriction_{j(z)}},(w,x)\right)=\hat{\Phi}_{h(w)}\left(I^g_{\mathcal{S}\restriction_{j(w)}},x\right)=I^g_{\mathcal{R}\restriction_w}(x)=I^g_{\mathcal{R}\restriction_z}(w,x),$$

and so $\hat{\Phi}_{f(z)}\left(g^{\langle \mathcal{S}\restriction_{j(z)}\rangle}\right)=g^{\langle \mathcal{R}\restriction_z\rangle}$ and $\hat{\Phi}_{h(z)}\left(I^g_{\mathcal{S}\restriction_{j(z)}}\right)=I^g_{\mathcal{R}\restriction_z}$ as required.

The details of the effective transfinite recursion, using the recursion theorem, are as in the proof of Lemma 2.25. □

It follows that if $\mathcal{R}$ and $\mathcal{S}$ are computably isomorphic, then for all g, $g^{\langle \mathcal{R}\rangle}\equiv_{\text{wtt}} g^{\langle \mathcal{S}\rangle}$ and $I^g_{\mathcal{R}}\equiv_{\text{wtt}} I^g_{\mathcal{S}}$. Hence, using canonical well-orderings, for $\alpha\leqslant\varepsilon_0$, we can unambiguously define $g^{\langle\alpha\rangle}$ and I^g_α for all g—these are unique up to weak truth-table degree, and in fact many-one degree if $\alpha>0$, and induce well-defined operations on the weak truth-table degrees. If $\alpha<\beta$ then $g^{\langle\alpha\rangle}<_{\text{wtt}} g^{\langle\beta\rangle}$ and $I^g_\alpha<_{\text{wtt}} I^g_\beta$.

Proposition 2.38. *Let $g\colon\omega\to\omega$ and let $\alpha\leqslant\varepsilon_0$. Then $g^{\langle\alpha\rangle}\leqslant_{\text{wtt}} I^g_\alpha$.*

Proof. By effective transfinite recursion on ε_0+1 we build a computable function R such that for all $\alpha\leqslant\varepsilon_0$ and all g, $\hat{\Phi}_{R(\alpha)}\left(I^g_\alpha\right)=g^{\langle\alpha\rangle}$. This is done by cases:

(1) Since $I^g_\alpha = g^{\langle 0\rangle} = g$, we let $R(0)$ be a number such that for all g, $\hat{\Phi}_{R(0)}(g) = g$.
(2) The proofs of Lemma 2.33(1) and Lemma 2.35(2) show that there is a computable function S such that for all $\alpha < \varepsilon_0$ and all $a < \omega$, if $\hat{\Phi}_a(g) = f$ then $\hat{\Phi}_{S(\alpha,a)}(I^g) = f^\dagger$. We then let, for all $\alpha < \varepsilon_0$, $R(\alpha+1) = S(\alpha, R(\alpha))$.
(3) For limit α, we string together the reductions for $\beta < \alpha$. The construction for part (3) of the proof of Proposition 2.37 shows that there is a computable function T such that for all limit ordinals $\alpha \leqslant \varepsilon_0$ and all $a < \omega$, for all sequences $\langle g_\beta \rangle_{\beta<\alpha}$ of functions,

$$\hat{\Phi}_{T(\alpha,a)}\left(\bigoplus_{\beta<\alpha} g_\beta\right) = \bigoplus_{\beta<\alpha} \hat{\Phi}_a(g_\beta).$$

We then let $R(\alpha) = T(\alpha, a)$, where a is an index such that $\varphi_a\restriction_\alpha = R\restriction_\alpha$.

Again to make things concrete, we show how to perform this recursion: we define a function F. For all $a < \omega$, we let $F(0,a) = R(0)$, $F(\alpha+1, a) = S(\alpha, a)$ and for limit α, $F(\alpha, a) = T(\alpha, a)$. By the recursion theorem, there is an index a such that $F(-, a) = \varphi_a$. Since $F(a, 0)$ is defined for all a, and since S and T are total, we have $\varepsilon_0 + 1 \subseteq \operatorname{dom} \varphi_a$. The function $R = \varphi_a\restriction_{\varepsilon_0+1}$ is as required. ☐

Proposition 2.39. *For any Δ^0_2 function g and any $\alpha \leqslant \varepsilon_0$, I^g_α is Δ^0_2.*

Proof. Fix a Δ^0_2 function g. By effective transfinite recursion we build a computable function R such that for all $\alpha \leqslant \varepsilon_0$, $\varphi_{R(\alpha)} = \langle g^\alpha_s \rangle$ is a computable approximation of I^g_α.

(1) We let $R(0)$ be an index for a computable approximation of g.
(2) The proof of Lemma 2.36 shows that there is a computable function S such that for all $a < \omega$, if φ_a is a computable approximation of a function h, then $\varphi_{S(a)}$ is a computable approximation of I^h. For any $\alpha < \varepsilon_0$, we let $R(\alpha+1) = S(R(\alpha))$.
(3) An argument similar to previous ones shows that there is a computable function T such that for all $a < \omega$ and all limit $\alpha \leqslant \varepsilon_0$, if for all $\beta < \alpha$, $\varphi_{\varphi_a(\beta)}$ is a computable approximation of a function h_β, then $\varphi_{T(\alpha,a)}$ is a computable approximation of $\bigoplus_{\beta<\alpha} h_\beta$. Then we let, for limit ordinals α, $R(\alpha) = T(\alpha, a)$, where a is an index for $R\restriction_\alpha$. ☐

The following theorem, a refinement of Proposition 2.39, is the goal of this section:

Theorem 2.40. *Let $\alpha \leqslant \varepsilon_0$.*

(1) A set A is ω^α-c.a. if and only if $A \leqslant_{\mathrm{wtt}} \varnothing^{\langle\alpha\rangle}$.
(2) A function g is ω^α-c.a. if and only if $g \leqslant_{\mathrm{wtt}} I^\varnothing_\alpha$.

This theorem generalises the fact that a function or a set is ω-c.a. if and only if it is weak truth-table reducible to $\varnothing'$. Anderson and Csima [3] proved part (1) of the theorem for $\alpha<\omega$.

We note that for $\alpha\geqslant 2$, we really do need to use the function jump I^g rather than the set jump $g^\dagger$:

Proposition 2.41. *There is an $(\omega+1)$-c.a. function which is not weak truth-table reducible to any set.*

Proof. We define an $(\omega+1)$-computable approximation $\langle f_s, o_s\rangle$ for a function f. For each $e<\omega$, we want to ensure that for any set A, $f(e)\neq\hat{\Phi}_e(A,e)$. Let $e<\omega$. We let V_e be the collection of all values $\hat{\Phi}_e(\sigma,e)$, as σ ranges over all binary strings of length $\varphi_e(e)$. For any set A, if $\hat{\Phi}_e(A,e)\downarrow$, then $\hat{\Phi}_e(A,e)\in V_e$. The sequence $\langle V_e\rangle$ is c.e., uniformly in e; if $\varphi_e(e)\uparrow$, then $|V_e|=0$, and otherwise $|V_e|\leqslant 2^{\varphi_e(e)}$. Let $\langle V_{e,s}\rangle$ be a uniformly computable enumeration of the sets V_e.

For all $s<\omega$, if $\varphi_{e,s}(e)\uparrow$, then we let $f_s(e)=0$ and $o_s(e)=\omega$. Otherwise, we let $f_s(e)$ be the least element of $\omega\setminus V_{e,s}$, and let $o_s(e)=2^{\varphi_e(e)}-|V_{e,s}|$. Then $f(e)\notin V_e$, which gives the desired diagonalisation. □

2.3.3 Commutative addition and powers of ω

We focus on ordinal powers of ω because these consist precisely of the ordinals which are closed under addition.

Proposition 2.42. *An ordinal $\alpha>0$ is closed under addition if and only if $\alpha=\omega^\beta$ for some β.*

Proof. Let β be any ordinal, and let $\gamma,\delta<\omega^\beta$. Let $\gamma=\omega^{\gamma_1}n_1+\dots\omega^{\gamma_k}n_k$ and $\delta=\omega^{\delta_1}m_1+\cdots+\omega^{\delta_l}m_l$ be the Cantor normal forms of γ and δ. Since $\omega^{\gamma_1}\leqslant\gamma<\omega^\beta$, we have $\gamma_1<\beta$; similarly, $\delta_1<\beta$. Hence

$$\gamma+\delta\leqslant\omega^{\gamma_1}(n_1+1)+\omega^{\delta_1}(m_1+1)\leqslant\omega^{\max\{\gamma_1,\delta_1\}}(n_1+m_1)<\omega^\beta,$$

so ω^β is closed under addition.

Let α be an ordinal which is not a power of ω. Let $\alpha=\omega^{\alpha_1}n_1+\cdots+\omega_{\alpha_k}n_k$ be the Cantor normal form of α. Since $\alpha\neq\omega^{\alpha_1}$, we have $\omega^{\alpha_1}<\alpha<\omega^{\alpha_1}(n_1+1)$. This shows that α is not closed under addition. □

While addition of ordinals is a natural and useful operation, it has a few shortcomings, in particular its lack of commutativity. Less well-used is the operation of "commutative addition" (as termed for instance in [6]), based on Cantor normal form.

Let $\alpha_1>\alpha_2>\cdots>\alpha_k$. Let $\beta=\omega^{\alpha_1}n_1+\omega^{\alpha_2}n_2+\cdots+\omega^{\alpha_k}n_k$, and $\gamma=\omega^{\alpha_1}m_1+\omega^{\alpha_2}m_2+\cdots+\omega^{\alpha_k}m_k$, where of course $n_i,m_i<\omega$, but we allow some

$n_i, m_i = 0$. We let

$$\beta \oplus \gamma = \omega^{\alpha_1}(n_1 + m_1) + \omega^{\alpha_2}(n_2 + m_2) + \cdots + \omega^{\alpha_k}(n_k + m_k).$$

Cantor normal form allows us to define $\beta \oplus \gamma$ for all ordinals β and γ: we extend their Cantor normal form to a presentation as above with a common sequence of decreasing exponents by adding zero coefficients; for any sequence of exponents, this presentation is unique.

Moreover, canonicity of our fixed computable well-orderings implies that the operation $\oplus$ for pairs of ordinals below ε_0 is computable.

Lemma 2.43. *Let α, β and γ be ordinals.*

(1) $\beta \oplus \gamma = \gamma \oplus \beta$.
(2) $\alpha \oplus (\beta \oplus \gamma) = (\alpha \oplus \beta) \oplus \gamma$.

Proof. This is quite straightforward, based on the commutativity and associativity of addition of natural numbers. For associativity, the point is that if $\alpha_1 > \alpha_2 > \dots \alpha_k$ mentions all exponents of ω in the Cantor normal forms of α, β and γ, and

$$\alpha = \omega^{\alpha_1} n_1 + \cdots + \omega^{\alpha_k} n_k,$$

$$\beta = \omega^{\alpha_1} m_1 + \cdots + \omega^{\alpha_k} m_k,$$

$$\gamma = \omega^{\alpha_1} l_1 + \cdots + \omega^{\alpha_k} l_k,$$

then

$$(\alpha \oplus \beta) \oplus \gamma = \alpha \oplus (\beta \oplus \gamma) = \omega^{\alpha_1}(n_1 + m_1 + l_1) + \cdots + \omega^{\alpha_k}(n_k + m_k + l_k). \quad \square$$

The associativity and commutativity of $\oplus$ allows us to unambiguously define $\bigoplus A$ for finite multisets of ordinals A.

Lemma 2.44. *Any power of ω is closed under $\oplus$.*

Proof. Let $\beta = \omega^{\alpha_1} n_1 + \cdots + \omega^{\alpha_k} n_k$ and $\gamma = \omega^{\alpha_1} m_1 + \cdots + \omega^{\alpha_k} m_k$ be smaller than ω^δ. Then for all $i \leqslant k$, $\omega^{\alpha_i} n_i, \omega^{\alpha_i} m_i < \omega^\delta$. Since ω^δ is closed under addition, it follows that $\beta \oplus \gamma < \omega^\delta$. $\square$

Lemma 2.45. *Let $\beta_1, \beta_2, \dots, \beta_n$ and $\gamma_1, \gamma_2, \dots, \gamma_n$ be two n-tuples of ordinals. Suppose that for all $i \leqslant n$, $\beta_i \leqslant \gamma_i$. Then $\bigoplus_{i \leqslant n} \beta_i \leqslant \bigoplus_{i \leqslant n} \gamma_i$, and $\bigoplus_{i \leqslant n} \beta_i < \bigoplus_{i \leqslant n} \gamma_i$ if and only if there is some $i \leqslant n$ such that $\beta_i < \gamma_i$.*

Proof. Again, this is known and quite straightforward, but we give details for completeness of our presentation. Let $\alpha_1 > \alpha_2 > \cdots > \alpha_k$ be the exponents of ω

appearing in the Cantor normal form of any of the β_i's and γ_i's; let $\beta_i = \sum_{j\leqslant k} \omega^{\alpha_j} n_{i,j}$ and $\gamma_i = \sum_{j\leqslant k} \omega^{\alpha_j} m_{i,j}$. So $\bigoplus_{i\leqslant n} \beta_i = \sum_{j\leqslant k} \left(\omega^{\alpha_j} \sum_{i\leqslant n} n_{i,j}\right)$, and $\bigoplus_{i\leqslant n} \gamma_i = \sum_{j\leqslant k} \left(\omega^{\alpha_j} \sum_{i\leqslant n} m_{i,j}\right)$.

If $\bigoplus_{i\leqslant n} \beta_i = \bigoplus_{i\leqslant n} \gamma_i$ then by the uniqueness of Cantor normal form, for all $j\leqslant k$, $\sum_{i\leqslant n} n_{i,j} = \sum_{i\leqslant n} m_{i,j}$. By induction on $j\leqslant k$, we show that for all i, $n_{i,j} = m_{i,j}$; it would follow that for all i, $\beta_i = \gamma_i$. Fix j, and suppose that for all $j' < j$, for all $i\leqslant n$, $n_{i,j'} = m_{i,j'}$. Since $\beta_i \leqslant \gamma_i$, the induction assumption implies that $n_{i,j} \leqslant m_{i,j}$. Now $\sum_{i\leqslant n} n_{i,j} = \sum_{i\leqslant n} m_{i,j}$ implies that for all $i\leqslant n$, $n_{i,j} = m_{i,j}$.

Suppose that $\bigoplus_{i\leqslant n} \beta_i \neq \bigoplus_{i\leqslant n} \gamma_i$. Let j be the least index such that $\sum_{i\leqslant n} n_{i,j} \neq \sum_{i\leqslant n} m_{i,j}$. An induction as in the previous paragraph shows that for all $j' < j$, for all $i\leqslant n$, $n_{i,j'} = m_{i,j'}$. This information, together with the fact that $\beta_i \leqslant \gamma_i$ for all i, shows that for all i, $n_{i,j} \leqslant m_{i,j}$, and so that $\sum_{i\leqslant n} n_{i,j} \leqslant \sum_{i\leqslant n} m_{i,j}$. Since $\sum_{i\leqslant n} n_{i,j} \neq \sum_{i\leqslant n} m_{i,j}$, we must have $\sum_{i\leqslant n} n_{i,j} < \sum_{i\leqslant n} m_{i,j}$. The choice of j now shows that $\bigoplus_{i\leqslant n} \beta_i < \bigoplus_{i\leqslant n} \gamma_i$. □

The operation of commutative addition allows us to show that if $\alpha \leqslant \varepsilon_0$ is closed under addition, then the α-c.a. functions induce an initial segment of the weak truth-table degrees.

Proposition 2.46. *Let $\alpha \leqslant \varepsilon_0$. If $f\colon \omega \to \omega$ is ω^α-c.a. and $g \leqslant_{\mathrm{wtt}} f$, then g is ω^α-c.a.*

Proof. Let $\langle f_s, o_s\rangle_{s<\omega}$ be an ω^α-computable approximation of f, and let (Φ, φ) be a weak truth-table functional such that $\hat{\Phi}(f) = g$. For any $x, s < \omega$, we recursively define a strictly increasing sequence $\langle t_s(x)\rangle_{s<\omega}$ of stages such that for all s, $\hat{\Phi}(f_{t_s(x)}, x)\downarrow$. Let $g_s(x) = \hat{\Phi}(f_{t_s(x)}, x)$, and let $m_s(x) = \bigoplus_{y<\varphi(x)} o_{t_s(x)}(y)$. Then $\langle g_s, m_s\rangle$ is an ω^α-computable approximation of g: by Lemma 2.44, for all x and s, $m_s(x) < \omega^\alpha$, and by Lemma 2.45, for all x and s, $m_{s+1}(x) \leqslant m_s(x)$ and if $g_{s+1}(x) \neq g_s(x)$ then $m_{s+1}(x) < m_s(x)$, because $f_{t_{s+1}(x)}\restriction_{\varphi(x)} \neq f_{t_s(x)}\restriction_{\varphi(x)}$. □

2.3.4 The complexity of the iterated bounded jump

We wish to establish the following:

Proposition 2.47. *For all $\alpha \leqslant \varepsilon_0$, $I^{\varnothing}_{\alpha}$ is ω^α-c.a.*

As a result, by Proposition 2.38 and Proposition 2.46, $\varnothing^{\langle\alpha\rangle}$ is also ω^α-c.a.

Proposition 2.47 is proved by effective transfinite recursion, which means that the ω^α-computable approximation for $I_\alpha(\varnothing)$ will be given uniformly in α. That is, by effective transfinite recursion on ε_0, we will show that there is a computable function $R\colon \varepsilon_0 \to \omega$ such that for all $\alpha < \varepsilon_0$, $\left\langle f_s^{R(\alpha),\omega^\alpha}, o_s^{R(\alpha),\omega^\alpha}\right\rangle$

is an ω^α-computable approximation of $I^{\varnothing}_{\alpha}$; recall that $\langle\langle f_s^{e,\alpha}, o_s^{e,\alpha}\rangle_{s<\omega}\rangle_{e<\omega}$ is an effective list, uniform in α, of all tidy $(\alpha+1)$-computable approximations (Proposition 2.7).

The following two lemmas correspond to two of the three cases in the definition of R.

Lemma 2.48. *Let $\alpha<\varepsilon_0$. If g is an ω^α-c.a. function, then I^g is an $(\omega^\alpha+1)$-c.a. function. From α, and an index of an ω^α-computable approximation of a function g, we can effectively obtain an index of an $(\omega^\alpha+1)$-computable approximation of I^g.*

Proof. Let $\langle g_s, o_s\rangle$ be an ω^α-computable approximation of g. For $e, x, s<\omega$, if $\hat{\Phi}_e(g_s, x)$ converges in s many steps, we let $h_s(e,x)=1+\hat{\Phi}_e(g_s,x)$; otherwise, we let $h_s(e,x)=0$. Then $\langle h_s\rangle$ is a computable approximation of I^g. We may assume that for all e and x, $h_0(e,x)=0$.

Fix $e, x<\omega$. For all $s<\omega$, let $r_s(e,x)$ be the least $r\leqslant s$ such that for all $t\in[r,s]$, $h_t(e,x)=h_s(x)$. We define a function $m_s(e,x)$:

- If $r_s(e,x)=0$, let $m_s(e,x)=\omega^\alpha$.
- If $r_s(e,x)>0$ then we know that $\varphi_e(x)\downarrow$. There are two sub-cases:
  - If $h_s(e,x)>0$, then we let
$$m_s(e,x)=\bigoplus_{y<\varphi_e(x)}\left(o_s(y)\oplus o_s(y)\right).$$
  - If $h_s(e,x)=0$, then we let
$$m_s(e,x)=\bigoplus_{y<\varphi_e(x)}\left(o_s(y)\oplus o_{r_s(e,x)-1}(y)\right).$$

By Lemma 2.44, for all e, x and s, $m_s(e,x)\leqslant\omega^\alpha$. We show that $\langle h_s, m_s\rangle$ is an $(\omega^\alpha+1)$-computable approximation. Fix $e, x, s<\omega$.

If $h_{s+1}(e,x)=h_s(e,x)$, then $r_s(e,x)=r_{s+1}(e,x)$. In the three cases for defining $m_s(e,x)$ and $m_{s+1}(e,x)$, Lemma 2.45, and the fact that $o_{s+1}(y)\leqslant o_s(y)$ for all y, implies that $m_{s+1}(e,x)\leqslant m_s(e,x)$.

Now suppose that $h_{s+1}(e,x)\neq h_s(e,x)$; we want to show that $m_{s+1}(e,x)<m_s(e,x)$. Note that $r_{s+1}(e,x)=s+1$; let $r=r_s(e,x)$. There are four cases.

(1) If $r=0$, then $m_s(e,x)=\omega^\alpha$ and $m_{s+1}(e,x)<\omega^\alpha$.
(2) Suppose that $r>0$ and that $h_{s+1}(e,x)=0$. Then $h_s(e,x)>0$. This means that $\hat{\Phi}_e(g_s,x)$ converges in s steps, but that $\hat{\Phi}_e(g_{s+1},x)$ does not converge in $s+1$ steps; so necessarily $g_{s+1}\restriction_{\varphi_e(x)}\neq g_s\restriction_{\varphi_e(x)}$. So there is some $y<\varphi_e(x)$ such that $o_{s+1}(y)<o_s(y)$. We have

$$m_s(e,x) = \bigoplus_{y<\varphi_e(x)} (o_s(y) \oplus o_s(y)),$$

and

$$m_{s+1}(e,x) = \bigoplus_{y<\varphi_e(x)} (o_{s+1}(y) \oplus o_s(y)).$$

The desired inequality then follows from Lemma 2.45.

(3) Suppose that $r > 0$, that $h_{s+1}(e,x) > 0$, and that $h_s(e,x) > 0$. Then $h_{s+1}(e,x) \neq h_s(e,x)$ implies that $g_{s+1}\restriction_{\varphi_e(x)} \neq g_s\restriction_{\varphi_e(x)}$, so again, there is some $y < \varphi_e(x)$ such that $o_{s+1}(y) < o_s(y)$. We have

$$m_s(e,x) = \bigoplus_{y<\varphi_e(x)} (o_s(y) \oplus o_s(y)),$$

and

$$m_{s+1}(e,x) = \bigoplus_{y<\varphi_e(x)} (o_{s+1}(y) \oplus o_{s+1}(y)),$$

so again $m_{s+1}(e,x) < m_s(e,x)$.

(4) The last case is that $r > 0$, $h_{s+1}(e,x) > 0$ and $h_s(e,x) = 0$. Now the point is that $h_{r-1}(e,x) > 0$, so the argument in case (2) shows that there is some $y < \varphi_e(x)$ such that $o_r(y) < o_{r-1}(y)$, whence $o_{s+1}(y) < o_{r-1}(y)$. We have

$$m_{s+1}(e,x) = \bigoplus_{y<\varphi_e(x)} (o_{s+1}(y) \oplus o_{s+1}(y))$$

and

$$m_s(e,x) = \bigoplus_{y<\varphi_e(x)} (o_s(y) \oplus o_{r-1}(y)),$$

so we get the required inequality in this case too. □

Lemma 2.49. *There is a computable function T such that for any limit ordinal $\alpha \leqslant \varepsilon_0$ and $a < \omega$, if for some g, for all $\beta < \alpha$, $\left\langle f_s^{\varphi_a(\beta),\omega^\beta}, o_s^{\varphi_a(\beta),\omega^\beta} \right\rangle_{s<\omega}$ is a total ω^β-computable approximation of I^g_β, then $\left\langle f_s^{T(\alpha,a),\omega^\alpha}, o_s^{T(\alpha,a),\omega^\alpha} \right\rangle_{s<\omega}$ is an ω^α-computable approximation of I^g_α.*

Proof. Given α and a, define $\langle h_s, m_s \rangle$ by letting $h_s(\beta,x) = f_s^{\varphi_a(\beta),\omega^\beta}(x)$ and $m_s(\beta,x) = o_s^{\varphi_a(\beta),\omega^\beta}(x)$, if $\varphi_a(\beta)\downarrow$ and $o_s^{\varphi_a(\beta),\omega^\beta}(x)\downarrow$, otherwise we let $h_s(\beta,x)$ and $m_s(\beta,x)$ diverge. For $z \neq \beta$ for any $\beta < \alpha$, we of course let $h_s(z,x) = m_s(z,x) = 0$. By the acceptability of the list of tidy $(\omega^\alpha + 1)$-computable approximations, we can define $T(\alpha, a)$ such that if φ_a is total, and for all $\beta < \alpha$, $\left\langle o_s^{\varphi_a(\beta),\omega^\beta} \right\rangle$ is total, then $\left\langle o_s^{T(\alpha,a),\omega^\alpha} \right\rangle$ is total, and $\lim_s f_s^{T(\alpha,a),\omega^\alpha} = \lim_s h_s$. □

Proposition 2.47 now follows by effective transfinite recursion.

2.3.5 Reducing ω^α-c.a. sets and functions to iterations of the wtt-jump

Let $\alpha < \varepsilon_0$. An *instance of an ω^α-computable approximation* is a pair (f, o) of computable functions $f\colon \omega \to \omega$ and $o\colon \omega \to \omega^\alpha$ such that for all $s < \omega$, $o(s+1) \leqslant o(s)$ and if $f(s+1) \neq f(s)$ then $o(s+1) < o(s)$.

As is done in the proof of Proposition 2.7, we can list, uniformly in α, tidy instances of $(\omega^\alpha + 1)$-computable approximations. In other words, there is an effective list $\langle f_e^\alpha, o_e^\alpha \rangle$ of pairs of computable functions with the following properties:

(1) For every $\alpha < \varepsilon_0$ and every $e < \omega$, (f_e^α, o_e^α) is an instance of an $(\omega^\alpha + 1)$-computable approximation with $f_e^\alpha(0) = 0$ and $o_e^\alpha(0) = \omega^\alpha$.
(2) The listing is *acceptable*: there is a (total) computable function $c(\alpha, d, e)$ such that for all $\alpha \leqslant \varepsilon_0$ and $d, e < \omega$, if (φ_d, φ_e) is an instance of an ω^α-computable approximation, then $\lim_s \varphi_d(s) = \lim_s f^\alpha_{c(\alpha,d,e)}$.

Again the idea is to convert any pair (g, m) of partial computable functions into a pair functions (f, o) as in (1). We enumerate the graphs of m and g until we see that both $m(0)$ and $g(0)$ converge. As long as we don't see convergence, both f and o are constant; otherwise we slowly copy the values that we see.

Given the lists $\langle f_e^\alpha, o_e^\alpha \rangle$, we define the following for $\alpha \leqslant \varepsilon_0$:

(1) $C_\alpha = \{e : \exists s\ (o_e^\alpha(s) < \omega^\alpha)\}$. The sets C_α are c.e., uniformly in α.
(2) Partial functions $F_\alpha\colon C_\alpha \to \omega$ by letting, for $e \in C_\alpha$, $F_\alpha(e) = \lim_s f_e^\alpha(s)$.

Lemma 2.50. *For every $\alpha \leqslant \varepsilon_0$, F_α is partial bounded $I_\alpha^\varnothing$-computable.*

Lemma 2.50 is proved by effective transfinite recursion on $\varepsilon_0 + 1$, so again it has to be uniform: we construct a computable function R such that for all $\alpha \leqslant \varepsilon_0$,

$$\hat{\Phi}_{R(\alpha)}\left(I_\alpha^\varnothing\right) = F_\alpha.$$

The following three lemmas explain how to define $R(\alpha)$ for the three kinds of ordinals α: $\alpha = 0$, successor α, and limit α.

Lemma 2.51. *F_0 is a partial computable function, and so is partial bounded $\varnothing$-computable.*

Proof. For each $e \in C_0$, $F_0(e) = f_e^0(n)$ for n such that $o_e^0(n) = 0$. □

Lemma 2.52. *There is a computable function S such that for all $\alpha < \varepsilon_0$ and all $a < \omega$, for any function $g\colon \omega \to \omega$, if $\hat{\Phi}_a(g) = F_\alpha$, then $\hat{\Phi}_{S(\alpha,a)}(I^g) = F_{\alpha+1}$.*

Proof. We show how to define, effectively from α and a, a weak truth-table functional (Ψ, ψ) such that for any function g, if $\hat{\Phi}_a(g) = F_\alpha$ then $\hat{\Psi}(I^g) = F_{\alpha+1}$. The acceptability of the enumeration of weak truth-table functionals then allows us to effectively find an index $S(\alpha, a)$ such that $(\Psi, \psi) = (\Phi_{S(\alpha,a)}, \varphi_{\alpha,a})$.

Let $e < \omega$. If $e \notin C_{\alpha+1}$, then we leave $\psi(e)\uparrow$, and for any oracle g, $\Psi(g, e)\uparrow$.

Suppose that $e \in C_{\alpha+1}$. For abbreviation, let $(f, o) = (f_e^{\alpha+1}, o_e^{\alpha+1})$. The idea now is to break up the instance (f, o) into a finite sequence of instances, each within a copy of ω^α sitting inside $\omega^{\alpha+1}$. Let s^* witness that $e \in C_{\alpha+1}$: $o(s^*) < \omega^{\alpha+1}$. Since $\omega^{\alpha+1} = \omega^\alpha \cdot \omega$, we can write, for $s \geqslant s^*$, $o(s) = \omega^\alpha n(s) + \beta(s)$ for unique $n(s) < \omega$ and $\beta(s) < \alpha$.

Let $M = n(s^*)$. For $m \leqslant M$ we define an instance (f^m, o^m) of an ω^α-computable approximation by copying $\beta(s)$ on stages on which $n(s) = m$. Namely let $J_m = \{s \geqslant s^* : n(s) = m\}$. Then $J_M < J_{M-1} < \cdots < J_k$ is a partition of $[s^*, \omega)$ for some $k \leqslant M$; let us assume that J_m for $m \geqslant k$ is nonempty (i.e., the approximation (f, o) does not skip over the mth copy of ω^α); this is easily arranged. For $m \geqslant k$ we define $f^m(s) = f(s)$ and $o^m(s) = \beta(s)$ for $s \in J_m$, and extend in a constant way otherwise (i.e., for $s < J_m$, $f^m(s) = f(\min J_m)$ and $o^m(s) = \beta(\min J_m)$; and if $m > k$ and $s > J_m$, we define similarly but with $\max J_m$ replacing $\min J_m$). For $m < k$ we leave $f^m(s)$ and $o^m(s)$ undefined for all s. The point is of course that $\lim_s f^k(s) = F_{\alpha+1}(e)$.

By the acceptability of the list $\langle f_d^\alpha, o_d^\alpha \rangle$, we can effectively get numbers d_m for $m \leqslant M$ such that for all $m \leqslant M$,

- $d_m \in C_\alpha$ if and only if $m \geqslant k$; and
- if $m \geqslant k$, then $F_\alpha(d_m) = \lim_s f^m(s)$.

Now the procedure Ψ queries the oracle on each pair (a, d_m). The use is bounded by $\max\{(a, d_m) : m < M\}$; this is revealed to us once we see that $e \in C_{\alpha+1}$, so this use is partial computable (uniformly in a, α and e). If indeed $\hat{\Phi}_a(g) = F_\alpha$ then $I^g(a, d_m) = 0$ if and only if $m < k$ and $I^g(a, d_k) = 1 + F_{\alpha+1}(e)$, so this is what Ψ outputs. ☐

Lemma 2.53. *There is a recursive function T such that for any limit ordinal α, $a < \omega$ and sequence $\langle g_\beta \rangle_{\beta<\alpha}$ of functions, if for all $\beta < \alpha$, $\hat{\Phi}_{\varphi_a(\beta)}(g_\beta) = F_\beta$, then $\hat{\Phi}_{T(\alpha,a)}\left(\bigoplus_{\beta<\alpha} g_\beta\right) = F_\alpha$.*

Proof. Of course now the point is that ω^α is the limit of the ordinals ω^β for $\beta < \alpha$. So given $e \in C_\alpha$ we can effectively find some $\beta < \alpha$ and some s^* such that $o_e^\alpha(s^*) < \omega^\beta$, and so can translate (f_e^α, o_e^α) to an instance of an ω^β-computable approximation; so we can find some $d \in C_\beta$ such that $\lim_s f_e^\alpha(s) = \lim_s f_d^\beta(s)$. We can then find some number $\psi(e)$, effectively computed from e, such that from $\left(\bigoplus_{\beta<\alpha} g_\beta\right)\upharpoonright_{\psi(e)}$ we can compute $g_\beta\upharpoonright_{\varphi_a(d)}$, and so using $\Phi_{\varphi_a(\beta)}$ can output

$$\hat{\Phi}_{\varphi_a(\beta)}(g_\beta, d) = F_\beta(d) = F_\alpha(e)$$

as required. Again, all this can be coded by a functional Ψ, and by acceptability we can effectively find an index $T(\alpha, a)$ such that $(\Phi_{T(\alpha,a)}, \varphi_{T(\alpha,a)}) = (\Psi, \psi)$. □

Now effective transfinite recursion on $\varepsilon_0 + 1$, using Lemmas 2.51, 2.52, and 2.53, builds a computable function R such that for all $\alpha \leqslant \varepsilon_0$, $\hat{\Phi}_{R(\alpha)}(I_\alpha^\varnothing) = F_\alpha$, and so proves Lemma 2.50.

Proof of part (2) of Theorem 2.40. Let $\alpha \leqslant \varepsilon_0$. Proposition 2.47 states that $I_\alpha^\varnothing$ is ω^α-c.a. By Proposition 2.46, every function $g \leqslant_{\mathrm{wtt}} I_\alpha^\varnothing$ is also ω^α-c.a.

For the converse, let g be an ω^α-c.a. function; let $\langle g_s, m_s\rangle$ be an ω^α-computable approximation for g. For every $x<\omega$, the sequence $\langle g_s(x), m_s(x)\rangle_{s<\omega}$ is an instance of an ω^α-computable approximation, and so by acceptability of the numbering of the partial instances of such approximations, there is a computable function h such that for all x, $h(x) \in C_\alpha$ and $g(x) = F_\alpha(h(x))$. By Lemma 2.50, there is a weak truth-table functional (Φ, φ) such that $F_\alpha = \hat{\Phi}(I_\alpha^\varnothing)$. Let $\psi(x) = \varphi(h(x))$, and for any oracle f, let $\Psi(f, x) = \Phi(f, h(x))$ with the same use. Then ψ is total (as $\operatorname{range} h \subseteq C_\alpha$), and $\hat{\Psi}(I_\alpha^\varnothing) = g$, so $g \leqslant_{\mathrm{wtt}} I_\alpha^\varnothing$. □

Proof of part (1) of Theorem 2.40. The proof of the backward direction is identical to the corresponding proof of part (2), because as mentioned after the statement of Proposition 2.47, Proposition 2.38 implies that $\varnothing^{\langle\alpha\rangle}$ is ω^α-c.a.

For $\alpha \leqslant \varepsilon_0$, define $D_\alpha\colon C_\alpha \to \{0,1\}$ by letting $D_\alpha(e) = F_\alpha(e) \mod 2$. If A is an ω^α-c.a. set, then there is a computable function $h\colon \omega \to C_\alpha$ such that for all x, $A(x) = D_\alpha(h(x))$. Hence to show that every ω^α-c.a. set is weak truth-table reducible to $\varnothing^{\langle\alpha\rangle}$, we show that D_α is a partial bounded $\varnothing^{\langle\alpha\rangle}$-computable function.

The proof follows the line of argument for Lemma 2.50. A computable function R such that for all α, $\hat{\Phi}_{R(\alpha)}\left(\varnothing^{\langle\alpha\rangle}\right) = D_\alpha$ is constructed by effective transfinite recursion on $\varepsilon_0 + 1$, once analogues of Lemmas 2.51, 2.52, and 2.53 are proved:

(1) D_0 is a partial computable function.
(2) There is a computable function S such that for all $\alpha < \varepsilon_0$ and all $a < \omega$, for any set $A \in 2^\omega$, if $\hat{\Phi}_a(A) = D_\alpha$, then $\hat{\Phi}_{S(\alpha,a)}(A^\dagger) = D_{\alpha+1}$.
(3) There is a recursive function T such that for any limit ordinal α, $a<\omega$ and sequence $\langle A_\beta\rangle_{\beta<\alpha}$ of sets, if for all $\beta<\alpha$, $\hat{\Phi}_{\varphi_a(\beta)}(A_\beta) = D_\beta$, then $\hat{\Phi}_{T(\alpha,a)}\left(\bigoplus_{\beta<\alpha} A_\beta\right) = D_\alpha$.

(1) follows immediately from the definition of D_0 and Lemma 2.51. For (3) we can simply take the function given by the proof of Lemma 2.53. The only

new ingredient is in the proof of (2). Again, given α, a and $e \in C_{\alpha+1}$, we let $(f, o) = (f_e^{\alpha+1}, o_e^{\alpha+1})$, and define the functions β and n, the number M, the pairs (f^m, o^m) for $m \leqslant M$ and the numbers k and d_m exactly as was done in the proof of Lemma 2.52. So $m \geqslant k$ if and only if $d_m \in C_\alpha$, and $D_{\alpha+1}(e) = D_\alpha(d_k)$.

The difficulty of course is that $A^\dagger$ does not tell us the value of $\hat{\Phi}_a(A, d_m) = D_\alpha(d_m)$, only whether $\hat{\Phi}_a(A, d_m)$ converges or not. But since the value is either 0 or 1, we can convert it to convergence or divergence of an auxiliary functional. That is, we can effectively calculate an index b and numbers c_m for $m \leqslant M$ such that for any oracle X, $\hat{\Phi}_b(X, c_m)\downarrow$ if and only if $d_m \in C_\alpha$ and $\hat{\Phi}_a(X, d_m)\downarrow = 1$; for the use we can let $\varphi_b(c_m) = \varphi_a(d_m)$. We then let

$$\psi(e) = 1 + \max\{(a, d_m), (b, c_m) : m \leqslant M\},$$

which is again partial computable; and for any oracle $Y \in 2^\omega$, we calculate, for $e \in C_{\alpha+1}$, $\hat{\Psi}(Y, e)$ by first finding the least m such that $(a, d_m) \in Y$ (we diverge if there is none), and then output $Y(b, c_m)$. If $\hat{\Phi}_a(A) = D_\alpha$ and $e \in C_{\alpha+1}$ then the least m such that $(a, d_m) \in A^\dagger$ is k, and

$$\Psi(A^\dagger, e) = A^\dagger(b, c_k) = \hat{\Phi}_a(A, d_k) = D_\alpha(d_k) = D_{\alpha+1}(e)$$

as required. □

Chapter Three

The hierarchy of totally α-c.a. degrees

THE FOLLOWING IS the central definition of this work. For $\alpha = \omega$, this definition was originally made by J. S. Miller (unpublished), and first investigated in detail in [25].

Definition. Let $\alpha \leqslant \varepsilon_0$. A Turing degree $\mathbf{d}$ is *totally α-c.a.* if every function $f \in \mathbf{d}$ is α-c.a.

3.1 TOTALLY $\mathcal{R}$-C.A. DEGREES

Basic properties of totally α-c.a. degrees are shared among totally $\mathcal{R}$-c.a. degrees, even when $\mathcal{R}$ is not canonical. For any computable well-ordering $\mathcal{R}$, we say that a Turing degree $\mathbf{d}$ is *totally $\mathcal{R}$-c.a.* if every function $f \in \mathbf{d}$ is $\mathcal{R}$-c.a.

We note the following:

Lemma 3.1. *Let $\mathcal{R}$ be a computable well-ordering. A degree $\mathbf{d}$ is totally $\mathcal{R}$-c.a. if and only if every $f \leqslant_{\mathrm{T}} \mathbf{d}$ is $\mathcal{R}$-c.a.*

Proof. Suppose that $\mathbf{d}$ is a totally $\mathcal{R}$-c.a. degree. Let $g \in \mathbf{d}$ be any function. Let $f \leqslant_{\mathrm{T}} \mathbf{d}$. Then $f \oplus g \in \mathbf{d}$, so $f \oplus g$ has an $\mathcal{R}$-computable approximation, from which we can get an $\mathcal{R}$-computable approximation for f. □

3.1.1 Totally $\mathcal{R}$-c.a. degrees and low$_2$ degrees

The following theorem shows that total $\mathcal{R}$-c.a.-ness is indeed a notion of lowness.

Theorem 3.2. *For any computable well-ordering $\mathcal{R}$, every totally $\mathcal{R}$-c.a. degree is* low$_2$.

Proof. Let $\mathcal{R}$ be a computable well-ordering. By Corollary 2.8, there is a $\mathbf{0}'$-computable sequence $\langle f^e \rangle_{e<\omega}$ consisting of all $\mathcal{R}$-c.a. functions. Using this sequence, it is easy to construct a $\mathbf{0}'$-computable function f which dominates every $\mathcal{R}$-c.a. function. Hence if $\mathbf{d}$ is a totally $\mathcal{R}$-c.a. degree, then f dominates all functions in $\mathbf{d}$. By a classic result of Martin's [70], $\mathbf{d}$ is low$_2$. □

Ershov's Theorem 2.13 can be extended to low_2 degrees.

Proposition 3.3. *Every Δ^0_2, low_2 degree is totally $\mathcal{R}$-c.a. for some computable well-ordering $\mathcal{R}$ of order-type ω.*

Proof. Let $\mathbf{d}$ be a Δ^0_2, low_2 degree. The proof of Theorem 2.13 can be adapted once we give a uniform $\mathbf{0}'$-enumeration of all the functions reducible to $\mathbf{d}$.

Let $D \in \mathbf{d}$, and let $\langle D_s \rangle$ be a computable approximation of D.

Since $\mathbf{d}$ is low_2, the collection of e such that $\Phi_e(D)$ is total is Σ^0_3; say $\Phi_e(D)$ is total if and only if $\exists x \forall y \exists z\, Q(e,x,y,z)$ where Q is computable. For e, x and $s < \omega$, let $y_s(e,x)$ be the greatest y such that for all $y' \leqslant y$ there is some $z < s$ such that $Q(e,x,y',z)$ holds. Now for all such e, x and s, for $n < \omega$, let $f^{e,x}_s(n) = \Phi_e(D,n)[y_s(e,x)]$ if $n < \operatorname{dom} \Phi_e(D)[y_s(e,x)]$, and $f^{e,x}_s(n) = 0$ for other n (recall that we write $\Phi_e(D)[s]$ for $\Phi_{e,s}(D_s)$).

If x witnesses that $\Phi_e(D)$ is total (it is an existential witness to the outermost quantifier in the Σ^0_3 property above, meaning that $y_s(e,x) \to \infty$), then $\lim_s f^{e,x}_s = \Phi_e(D)$; if not, then the sequence $\langle f^{e,x}_s \rangle_{s<\omega}$ is eventually constant. Hence, renumbering, we get a uniformly computable sequence $\left\langle \langle f^d_s \rangle_{s<\omega} \right\rangle_{d<\omega}$ of computable approximations, with the collection of limits $\{f^d : d < \omega\}$ (where $f^d = \lim_s f^d_s$) consisting precisely of all the functions computable from $\mathbf{d}$.

Now we let $\mathcal{R}$ be the interspersed union of the well-orderings built in the proof of Theorem 2.13 for the approximations $\langle f^d_s \rangle$. We let

$$R = \left\{ (d,x,s) \in \omega \times \omega \times \omega : s = 0 \text{ or } f^d_s(x) \neq f^d_{s-1}(x) \right\},$$

and for $(d,x,s), (e,y,t) \in R$, let $(d,x,s) <_{\mathcal{R}} (e,y,t)$ if $\langle d,x \rangle < \langle e,y \rangle$ or if $(d,x) = (e,y)$ and $t < s$. The argument of the proof of Theorem 2.13 shows that $\mathcal{R} = (R, <_{\mathcal{R}})$ has order-type ω and that for every $d < \omega$, $\langle f^d_s \rangle_{s<\omega}$ can be extended to an $\mathcal{R}$-computable approximation. Hence $\mathbf{d}$ is totally $\mathcal{R}$-c.a. □

The argument for Corollary 2.29 now shows:

Corollary 3.4. *Every Δ^0_2, low_2 degree is totally $\mathcal{R}_o$-c.a. for some notation $o \in \mathcal{O}$ for ω^2.*

3.1.2 C.e. degrees

In this work we focus on totally α-c.a. c.e. degrees, namely those totally α-c.a. Turing degrees which contain a c.e. set.

The following result shows that for c.e. degrees, sets (elements of Cantor space) capture everything expressed by functions (elements of Baire space) as far as approximations are concerned. Technically, this is the first application in this monograph of the *permitting* method, calibrated at the level of total $\mathcal{R}$-c.a.-ness.

Proposition 3.5. *Let $\mathcal{R}$ be a computable well-ordering. A c.e. degree* $\mathbf{d}$ *is totally $\mathcal{R}$-c.a. if and only if every set $Z \leqslant_{\mathrm{T}} \mathbf{d}$ is $\mathcal{R}$-c.a.*

The argument of Lemma 3.1 shows now that a c.e. degree is totally $\mathcal{R}$-c.a. if and only if every set $Z \in \mathbf{d}$ is $\mathcal{R}$-c.a.

Proof. Let $\mathbf{d}$ be a c.e. degree, and suppose that some $g \leqslant_{\mathrm{T}} \mathbf{d}$ is not $\mathcal{R}$-c.a. Since $\mathbf{d}$ is c.e., there is some computable approximation $\langle g_s \rangle$ of g such that $\mathbf{d}$ computes the modulus of this approximation.

We construct Z by giving a computable approximation $\langle Z_s \rangle$ for Z. Let $\langle \langle Z_s^e, o_s^e \rangle_{s<\omega} \rangle_{e<\omega}$ be an effective enumeration of tidy $(\mathcal{R}+1)$-computable approximations such that letting $Z^e = \lim_s Z_s^e$, the sequence $\langle Z^e \rangle$ enumerates the $\mathcal{R}$-c.a. sets. Further, as is clear from the construction in Proposition 2.7, every $\mathcal{R}$-c.a. set appears as Z^e for some e such that the approximation $\langle Z_s^e, o_s^e \rangle$ is eventually $\mathcal{R}$-computable: for all n there is some s such that $o_s^e(n) \in R$.

To defeat the threat that $Z = Z^e$, we pick potential witnesses x for this e^{th} requirement, and try to ensure that $Z(x) \neq Z^e(x)$. Naturally, we examine the sequence $\langle Z_s^e(x) \rangle_{s<\omega}$, and if there is equality between $Z_s(x)$ and $Z_{s+1}^e(x)$, we will want to change the value of $Z(x)$. To keep Z being computable from D, each such change must be permitted by g. We prompt g to give us such a permission by making a threat of our own, of giving an $\mathcal{R}$-computable approximation for g.

Since permission will only be granted eventually, we need to attempt to ensure that $Z(x) \neq Z^e(x)$ for infinitely many numbers x. To avoid unnecessary interaction between requirements, these all have to be distinct. Recall that $\langle \omega^{[e]} \rangle_{e<\omega}$ is a partition of ω into uniformly computable sets (which we often refer to as "columns").

We start by defining $Z_0 = \varnothing$. At stage s we *wish to flip* $x \in \omega^{[e]}$ if $o_s^e(x) \in R$ and $Z_s^e(x) = Z_s(x)$. We are *allowed to flip* x at stage s if $g_{s+1} \upharpoonright_x \neq g_s \upharpoonright_x$. If we both wish to flip and are allowed to flip some x, then we flip it: we set $Z_{s+1}(x) = 1 - Z_s(x)$. Otherwise, we set $Z_{s+1}(x) = Z_s(x)$. This defines the sequence $\langle Z_s \rangle$.

Let $x < \omega$. If $g_s \upharpoonright_x = g_t \upharpoonright_x$ for all $s \geqslant t$ then $Z_s(x) = Z_t(x)$ for all $s \geqslant t$. Hence $\langle Z_s \rangle$ is a computable approximation of a set Z. In fact, since $\mathbf{d}$ computes the modulus for $\langle g_s \rangle$, $Z \leqslant_{\mathrm{T}} \mathbf{d}$.

To show that Z is not $\mathcal{R}$-c.a. we show that if the approximation $\langle Z_s^e, o_s^e \rangle$ is eventually $\mathcal{R}$-computable then $Z \neq Z^e$. Fix such e and suppose for a contradiction that $Z^e = Z$. We define a sequence $\langle h_s, m_s \rangle$ by recursion. For $y < \omega$ let x be the least element of $\omega^{[e]}$ greater than y. For all s let $m_s(y) = o_s^e(x)$. Start with $h_s(y) = 0$ for all $s < y$. Now if $m_s(y) = m_{s-1}(y)$ then let $h_s(y) = h_{s-1}(y)$; otherwise let $h_s(y) = g_s(y)$. Then $\langle h_s, m_s \rangle$ is an eventually $\mathcal{R}$-computable approximation for $h = \lim_s h_s$ (which is therefore $\mathcal{R}$-c.a.); we show that $h = g$.

Suppose not. Again let $y < \omega$ and let x be the least element of $\omega^{[e]}$ greater than y. Let t be the stage at which the sequence $\langle m_s(y) \rangle$ stabilizes. So $h(y) = h_t(y) = g_t(y)$ (by minimality of t) and for all $s \geqslant t$, $Z_s^e(x) = Z_t^e(x) = Z^e(x)$. Suppose that $g(y) \neq g_t(y)$. Let s be the least stage $s > t$ at which we see that

$g_{s+1}\restriction_x \neq g_s\restriction_x$. We are permitted to flip $Z(x)$ at stage s, so $Z_{s+1}(x) \neq Z^e_{s+1}(x)$ (either because we flipped it at stage s, or we did not need to). By induction, at no later stage will we want to flip x, so $Z(x) \neq Z^e_{s+1}(x) = Z^e(x)$, contradicting the assumption that $Z = Z^e$. □

The fact that **d** is a c.e. degree is heavily used in the proof of Proposition 3.5. Barmpalias (unpublished) constructed a degree **d** such that every set $Z \in \mathbf{d}$ is ω-c.a. (in fact, **d** is superlow), but some function $f \in \mathbf{d}$ is not ω-c.a.

3.2 THE FIRST HIERARCHY THEOREM: TOTALLY ω^α-C.A. DEGREES

Let $\gamma < \alpha \leqslant \varepsilon_0$. Since every γ-c.a. function is also α-c.a. (see Section 2.2), every totally γ-c.a. degree is also totally α-c.a. The question is when does this hierarchy collapse?

Theorem 3.6. *Let $\alpha \leqslant \varepsilon_0$. There is a totally α-c.a. degree which is not totally γ-c.a. for any $\gamma < \alpha$ if and only if α is a power of ω. If α is a power of ω, then in fact there is a c.e. degree which is totally α-c.a. but not totally γ-c.a. for any $\gamma < \alpha$.*

The first $\omega \cdot 2$ many levels of the hierarchy of totally α-c.a. degrees are depicted in Figure 3.1.

For the forward direction of the first hierarchy theorem, we prove the following lemma. It is proved in generality greater than is currently necessary, but which will be useful later.

Lemma 3.7. *Let $\gamma < \varepsilon_0$, and let **d** be a Turing degree such that every $g \in \mathbf{d}$ is γm-c.a. for some $m < \omega$. Then **d** is totally γ-c.a.*

Proof. Let $f \in \mathbf{d}$. Define $g(x) = f\restriction_x$; then $g \in \mathbf{d}$. By assumption, there is some $m < \omega$ such that g is γm-c.a. Let $\langle g_s, o_s \rangle$ be a γm-computable approximation for g. By speeding up this approximation, we may assume that for all x and s, $g_s(x)$ is a string of length x.

For every x and s there is some unique $k < m$ such that $o_s(x) \in [\gamma \cdot k, \gamma \cdot (k+1))$; we denote this k by $k_s(x)$. We have $o_s(x) = \gamma \cdot k_s(x) + \beta_s(x)$ for some $\beta_s(x) < \gamma$. For every x and s, $k_{s+1}(x) \leqslant k_s(x)$, and so $k_\omega(x) = \lim_s k_s(x)$ is well-defined. We let $k^* = \liminf_x k_\omega(x)$.

We can now give a γ-computable approximation $\langle f_s, m_s \rangle$ for f. Fix x^* such that for all $x \geqslant x^*$, $k_\omega(x) \geqslant k^*$; so for all s and all $x \geqslant x^*$, $k_s(x) \geqslant k^*$. For any $y < \omega$ we can effectively find some $x = h(y) > y$ such that $k_\omega(x) = k^*$, by insisting that $x \geqslant x^*$ and waiting until we see some stage s such that $k_s(x) = k^*$. We let $t(y)$ be some stage t such that $k_t(h(y)) = k^*$. Fix y, and let $x = h(y)$; we then let

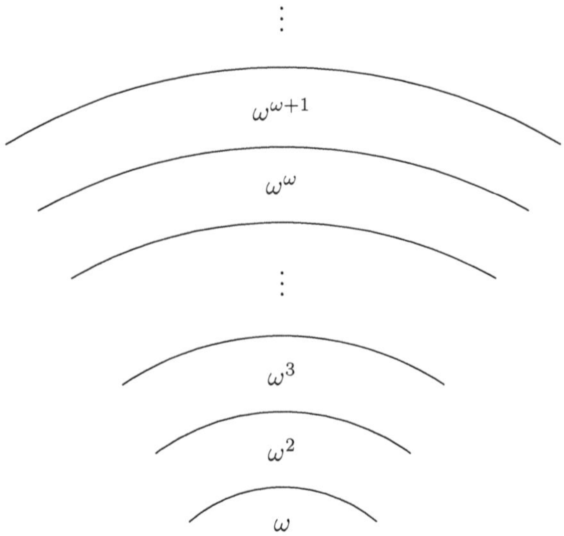

Figure 3.1. The first hierarchy theorem. "ω^α" denotes the collection of totally ω^α-c.a. degrees.

$$m_s(y) = \beta_{\max\{s,t(y)\}}(x)$$

and

$$f_s(y) = \left(g_{\max\{s,t(y)\}}(x)\right)(y).$$

If $f_{s+1}(y) \neq f_s(y)$ then $s \geqslant t(y)$ and $g_{s+1}(x) \neq g_s(x)$, and so $o_{s+1}(x) < o_s(x)$. Since $s \geqslant t(y)$, we have $o_s(x) = \gamma \cdot k^* + \beta_s(x)$ and $o_{s+1}(x) = \gamma \cdot k^* + \beta_{s+1}(x)$, and so $m_{s+1}(y) = \beta_{s+1}(x) < \beta_s(x) = m_s(y)$. Hence $\langle f_s, m_s \rangle$ is indeed a γ-computable approximation. If $g_s(x) = g(x)$ then $f_s(y) = (g(x))(y) = f(y)$, so $\lim_s f_s = f$. $\square$

The forward direction of Theorem 3.6 follows: if α is not a power of ω, then there is some $\gamma < \alpha$ and some m such that $\alpha \leqslant \gamma m$, and so every totally α-c.a. degree is totally γm-c.a., and so by Lemma 3.7 is actually totally γ-c.a.

The rest of this section is devoted to the proof of the backward direction of Theorem 3.6: given some $\alpha \leqslant \varepsilon_0$ which is a power of ω, the construction of a c.e. degree which is totally α-c.a. but not totally γ-c.a. for any $\gamma < \alpha$. Fix such α. The key property of α, which makes the construction work, is that α is closed under addition (Proposition 2.42). We define a computable enumeration $\langle D_s \rangle$ of a c.e. set D, and ensure that $\deg_{\mathrm{T}}(D)$ is totally α-c.a. but not totally γ-c.a. for any $\gamma < \alpha$.

To witness the properness, we enumerate a Turing functional Λ and ensure that $\Lambda(D)$ is not γ-c.e. for any $\gamma < \alpha$. We fix, for each $\gamma < \alpha$, an enumeration

$\langle\langle f_s^{e,\gamma}, o_s^{e,\gamma}\rangle_{s<\omega}\rangle_{e<\omega}$ of tidy $(\gamma+1)$-computable approximations whose limits $f^{e,\gamma} = \lim_s f_s^{e,\gamma}$ consist of all γ-c.a. functions (Proposition 2.7). To show that $\Lambda(D)$ is not γ-c.a. for any $\gamma < \alpha$, it is sufficient to meet, for all $\gamma < \alpha$ and $e < \omega$, the requirement

$P^{e,\gamma}$: There is some p such that $\Lambda(D,p) \neq f^{e,\gamma}(p)$.

Of course, we also need to ensure that $\Lambda(D)$ is total. To show that $\deg_T(D)$ is totally α-c.a., we need to meet, for all $e < \omega$, the requirement

Q_e: If $\Phi_e(D)$ is total, then it is α-c.a.

Discussion

Perhaps surprisingly, the simplest construction one would hope to work, does work. We give full details because several other constructions we present later are elaborations on this one. We use the terminology discussed in Subsection 1.6.3.

First, independently consider the strategies for meeting each requirement. To meet $P^{e,\gamma}$, we pick a witness p (which recall is also called a *follower*), and whenever we observe that $f_s^{e,\gamma}(p) = \Lambda_s(D_s, p)$, we change the value of $\Lambda(D,p)$ by enumerating the use $\lambda_s(p) = \lambda_s(D_s, p)$ into D_{s+1}.[1] If this is performed without interruption, success is guaranteed, because our opponent can change the value of $f^{e,\gamma}(p)$ only finitely many times.

To meet Q_e, the only thing we can do is observe, for each input x, the value of $\Phi_{e,s}(D_s, x)$, and at various stages s declare that we believe that $\Phi_e(D,x) = \Phi_{e,s}(D_s, x)$. If $\Phi_e(D)$ is total then we will eventually be right; we need to ensure, informally speaking, that the "number of times" we change our mind about the value of $\Phi_e(D,x)$ is bounded by α. (Of course, technically we mean that we need to define a decreasing sequence of ordinals below α which is associated with the mind-changes. However, it is useful to think of α as bounding the number of mind-changes, in an analogy with the situation $\alpha = \omega$.) There is one possible action Q_e can take, and that is to impose restraint: if we freeze D below the use $\varphi_{d,s}(D_s, x)$, then our guess is correct.

The conflict between different requirements is now clear: when a requirement $P^{e,\gamma}$ enumerates $\lambda_s(p)$ into D, this may destroy a computation $\Phi_{d,s}(D,x)$ for some $d \leqslant e$ say, which Q_d has earlier declared it believed. The requirement Q_d can tolerate some injury; after all, it is not trying to make $\Phi_d(D)$ computable. It needs to limit the "amount of injury" to be below α. This is possible because once a follower p is chosen, we can tell "how many times" the requirement $P^{e,\gamma}$ will act: the bound is $o_0^{e,\gamma}(p)$. Before starting to make guesses about $\Phi_d(D,x)$, the requirement Q_d will observe which requirements will bother it and take

[1]Recall our convention that since both D and Γ are defined by us, the use $\lambda(p)$ is the largest number actually queried during the computation.

their bounds $o_0^{e,\gamma}(p)$ into consideration. The fact that α is closed under addition means it can deal with injury from more than one other requirement.

This plan will not succeed if we allow requirements $P^{e,\gamma}$ to "gang up" on Q_d. Suppose that at some stage s, Q_d starts making guesses about $\Phi_d(D,x)$, and declares an ordinal $\beta < \alpha$ bounding the "number of times" it will change its mind about this value. This bound β is calculated on the basis of which followers p for requirements $P^{e,\gamma}$ it is observing at stage s. It would be bad if we allow a different requirement $P^{e,\gamma}$ (say for some $e > s$) to also destroy $\Phi_d(D,x)$: the bound on the action of such a requirement cannot be comprehended by Q_d at stage s. Such requirements need to be restrained by Q_d: the numbers $\lambda_t(p)$ which they enumerate into D must be greater than the use $\varphi_{d,s}(x)$.

On the face of it, this can be arranged using only finite injury: when Q_d observes a new $\Phi_d(D,x)$ computation, it initialises all requirements $P^{e,\gamma}$ which are not allowed to injure this computation. The use $\lambda_t(p)$ for followers picked by these requirements later will be greater than $\varphi_{d,s}(x)$ as required. The reason that the injury will be finite is that it is guaranteed that the finitely many requirements which do have the right to injure $\Phi_d(D,x)$ only act at finitely many stages. Thus, it would seem, we would eventually either see a final computation $\Phi_d(D,x)$ and injury to weaker $P^{e,\gamma}$ on behalf of this computation will cease; or the computation $\Phi_d(D,x)$ never recovers, in which case, also, eventually initialisation of weaker requirements will stop.

However, a complication arises from the combined influence of several negative requirements on some positive requirement. To see this, we first note that the permission to injure a computation that some Q_d is monitoring is *follower-based* rather than *requirement-based.* Say that a positive requirement $P^{e,\gamma}$ picks a follower p. Then we see a computation $\Phi_d(D,x)$. Since p is already chosen, Q_d can observe $o_0^{e,\gamma}(p)$ and allow $P^{e,\gamma}$ to injure the computation. However, if for some reason later, $P^{e,\gamma}$ abandons the follower p and replaces it by a new follower p', the requirement Q_d can no longer tolerate any action by $P^{e,\gamma}$: the ordinal $o_s^{e,\gamma}(p')$ may be much larger than $o_0^{e,\gamma}(p)$, and could not have been observed by Q_d at the stage it first started copying $\Phi_d(D,x)$. In a sense, the requirement $P^{e,\gamma}$ is demoted (it loses priority) relative to the pair (d,x).

Now consider such a positive requirement $P = P^{e,\gamma}$ and two negative requirements Q_c and Q_d. Suppose that, by an action of a positive requirement stronger than P, P is no longer allowed to destroy $\Phi_c(D,0)$, but that currently, $\Phi_c(D,0)\uparrow$. Meanwhile, P has a follower p_0, and we observe $\Phi_d(D,0)$ for the first time. The follower p_0 is allowed to injure that computation, and that computation is indeed destroyed (by P or by some weaker positive requirement). Then, we see that $\Phi_c(D,0)\downarrow$ with large use; this forces P to cancel p_0 and appoint a new follower p_1. In turn, this means that $\Phi_d(D,0)$ no longer tolerates P-action. While $\Phi_d(D,0)\uparrow$, we see that $\Phi_c(D,1)\downarrow$, and it observes p_1; some action destroys the computation. We then see that $\Phi_d(D,0)\downarrow$, and p_1 is abandoned and replaced by a new follower p_2, and so $\Phi_c(D,1)$ can no longer tolerate P. The see-saw between Q_c and Q_d eventually causes infinitely much injury to P. Note that one negative requirement is not sufficient for this argument, as we assume that $\operatorname{dom} \Phi_d(D)$ is an initial segment of ω.

The source of this problem is P's haste in appointing a replacement follower. If it waited until $\Phi_c(D, 0)$ converged before it appointed p_0, no injury would be necessary. For this to be possible, P needs to guess whether $\Phi_c(D, 0)$ will indeed converge in the future; if not, it will not wait. This necessitates the use of a tree of strategies in the construction.

The tree of strategies

As mentioned above (Section 1.6), to define the tree, we specify recursively the association of nodes to requirements, and specify the outcomes of nodes working for particular requirements. To specify the priority ordering of nodes, we specify the ordering between outcomes of any node.

We order all the requirements, Q_e and $P^{e,\gamma}$, in order-type ω; all nodes of length k work for the kth requirement on the list. The outcomes of a node working for Q_e are ∞ and $\mathtt{fin}$, with $\infty < \mathtt{fin}$; a node working for $P^{e,\gamma}$ has only one outcome.

Construction

At stage s, we let the collection of accessible nodes δ_s be an initial segment of the tree of strategies.

Let σ be a node which is accessible at stage s. We describe the action that σ takes, and if it does not end the stage, then we specify which immediate successor of σ is also accessible at stage s. Both of these depend, of course, on the requirement for which σ works.

Suppose first that σ works for Q_e. Then σ takes no action beyond determining which successor is accessible. If s is the least stage at which σ is accessible, we let $\sigma\hat{}\infty \in \delta_s$. If not, let t be the last stage before stage s at which $\sigma\hat{}\infty$ was accessible. If $t < \operatorname{dom} \Phi_{e,s}(D_s)$, let $\sigma\hat{}\infty \in \delta_s$.[2] Otherwise, we let $\sigma\hat{}\mathtt{fin} \in \delta_s$.

Now suppose that σ works for $P^{e,\gamma}$. As σ has but one outcome, the determination of the next element of δ_s is immediate, unless σ acts and ends the stage, in which case σ is the last element of δ_s. We let σ act as follows:

(1) If σ has no follower, then σ appoints a new, large follower p for itself.
(2) If σ has a follower p, and $\Lambda_s(D_s, p) = f_s^{e,\gamma}(p)$, then σ enumerates $\lambda_s(p)$ into D_{s+1}. We will later verify that $\lambda_s(p) \notin D_s$.

In either case, we set $\Lambda_{s+1}(D_{s+1}, p) = s + 1$ with large use. Technically, this means that we pick a large number u, and enumerate the axiom $D_{s+1}\restriction_u \mapsto (p, s)$ into Λ_{s+1}. The point of the value $s+1$ is that $\Lambda_{s+1}(D_{s+1}, p) \neq f_{s+1}^{e,\gamma}(p)$, since by convention, for all t, $f_t^{e,\gamma}(p) < t$.

[2] Again recall that we assume that $\operatorname{dom} \Phi_{e,s}(D_s)$ is an initial segment of ω (Convention 1.10), and that we use von Neumann natural number notation (Convention 1.11).

Also, in either case, we end the stage. If neither case (1) nor case (2) holds, then σ does not act, and the unique immediate successor of σ on the tree of strategies is accessible at stage s.

If σ ended the stage, then all nodes that are weaker than σ are *initialised.* For positive requirements $P^{e,\gamma}$, being initialised means that their followers are cancelled, and so at the next time they are visited, they have no follower and need to appoint a new one.

At the end of the stage, for each $p < s$ which is not at that moment a follower for some node on the tree, if $\Lambda_s(D_s, p)\uparrow$ then we set $\Lambda_{s+1}(D_{s+1}, p) = 0$ with use -1. That is, we enumerate the axiom $\langle\rangle \mapsto (p, 0)$ into Λ_{s+1}.

Verification

The following lemma will be familiar to experts in effective constructions, indeed, it is usually taken for granted and not mentioned explicitly. We give a careful and detailed presentation here, but will subsequently only sketch such proofs. For the following lemma, we first note that if p is a follower for some node σ at the beginning of stages $t < s$, and $p \in \operatorname{dom} \Lambda_t(D_t)$ and $p \in \operatorname{dom} \Lambda_s(D_s)$ (as we shall soon verify), then $p < \lambda_t(p) \leqslant \lambda_s(p)$, since $\lambda_r(p)$ is always chosen to be large.

Lemma 3.8. *The functional Λ is consistent for D. Further, at every stage s:*

(a) Λ_s is consistent for D_s.

Let σ be a node which works for a positive requirement, and suppose that at the beginning of stage s, σ has a follower p.

(b) $\Lambda_s(D_s, p)\downarrow$ and $\lambda_s(p) \notin D_s$.
(c) If p' is, at the beginning of stage s, a follower for a node σ' weaker than σ, then $\lambda_s(p) < p'$. And so $\lambda_s(p) < \lambda_s(p')$.
(d) Let $t < s$, and suppose that p was a follower for σ at the beginning of stage t. So σ was not initialised at any stage $r \in [t, s)$. Then $D_t\restriction_{\lambda_t(p)} = D_s\restriction_{\lambda_t(p)}$. If, further, σ does not act at any stage $r \in [t, s)$, then $D_t\restriction_{\lambda_t(p)+1} = D_s\restriction_{\lambda_t(p)+1}$ (this implies that $\lambda_s(p) = \lambda_t(p)$).

Proof. We prove (a), (b), (c), and (d) simultaneously by induction on s. Assume the lemma holds for $s-1$; we consider the action taken at stage $s-1$.

For (a) at stage s, we invoke Lemma 1.9. Condition (1) of that lemma certainly holds at every stage of the construction. Condition (2) also holds: at stage $s-1$, at most one node σ enumerates a new axiom into Λ_s which pertains to its follower p; at the end of the stage we may enumerate further axioms, but only for numbers which are no longer followers, and so for numbers other than p. For condition (3), suppose that a new axiom pertaining to some number p is added to Λ_s during stage $s-1$, but that $p \in \operatorname{dom} \Lambda_{s-1}(D_{s-1})$; we need to show that $\lambda_{s-1}(p)$ is enumerated into D_s. The assumption on p implies that p is not chosen as a new follower at stage $s-1$. At the end of the stage we add axioms only for numbers $p \notin \operatorname{dom} \Lambda_{s-1}(D_{s-1})$; so it must be that p is a follower for some node σ at the beginning of stage $s-1$. Thus, σ acts at stage $s-1$ and enumerates $\lambda_{s-1}(p)$ into D_s (we use (b) at stage $s-1$); this shows condition (3) of Lemma 1.9 holds. This shows that (a) holds at stage s as well.

We next prove (d). Let t, σ and p be as described. Suppose that a number y enters D_{r+1} for some $r \in [t,s)$. Then $y = \lambda_r(p')$ for some follower p' for some node σ'. Since σ is not initialised at stage r, either σ' is weaker than σ, in which case by (c) at stage r we have $y > \lambda_r(p) \geqslant \lambda_t(p)$; or $\sigma' = \sigma$, in which case of course $y = \lambda_r(p) \geqslant \lambda_t(p)$. In either case, $D_t \restriction_{\lambda_t(p)} = D_s \restriction_{\lambda_t(p)}$. If σ does not act at any stage $r \in [t,s)$ then we always have $y > \lambda_t(p)$ and so $D_t \restriction_{\lambda_t(p)+1} = D_s \restriction_{\lambda_t(p)+1}$.

To show (b) at stage s, let p be a follower for a node σ at the beginning of stage s. If σ acts at stage $s-1$, then at that stage we define $\Lambda_s(D_s, p)\downarrow$ with large use $\lambda_s(p)$; since it is large, we have $\lambda_s(p) \notin D_s$. Otherwise, p is a follower for σ at the beginning of stage $s-1$, and p is not cancelled at that stage. By (b) at stage $s-1$, $\Lambda_{s-1}(D_{s-1}, p)\downarrow$. By (d) at stage s, with $t = s-1$, we have $D_s \restriction_{\lambda_{s-1}(p)+1} = D_{s-1} \restriction_{\lambda_{s-1}(p)+1}$. This implies that the axiom making $\Lambda_{s-1}(D_{s-1}, p)\downarrow$ applies at stage s as well, and in fact $\lambda_s(p) = \lambda_{s-1}(p)$. By (b) at stage $s-1$ we have $\lambda_{s-1}(p) \notin D_{s-1}$, and the agreement between D_{s-1} and D_s just observed shows that $\lambda_s(p) = \lambda_{s-1}(p) \notin D_s$ as well.

For (c), let p' and σ' be as described. Let $t \leqslant s-1$ be the stage at which p' was chosen as a follower for σ'. The fact that the follower p' is kept from stage $t+1$ up to stage s shows that σ' was not initialised at any stage $r \in [t,s)$. Since σ is stronger than σ', this shows that σ was not initialised and did not act at any such stage. Thus, p must have been appointed by σ at a stage prior to stage t, and so p is a follower for σ at the beginning of stage t. At stage t, p' is chosen to be large, and so $p' > \lambda_t(p)$ (the latter exists by (b) at stage t). By (d) (at stage s, applied to stage t), we see that $D_s \restriction_{\lambda_t(p)+1} = D_t \restriction_{\lambda_t(p)+1}$, whence $\lambda_s(p) = \lambda_t(p)$. □

We start by working toward showing that the construction is fair.

Lemma 3.9. *Let σ be a node which works for requirement $P^{e,\gamma}$. Let $s < t$ be stages, and suppose that σ acts at both stages s and t, and is not initialised at any stage $r \in (s,t)$. Let p be the follower for σ at the end of stage s. Then $o_t^{e,\gamma}(p) < o_s^{e,\gamma}(p)$.*

Proof. The follower p is not cancelled at any stage $r \in (s,t]$. In particular, σ's action at stage t is not appointing a new follower, and so this action is prompted by the equality $f_t^{e,\gamma}(p) = \Lambda_t(D_t, p)$.

We observe that $\Lambda_t(D_t, p) > s$. This follows from the fact that at stage s, we set $\Lambda_{s+1}(D_{s+1}, p) = s+1$, and that at no later stage do we decrease the value of $\Lambda_r(D_r, p)$.

Now we have $f^{e,\gamma}_t(p) = \Lambda_t(D_t, p) > s$ and by convention, $f^{e,\gamma}_s(p) < s$. So $f^{e,\gamma}_s(p) \neq f^{e,\gamma}_t(p)$. Since $\langle f^{e,\gamma}_s, o^{e,\gamma}_s \rangle_{s<\omega}$ is a $(\gamma+1)$-computable approximation, and $f^{e,\gamma}_r(p)$ is not constant on $r \in [s,t]$, we must have $o^{e,\gamma}_t(p) < o^{e,\gamma}_s(p)$. □

Since for all s, δ_s is an initial segment of the tree of strategies, the true path δ_ω is an initial segment of the tree. Since every node on the tree of strategies has but finitely many outcomes, the only thing that could stop the true path from being infinite is that some node on the true path acts and ends the stage at almost every stage it is accessible.

Lemma 3.10. *Suppose that σ is a node on the true path working for some positive requirement $P^{e,\gamma}$, and that the construction is fair to σ. Then σ acts only finitely many times.*

Proof. Let s_0 be the last stage at which σ is initialised. Let s_1 be the least stage beyond s_0 at which σ is accessible. At stage s_1, σ appoints a follower p. Since $s_1 > s_0$, this follower is never cancelled.

The fact that σ acts only finitely many times beyond stage s_1 now follows from Lemma 3.9. Since $\langle f^{e,\gamma}_s, o^{e,\gamma}_s \rangle$ is a $(\gamma+1)$-computable approximation, there is some stage $t \geqslant s_1$ after which $o^{e,\gamma}_u(p)$ is constant. Then σ can act at most once after stage t. □

By induction on the length of nodes, we see that the construction is fair to every node on the true path, and so that no node can be the last node on the true path.

Corollary 3.11. *The true path δ_ω is infinite, and the construction is fair to every node on the true path.*

Next, we show that the positive requirements are met, and so that $\Lambda(D)$ witnesses that $\deg_{\mathrm{T}}(D)$ is not totally γ-c.a. for any $\gamma < \alpha$.

Lemma 3.12. *$\Lambda(D)$ is total.*

Proof. Let $p < \omega$. Suppose that there is some stage $s_0 > p$ at which p is not a follower for any node. After stage s_0, we enumerate an axiom into Λ regarding p at most once, because such an axiom has use -1 and so defines a computation that cannot be destroyed. So overall, only finitely many axioms in Λ are made for p. Thus, if $p \notin \operatorname{dom} \Lambda(D)$, then at almost every stage s we have $p \notin \operatorname{dom} \Lambda_s(D_s)$. But then at some such stage $s > s_0$ we would define $\Lambda_{s+1}(D_{s+1}, p)\downarrow$ with use -1, which would imply that $p \in \operatorname{dom} \Lambda(D)$ after all—contradiction.

Now suppose that p is picked as a follower for some node σ, and that p is never cancelled. The construction is fair to σ, and so either σ lies to the left of the true path, or lies on the true path. In either case, σ acts at most finitely many times (Lemma 3.10). Let $s-1$ be the last stage at which σ acts. Lemma 3.8(d) now shows that $D\upharpoonright_{\lambda_s(p)+1} = D_s\upharpoonright_{\lambda_s(p)+1}$ and so $p \in \operatorname{dom}\Lambda(D)$. □

Lemma 3.13. *Every positive requirement is met.*

Proof. Let $P^{e,\gamma}$ be a positive requirement. Let σ be a node on the true path which works for $P^{e,\gamma}$. As in the proof of Lemma 3.12 there is a last stage $s-1$ at which σ acts, and at that stage we define a D-correct computation $\Lambda_s(D_s, p)$. If $\Lambda(D,p) = f^{e,\gamma}(p)$, then for almost all stages $t > s$ we would have $\Lambda_t(D_t, p) = f_t^{e,\gamma}(p)$. There is such a stage $t > s$ at which σ is accessible. At such a stage, σ would act—contradiction. □

We now need to show that $\deg_{\mathrm{T}}(D)$ is totally α-c.a., that is, that every requirement Q_e is met. Fix $e < \omega$, and suppose that $\Phi_e(D)$ is total; we give $\Phi_e(D)$ an α-computable approximation.

Since the true path δ_ω is infinite, there is some node $\tau \in \delta_\omega$ that works for the requirement Q_e. Let s^* be the last stage at which the node τ is initialised (this is the same as the last stage at which the node $\tau\hat{\ }\infty$ is initialised). We let

$$S = \{s > s^* : \tau\hat{\ }\infty \in \delta_s\}.$$

Since $\Phi_e(D)$ is total, S is infinite (so $\tau\hat{\ }\infty$ is on the true path)—a greatest stage in S would yield a contradiction. Let $s_0, s_1, \ldots$ be the increasing enumeration of the (computable) set S.

For $x < \omega$, we let $i(x)$ be the least index i such that $x < \operatorname{dom}\Phi_e(D)[s_i]$. For $j \geqslant i(x)$, we let $a_j(x)$ be the collection of nodes $\sigma \succcurlyeq \tau\hat{\ }\infty$ which at the beginning of stage s_j have a follower $p = p(\sigma, x)$ which was chosen before stage $s_{i(x)}$. Note that for all $j \geqslant i(x)$, $a_{j+1}(x) \subseteq a_j(x)$. The next lemma says that only nodes in $a_j(x)$ can injure the computation $\Phi_e(D,x)[s_j]$.

Lemma 3.14. *Let $j \geqslant i(x)$. Suppose that $\Phi_e(D,x)[s_{j+1}] \neq \Phi_e(D,x)[s_j]$. Then the weakest node in $a_{j+1}(x)$ acts at stage s_j.*

Proof. There is some stage $r \in [s_j, s_{j+1})$ at which some node σ enumerates a number smaller than $\varphi_{e,s_j}(x)$ into D_{r+1}, destroying the computation $\Phi_e(D,x)[s_j]$.[3] We show that $r = s_j$ and that σ is the weakest node in $a_{j+1}(x)$.

Let p be σ's follower at stage r. Let t be the stage at which p was appointed. We have $\lambda_r(p) < \varphi_{e,s_j}(x)$, and so $t < s_j$.

Certainly, σ cannot be stronger than $\tau\hat{\ }\infty$, since $r > s^*$. On the other hand, $\tau\hat{\ }\infty$ is accessible at stage s_j, and σ is not initialised at stage s_j (this would

[3]Recall that since Φ_e is not enumerated by us, our convention is that $\varphi_e(x)$ is not the largest number queried but one greater, the length of the string in the axiom defining the computation.

cancel p), whence σ must extend $\tau\hat{}\infty$. From this we already conclude that $r = s_j$, as σ is not accessible at any stage in the interval (s_j, s_{j+1}).

Since σ acts at stage s_j, all nodes weaker than σ are initialised at stage s_j, and so no node weaker than σ can have, at stage s_{j+1}, a follower chosen prior to stage $s_{i(x)}$. I.e., no node weaker than σ can be an element of $a_{j+1}(x)$. To finish the proof of the lemma, it remains to show that $\sigma \in a_{j+1}(x)$, i.e., to show that $t < s_{i(x)}$, and that σ is not initialised at some stage $r \in [s_j, s_{j+1})$. The latter is immediate: at stage s_j, σ acts and so is not initialised; and at stage $r \in (s_j, s_{j+1})$, $\tau\hat{}\infty$ is not accessible, and so the fact that $\tau\hat{}\infty$ is not initialised at stage r implies that neither is σ.

Suppose, for a contradiction, that $t \geqslant s_{i(x)}$. Since σ extends $\tau\hat{}\infty$, we see that $t \in S$ and so that $x < \operatorname{dom} \Phi_e(D)[t]$. This is the crucial point for the entire construction: in this case every time we define $\lambda(p)$ we observe $\Phi_e(D, x)$, and so the former is larger than the use of the latter.

Let $u = \varphi_{e,t}(x)$. At stage t we pick $\lambda_t(p) > u$. Since σ is not initialised at any stage $r \in [t, s_j)$, Lemma 3.8(d) shows that $D_t \restriction_u = D_{s_j} \restriction_u$, which in turn implies that $\varphi_{e,s_j}(x) = u$. This contradicts $\lambda_{s_j}(p) < \varphi_{e,s_j}(x)$. □

Fix $x < \omega$. For $j \geqslant i(x)$ and $\sigma \in a_j(x)$ we let $t_j(\sigma)$ be the greatest stage $t < s_j$ at which σ acts. Such a stage t exists, because σ acts when it appoints the follower $p(\sigma, x)$. We note for later that σ is not initialised between stage $t_j(\sigma)$ and stage s_j. In fact, $t_j(\sigma) = s_i$ for some $i < j$, but this is not material.

For $j \geqslant i(x)$ and $\sigma \in a_j(x)$ we let $\beta_j(\sigma) = o^{i,\gamma}_{t_j(\sigma)}(p(\sigma, x))$, where σ works for the requirement $P^{i,\gamma}$. We order the set $a_j(x)$ by descending priority to obtain a sequence, and let

$$m_j(x) = \sum_{\sigma \in a_j(x)} \beta_j(\sigma),$$

with the addition performed along the order of $a_j(x)$: if $a_j(x) = \langle \sigma_1, \sigma_2, \ldots, \sigma_k \rangle$ then $m_j(x) = \beta_j(\sigma_1) + \beta_j(\sigma_2) + \cdots + \beta_j(\sigma_k)$. We let $g_j(x) = \Phi_e(D, x)[s_j]$. Certainly $\lim_{j\to\infty} g_j(x) = \Phi_e(D, x)$.

Lemma 3.15. *Let $j \geqslant i(x)$. Then $m_{j+1}(x) \leqslant m_j(x) < \alpha$, and if $g_{j+1}(x) \neq g_j(x)$ then $m_{j+1}(x) < m_j(x)$.*

We then let $m_j(x) = m_{i(x)}(x)$ and $g_j(x) = g_{i(x)}(x)$ for all $j < i(x)$, and see that $\langle g_j, m_j \rangle$ is an α-computable approximation for $\Phi_e(D)$.

Proof. First note that for each $j \geqslant i(x)$, for each $\sigma \in a_j(x)$, if σ works for $P^{i,\gamma}$ then $\beta_j(\sigma) \leqslant \gamma < \alpha$; as α is closed under addition (here is where we use the assumption), $m_j(x) < \alpha$ for all j.

Next, we observe that thought of as sequences, $a_{j+1}(x)$ is an *initial segment* of $a_j(x)$. This is because if $\sigma \in a_j(x) \setminus a_{j+1}(x)$, then σ is initialised at some stage

$r \in [s_j, s_{j+1})$; at that stage r, every node weaker than σ is also initialised and extracted from $a_{j+1}(x)$.

Now for each $\sigma \in a_{j+1}(x)$, $t_{j+1}(\sigma) \geqslant t_j(\sigma)$ and so $\beta_{j+1}(\sigma) \leqslant \beta_j(\sigma)$. Altogether, we see that $m_{j+1}(x) \leqslant m_j(x)$.

Suppose that $g_{j+1}(x) \neq g_j(x)$. Let σ be the weakest node in $a_{j+1}(x)$. We know (Lemma 3.14) that σ acts at stage s_j. Thus, $t_j(\sigma) < s_j = t_{j+1}(\sigma)$. Since σ acts at both stage $t_j(\sigma)$ and stage $t_{j+1}(\sigma)$, and is not initialised between these stages, Lemma 3.9 says that $\beta_{j+1}(\sigma) < \beta_j(\sigma)$. Together with $\beta_{j+1}(\tau) \leqslant \beta_j(\tau)$ for all other $\tau \in a_{j+1}(x)$, and since $\beta_{j+1}(\sigma)$ is the last summand in $m_{j+1}(x)$, we see that $m_{j+1}(x) < m_j(x)$. □

3.3 A REFINEMENT OF THE HIERARCHY: UNIFORMLY TOTALLY ω^α-C.A. DEGREES

Downey, Jockusch, and Stob [31] have shown that the following are equivalent for a c.e. degree $\mathbf{d}$:

(1) $\mathbf{d}$ is array computable;
(2) for every increasing computable function h, every function $f \in \mathbf{d}$ has an h-bounded computable approximation;
(3) there is some increasing computable function h such that every function $f \in \mathbf{d}$ has an h-bounded computable approximation.

By Proposition 2.12, every c.e., array computable degree is totally ω-c.a. Note that the computable enumerablility of $\mathbf{d}$ is necessary here, as there are uncountably many array computable degrees.

The converse does not hold: there is a c.e. degree which is totally ω-c.a. but not array computable. An indirect argument for the existence of such a degree is given by a conjunction of work by Walk [101] and Downey, Greenberg, and Weber [25]. Walk constructed a c.e. degree which is not array computable, but does not bound a critical triple. Downey, Greenberg, and Weber showed that such a degree must be totally ω-c.a.

Theorem 3.20 gives a direct construction of a c.e. degree which is totally ω-c.a. and not array computable, by finding a generalisation of the notion of array computability to all levels of the hierarchy of totally ω^α-c.a. degrees. We call this generalisation the *uniform version* of total ω^α-computable approximability. The key idea is the observation, mentioned above, that for ordinals $\alpha > \omega$, the first value $o_0(x)$ of an α-computable approximation $\langle f_s, o_s \rangle$ is the correct measure of "how many times" the approximation $\langle f_s(x) \rangle_{s<\omega}$ changes, rather than the natural number $m_{\langle f_s \rangle}(x)$, the value of the mind-change function.

Definition 3.16. Let $\alpha \leqslant \varepsilon_0$.

An *α-order function* is a non-decreasing computable function $h\colon \omega \to \alpha$ whose range is unbounded in α.

Let h be an α-order function. An α-computable approximation $\langle f_s, o_s\rangle$ is an *h-computable approximation* if for all x, $o_0(x) < h(x)$.[4]

We say that a function $f\colon \omega \to \omega$ is *h-computably approximable* (or *h-c.a.*) if there is an h-computable approximation $\langle f_s, o_s\rangle$ such that $\lim_s f_s = f$.

Note that for all $\alpha \leqslant \varepsilon_0$, α-order functions exist; in fact, there is a computable, strictly increasing and unbounded function from ω to α (see Lemma 2.24). This shows that a function is α-c.a. if and only if it is h-c.a. for some α-order function h.

The following uses an argument used by Terwijn and Zambella [95] in the context of computable traceability, and earlier by Downey, Jockusch, and Stob [31].

Lemma 3.17. *The following are equivalent for a Turing degree* $\mathbf{d}$ *and* $\alpha \leqslant \varepsilon_0$.

(1) There is some α-order function h such that every $f \in \mathbf{d}$ is h-c.a.
(2) For every α-order function h, every $f \in \mathbf{d}$ is h-c.a.

Proof. Let h and $\bar{h}$ be α-order functions. We show that for all $f\colon \omega \to \omega$ there is some $g \equiv_{\mathrm{T}} f$ such that if g is $\bar{h}$-c.a., then f is h-c.a.

The function g is obtained by "stretching" f along the composition of the "discrete inverse" of h with $\bar{h}$. Namely, we (computably) partition ω into an increasing sequence of finite intervals $I^* < I_0 < I_1 < I_2 < \dots$ so that for all n, for all $x \in I_n$, $h(x) \geqslant \bar{h}(n)$. Some intervals I_n are allowed to be empty (this is used when $\bar{h}$ is not injective). We simply let I^* be the set of x such that $h(x) < \bar{h}(0)$; and if

$$\bar{h}(n-1) < \bar{h}(n) = \dots = \bar{h}(m) < \bar{h}(m+1)$$

(possibly $m = n$) then we let I_n be the set of x such that $\bar{h}(n) \leqslant h(x) < \bar{h}(m+1)$; this is finite since h is unbounded in α. For k between n and $m+1$ we let I_k be empty.

We then define $g(n) = f\restriction_{I_n}$. Let $\langle g_s, o_s\rangle$ be an $\bar{h}$-computable approximation for g. By speeding up this approximation we may assume that for all s and n, $g_s(n)$ is a function from I_n to ω. We can then define $f_s(x) = (g_s(n))(x)$ for $x \in I_n$ (and let $m_s(x) = o_s(n)$); for $x \in I^*$ we let $f_s(x) = f(x)$ and $m_s(x) = 0$. Then $\langle f_s, m_s\rangle$ is an h-computable approximation for f. □

Definition 3.18. A Turing degree $\mathbf{d}$ is *uniformly totally α-c.a.* if for some (all) α-order function(s) h, every $f \in \mathbf{d}$ is h-c.a.

The Downey, Jockusch, and Stob characterisation shows that a c.e. degree is array computable if and only if it is uniformly totally ω-c.a.

[4]In the language of Section 2.3, for each x, the sequence $\langle f_s(x), o_s(x)\rangle$ is an instance of an $h(x)$-computable approximation.

Lemma 3.19. *A Turing degree* $\mathbf{d}$ *is uniformly totally* α*-c.a. if and only if for some (all)* α*-order function* h*, every* $f \leqslant_{\mathrm{T}} \mathbf{d}$ *is* h*-c.a.*

Proof. Suppose that $\mathbf{d}$ is uniformly totally α-c.a., and let h be an α-order function. Let $f \leqslant_{\mathrm{T}} \mathbf{d}$ and let $g \in \mathbf{d}$; so $f \oplus g \in \mathbf{d}$. Then $f \oplus g$ is $h \oplus h$-c.a.; it follows that f is h-c.a. □

The argument of Proposition 3.5 shows that a c.e. degree is uniformly totally α-c.a. if and only if for some (all) α-order function h, every set in $\mathbf{d}$ is h-c.a.

We turn to investigate the distribution of uniformly totally α-c.a. degrees in the hierarchy of totally α-c.a. degrees. An immediate fact, using the constant function with value α, is that for all $\alpha < \varepsilon_0$, every totally α-c.a. degree is uniformly totally $(\alpha+1)$-c.a.

It follows from the easy direction of Theorem 3.6 that if $\beta \in (\omega^\alpha, \omega^{\alpha+1})$ (that is, if β is not a power of ω), then every uniformly totally β-c.a. degree is totally ω^α-c.a. Hence, if β is not a power of ω, then there is an ordinal α which is a power of ω such that the collection of uniformly totally β-c.a. degrees is the same as the collection of totally α-c.a. degrees.

Thus, the only ordinals α for which the class of uniformly totally α-c.a. degrees does not necessarily coincide with the class of totally β-c.a. degrees for some ordinal β are the powers of ω. Theorem 3.20 shows that for ordinals $\alpha \leqslant \varepsilon_0$ which are powers of ω, the uniformly totally α-c.a. degrees indeed form a distinct level of the hierarchy.

Theorem 3.20. *Let* $\alpha \leqslant \varepsilon_0$ *be a power of* ω*.*

(1) *There is a uniformly totally* α*-c.a. c.e. degree which is not totally* γ*-c.a. for any* $\gamma < \alpha$*.*
(2) *There is a totally* α*-c.a. c.e. degree which is not uniformly totally* α*-c.a.*

The first $\omega \cdot 2$ many levels of the hierarchy of totally and uniformly totally α-c.a. degrees are depicted in Figure 3.2.

3.3.1 Proof of Theorem 3.20(1)

We show that the first part of Theorem 3.20 is actually already proved using the construction used for proving Theorem 3.6. Given $\alpha \leqslant \varepsilon_0$ which is a power of ω, that construction produces a c.e. set D whose Turing degree is totally α-c.a., but such that there is some $f \leqslant_{\mathrm{T}} D$ that is not γ-c.a. for any $\gamma < \alpha$. We show that $\deg_{\mathrm{T}}(D)$ is actually uniformly totally α-c.a. The reason for this is the long delay between expansionary stages that was already incorporated into the construction.

For concreteness, let $P^{e_0,\gamma_0}, P^{e_1,\gamma_1}, \ldots$ effectively enumerate all the positive requirements $P^{e,\gamma}$, and suppose that for all $k < \omega$, all nodes of length $2k$ work

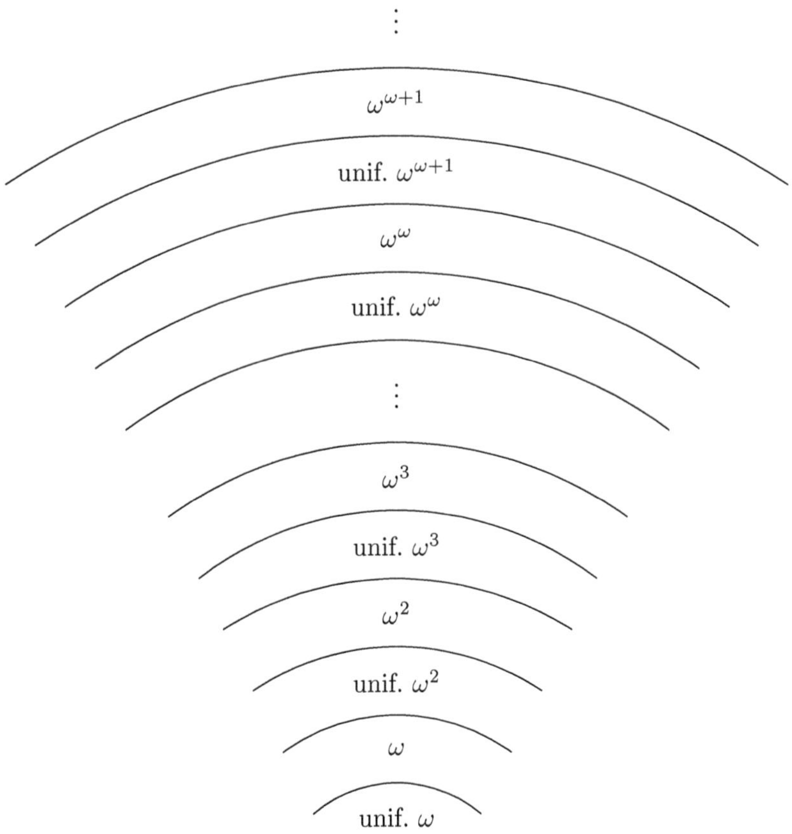

Figure 3.2. The first refinement of the hierarchy. "ω^α" denotes the collection of totally ω^α-c.a. degrees. "unif. ω^α" denotes the class of uniformly totally ω^α-c.a. degrees.

for the requirement P^{e_k,γ_k}. In particular, all nodes of even length work for some positive requirement.

Lemma 3.21. *For all stages s, for all $\sigma \in \delta_s$, $|\sigma| \leqslant 2s$.*

Proof. By induction on s. If this holds for all stages $t < s$, and if at stage s, some node σ of length $2s$ is accessible, then since it works for a positive requirement, and was not accessible at any stage before s, at stage s, the node σ acts by appointing a follower, and ends the stage. □

For all $n < \omega$, let

$$h(n) = \left(\max_{k \leqslant n} \gamma_k\right) \cdot 2^{2n}.$$

Since every ordinal below α appears as some γ_k, the function h is an α-order function. The combinatorial point is that if $\sigma_1, \dots, \sigma_l$ is a sequence of *distinct* nodes on the tree, each of length at most $2n$, with σ_i working for $P^{e_{k_i}, \gamma_{k_i}}$ (so $k_i \leqslant n$), then as the tree of strategies is (at most) binary branching, we have $l \leqslant 2^{2n}$, and so

$$\sum_{i \leqslant l} \gamma_{k_i} \leqslant h(n).$$

We show that every $f \leqslant_{\mathrm{T}} D$ is $(h+1)$-c.a. To this end, fix some $e < \omega$ such that $\Phi_e(D)$ is total, and let τ be a node on the true path which works for requirement Q_e. Recall the construction, during the proof of Theorem 3.20, of an α-computable approximation $\langle g_j, m_j \rangle$ for $\Phi_e(D)$. We let s^* be the last stage at which τ was initialised, and

$$S = \{s > s^* : \tau\hat{\ }\infty \in \delta_s\} = \{s_0, s_1, \dots\}.$$

For all $x < \omega$, $i(x)$ was the least index i such that $x < \operatorname{dom} \Phi_e(D)[s_i]$. For $j \geqslant i(x)$ we observed the set $a_j(x)$ of nodes $\sigma \succcurlyeq \tau\hat{\ }\infty$ that have followers at the beginning of stage $s_{i(x)}$, and are not initialised between stages $s_{i(x)}$ and s_j; we focus on $a(x) = a_{i(x)}(x)$. The ordinal $m_0(x) = m_{i(x)}(x)$ was defined to be the sum of ordinals of the form $o_t^{i,\gamma}(p)$, where $t < s_{i(x)}$ is some stage, and p is a follower at stage $s_{i(x)}$ for $\sigma \in a(x)$, working for $P^{i,\gamma}$. Certainly $o_t^{i,\gamma}(p) \leqslant \gamma$. And so, if $2n$ is a bound on the lengths of nodes in $a(x)$, then $m_0(x) \leqslant h(n)$. The proof will be complete when we show that for almost all x, $2x$ is a bound on the lengths of nodes in $a(x)$, and so $m_0(x) \leqslant h(x)$; so a modification of the approximation $\langle g_j, m_j \rangle$ on finitely many inputs yields an h-computable approximation for $\Phi_e(D)$.

Lemma 3.22. *For all $x \geqslant \operatorname{dom} \Phi_e(D)[s_1]$, for all $\sigma \in a(x)$, $|\sigma| \leqslant 2x$.*

Proof. Let $x \geqslant \operatorname{dom} \Phi_e(D)[s_1]$. So $i(x) \geqslant 2$; for brevity, we let $u_0 = s_{i(x)-2}$ and $u_1 = s_{i(x)-1}$. By the instructions for τ, $u_0 < \operatorname{dom} \Phi_e(D)[u_1]$; by minimality of $i(x)$, $\operatorname{dom} \Phi_e(D)[u_1] \leqslant x$; so $x > u_0$. By Lemma 3.21, all nodes accessible at any stage $t \leqslant u_0$ have length at most $2u_0$.

Let $\sigma \in \delta_{u_1}$ be a node working for some positive requirement $P^{i,\gamma}$ which has not been accessible at any stage $s \leqslant u_0$ (if there is such a node). Since τ and all of its predecessors are accessible at stage u_0, we have $\sigma \succcurlyeq \tau\hat{\ }\infty$. But since u_1 is the immediate successor of u_0 in S, σ was not accessible at any stage $s \in (u_0, u_1)$; so u_1 is the least stage at which σ is accessible, and so σ ends the stage u_1. It follows that for such σ we must have $|\sigma| \leqslant 2u_0 + 2$.

In total, if $\sigma \succcurlyeq \tau\hat{\ }\infty$ is accessible at any stage $s \leqslant u_1$, then $|\sigma| \leqslant 2(u_0+1) \leqslant 2x$.

Let $\sigma \in a(x)$. The node σ extends $\tau\hat{\ }\infty$, and was accessible at some stage $t \in S$, smaller than $s_{i(x)}$; so $t \leqslant u_1$. Hence $|\sigma| \leqslant 2x$ as required. ☐

3.3.2 Proof of Theorem 3.20(2)

A minor modification of the construction for Theorem 3.6 gives the proof of the second part of Theorem 3.20. Again, we are given an ordinal α which is a power of ω, so is closed under addition; and we enumerate a c.e. set D whose Turing degree will be totally α-c.a. but not uniformly so. By Lemma 3.19, it is sufficient to fix an α-order function h and enumerate a functional Λ such that $\Lambda(D)$ is total and is not h-c.a. What makes this construction work is that we can enumerate tidy $(h+1)$-computable approximations. The definition is the expected modification of Definition 2.5. A simpler version of the proof of Proposition 2.7 yields:

Lemma 3.23. *Let $\alpha \leqslant \varepsilon_0$ and let h be an α-order function. Then there is an effective enumeration $\langle f_s^e, o_s^e \rangle$ of tidy $(h+1)$-computable approximations such that letting $f^e = \lim f_s^e$, the sequence $\langle f^e \rangle_{e<\omega}$ contains all h-c.a. functions.*

Fixing h, we get an enumeration of $(h+1)$-computable approximations $\langle \langle f_s^e, o_s^e \rangle_{s<\omega} \rangle_{e<\omega}$ as in Lemma 3.23, and repeat the construction for Theorem 3.6 where the positive requirements are now:

P^e: There is some p such that $\Lambda(D,p) \neq f^e(p)$.

The rest of the construction is identical, as are the verifications, and so we omit them. The critical reader would ask, though: as was shown in the previous subsection, the construction for Theorem 3.6 actually produces a uniformly totally α-c.a. degree. Why can we not replicate the argument now to get a contradiction?

We recall the argument proving the first part of Theorem 3.20. Let $e<\omega$ such that $\Phi_e(D)$ is total. A uniform bound for $m_0(x)$, where $\langle g_j, m_j \rangle_{j<\omega}$ is the α-computable approximation for $\Phi_e(D)$, was given by seeing that for almost all x, the nodes in $a(x)$ all had length at most $2x$, a fact which is preserved in the current, modified construction. In the previous construction, this was sufficient to give the bound, since for any follower p for some node $\sigma \in a(x)$, working for some $P^{i,\gamma}$, we had $o_0^{i,\gamma}(x) \leqslant \gamma$. In the current construction, of course, we just have $o_0^i(p) \leqslant h(p)+1$, so the size of p plays a role.

Can we not use the argument showing that $\sigma \in a(x)$ has length at most $2x$ to also bound the size of followers for such σ? After all, these followers are chosen at some τ-expansionary stage t smaller than $s_{i(x)}$, and, roughly speaking, a follower chosen at stage t has size "close to t." As in the proof of Lemma 3.22, let $u_0 < u_1$ be the immediate predecessors of $s_{i(x)}$ in S. Then u_0 is bounded by x, but u_1 may be much larger than x; and one element σ of $a(x)$ may pick its follower at stage u_1. So even though the *length* of that σ is bounded by $2x$, the *size of its follower* cannot be computably bounded in x, and it is this single element of $a(x)$ that chooses a follower late, which prevents us from giving an approximation $\Phi_e(D,x)$ with some ordinal bound which depends on x but not on σ and p (and so not on τ).

3.4 ANOTHER REFINEMENT OF THE HIERARCHY: TOTALLY $<\omega^\alpha$-C.A. DEGREES

The hierarchy of totally α-c.a. degree is not, a priori, the finest one could devise. For a limit ordinal α, one could conceive of a totally α-c.a. degree $\mathbf{d}$ such that every $f \in \mathbf{d}$ is γ-c.a. for some $\gamma < \alpha$, but such that $\mathbf{d}$ is not totally γ-c.a. for any $\gamma < \alpha$.

Definition 3.24. Let $\alpha \leqslant \varepsilon_0$. A Turing degree $\mathbf{d}$ is *totally* $<\alpha$*-c.a.* if every $f \in \mathbf{d}$ is γ-c.a. for some $\gamma < \alpha$.

As is the case with totally α-c.a. degrees and with uniformly totally α-c.a. degrees, a Turing degree $\mathbf{d}$ is totally $<\alpha$-c.a. if and only if every $f \leqslant_{\mathrm{T}} \mathbf{d}$ is γ-c.a. for some $\gamma < \alpha$.

As was indicated in the introduction, the class of totally $<\omega^\omega$-c.a. degrees is the main class investigated in this work.

As we did for uniformly totally α-c.a. degrees, we now examine how the classes of totally $<\alpha$-c.a. degrees fit in the hierarchy of totally α-c.a. degrees. Of course, if $\gamma < \alpha$, then every totally γ-c.a. degree is totally $<\alpha$-c.a., and every totally $<\alpha$-c.a. degree is totally α-c.a. In fact, slightly more holds: for any ordinal α, every totally $<\alpha$-c.a. degree is uniformly totally α-c.a., because for any α-order function h and all $\gamma < \alpha$, $h(x) \geqslant \gamma$ for almost all x, so any γ-computable approximation can easily be converted into an h-computable approximation.

Lemma 3.7 shows that if $\beta \in (\omega^\alpha, \omega^{\alpha+1}]$, then every totally $<\beta$-c.a. degree is totally ω^α-c.a.; in particular, note that this holds even if $\beta = \omega^{\alpha+1}$. Hence, if β is not a *limit of powers of* ω, then there is some $\alpha < \beta$, a power of ω, such that the class of totally $<\beta$-c.a. degrees coincides with the class of totally α-c.a. degrees.

Also note that the construction proving Theorem 3.6 and Theorem 3.20(1) produces a degree that is uniformly totally α-c.a. but not totally $<\alpha$-c.a.; to show that the degree constructed was not totally γ-c.a. for any $\gamma < \alpha$, we constructed a single function $\Lambda(D)$ which was not γ-c.a. for any $\gamma < \alpha$.

The following theorem then completely determines the new levels of our hierarchy, the first $\omega \cdot 2$ levels of which are depicted in Figure 3.3.

Theorem 3.25. *If $\alpha \leqslant \varepsilon_0$ is a limit of powers of ω, then there is a c.e. degree which is totally $<\alpha$-c.a. but not totally γ-c.a. for any $\gamma < \alpha$.*

The rest of this section is devoted to the proof of Theorem 3.25. We are given an ordinal $\alpha \leqslant \varepsilon_0$, a limit of powers of ω, and give a computable enumeration $\langle D_s \rangle$ of a c.e. set D such that $\deg_{\mathrm{T}}(D)$ is totally $<\alpha$-c.a. but not totally γ-c.a. for any $\gamma < \alpha$.

For every $\gamma < \alpha$ we must meet the requirements

P^γ: There is a function $f \leqslant_{\mathrm{T}} D$ which is not γ-c.a.

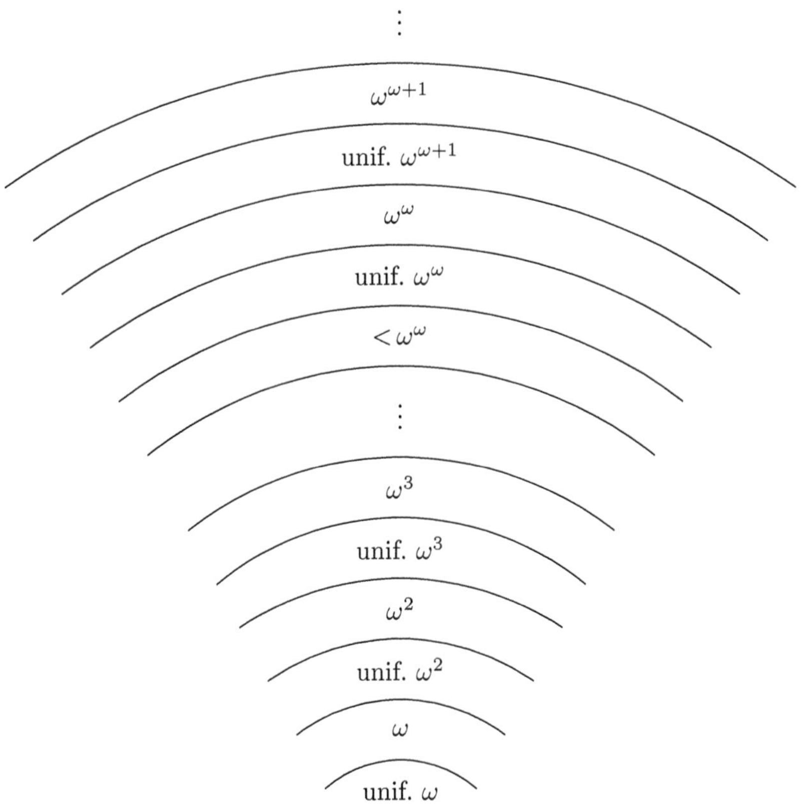

Figure 3.3. The second refinement of the hierarchy. "ω^α" denotes the collection of totally ω^α-c.a. degrees. "$<\omega^\alpha$" denotes the collection of totally $<\omega^\alpha$-c.a. degrees. "unif. ω^α" denotes the class of uniformly totally ω^α-c.a. degrees.

and

Q_e: If $\Phi_e(D)$ is total, then $\Phi_e(D)$ is γ-c.a. for some $\gamma < \alpha$.

Discussion

The first thing to notice is that we cannot, uniformly in γ, compute from D a function f which is not γ-c.a.; for we could string these functions together to get a single function which is not γ-c.a. for any $\gamma < \alpha$, and so fail to make $\deg_{\mathrm{T}}(D)$ totally $<\alpha$-c.a.

It is also fairly easy to see how the construction necessitates this nonuniformity. For suppose we tried to copy the construction proving Theorem 3.6. A node τ, working for Q_e, is now trying to make $\Phi_e(D)$ a γ-c.a. function for some $\gamma < \alpha$. But extending $\tau\hat{}\infty$ are nodes σ, working for P^β for ordinals β which are unbounded in α; their action would cause changes to τ's approximation of

$\Phi_e(D)$, and so force τ to have its γ larger than all of these β's, i.e., to be at least α.

The solution concerns that basic staple of both comedy and computability theory, namely timing. Remember that in a situation as above, a node σ extending $\tau\hat{}\infty$ can injure a computation $\Phi_e(D,x)[s]$ only if the follower p for σ at stage s was appointed before the τ-expansionary stage $t = s_{i(x)}$ at which we first observed and certified a computation $\Phi_e(D,x)[t]$. On the other hand, regardless of when p was appointed, upon enumerating $\lambda_s(p)$ into D_{s+1}, we need to *immediately* appoint a new use $\lambda_{s+1}(p)$, without waiting for a new $\Phi_e(D,x)[u]$ computation to recover; this, because we need to make $\Lambda(D)$ total. Even though σ guesses that $\Phi_e(D)$ is total, it is participating in the construction of the global functional Λ, and is responsible for making $p \in \operatorname{dom} \Lambda(D)$, even if its guess is incorrect. Inevitably, the new marker $\lambda_{s+1}(p)$ will be smaller than the use $\varphi_{e,r}(x)$ at the next τ-expansionary stage, and so further action with p will injure $\Phi_e(D,x)$ again.

In the previous construction this was fine, because σ provided τ with a bound $o_0^{i,\gamma}(p)$ on the "number of times" it will act for p, and the sum of these bounds was smaller than α. As mentioned above, this is insufficient when we want to show that the function $\Phi_e(D)$ is γ-c.a. for some $\gamma < \alpha$. Once we determined γ, what we need to do is break the cycle of repeated injury by the same follower p, when the bound for the follower is greater than γ. This is possible if we delay defining $\Lambda(D,p)$ until we see the computation $\Phi_e(D,x)$ recover. To do this, we distribute in a tree of strategies nodes η, working for P^β, which are responsible for a local version $\Lambda_\eta(D)$ of $\Lambda(D)$. Only nodes extending η contribute to the definition of $\Lambda_\eta(D)$, and the function $\Lambda_\eta(D)$ is required to be total only if η lies on the true path. If such η extends $\tau\hat{}\infty$, then indeed definitions of $\Lambda_\eta(D,p)$ can wait until $\Phi_e(D,x)$ recovers at the next τ-expansionary stage. We see how this gives the nonuniformity in defining the function witnessing P^β: we need the true path to find it.

How do we find γ? The approach of waiting to define Λ_η cannot be employed if τ extends η. If σ is a "child" node of such η with $\tau\hat{}\infty \preccurlyeq \sigma$, then we are back at the situation of the original construction: repeated action for a follower p for σ will keep injuring a computation $\Phi_e(D,x)$. Again, σ provides a bound for its action, and that bound is itself bounded by β, where η works for P^β. And $\beta < \alpha$. Since there are only finitely many "mother" nodes $\eta \prec \tau$, the bound γ will be any ordinal, closed under addition, which bounds the ordinals β for these nodes η. That such an ordinal $\gamma < \alpha$ can be found follows from the fact that α is a limit of ordinals closed under addition.

The tree of strategies

Let $\gamma < \alpha$. In order to meet the requirement P^γ, for each $e < \omega$, we need to meet the subrequirements $P^{e,\gamma}$ which diagonalise against $f^{e,\gamma}$. We arrange all of the requirements and subrequirements—Q_e, P^γ and $P^{e,\gamma}$—effectively, in a list of order-type ω, but ensuring that for each γ and e, P^{γ} appears before $P^{e,\gamma}$. We let all strategies on the tree of length k work for the kth requirement on the list.

Nodes working for requirements P^{γ} and $P^{e,\gamma}$ have only one outcome. Nodes working for Q_e have two outcomes, ∞ and fin, the former stronger than the latter.

Nodes η working for P^{γ} enumerate a functional Λ_{η}. For any node σ working for $P^{\gamma,e}$ there is a unique node $\eta \prec \sigma$ working for P^{γ}. We denote this node, the "mother" of σ, by $\eta(\sigma)$.

Construction

At stage s, we let δ_s, the collection of nodes accessible at stage s, be an initial segment of the tree of strategies.

Suppose that a node τ that works for requirement Q_e is accessible at stage s. If s is the least stage at which τ is accessible, then we let $\tau\hat{\ }\infty \in \delta_s$. Otherwise, we let t be the last stage before s at which $\tau\hat{\ }\infty$ was accessible. If $t < \operatorname{dom} \Phi_{e,s}(D_s)$, then we let $\tau\hat{\ }\infty \in \delta_s$. Otherwise, we let $\tau\hat{\ }\mathtt{fin} \in \delta_s$.

Suppose that a node η that works for requirement P^{γ} is accessible at stage s. If there is some p which is a follower for some child $\sigma \succ \eta$ of η (an extension of η working for some subrequirement $P^{e,\gamma}$) such that $p \notin \operatorname{dom} \Lambda_{\eta,s}(D_s)$, then we define $\Lambda_{\eta,s+1}(D_s, p) = s+1$ with large use, and end the stage (in this case, we do not initialise all nodes weaker than η; but as usual, we do initialise all nodes which lie to the right of η).

Otherwise, for all $p < s$ which are not in $\operatorname{dom} \Lambda_{\eta,s}(D_s)$, we define $\Lambda_{\eta,s+1}(D_s, p) = 0$ with use -1; the unique immediate successor of η on the tree of strategies is accessible next.

Suppose that a node σ that works for a subrequirement $P^{e,\gamma}$ is accessible at stage s.

(1) If σ has no follower, then σ appoints a new, large follower for itself.
(2) If σ has a follower p, and $\Lambda_{\eta(\sigma),s}(D_s, p)\downarrow = f_s^{e,\gamma}(p)$, then we enumerate $\lambda_{\eta(\sigma),s}(p)$ into D_{s+1}.

Note that in either case, we do not define a new computation $\Lambda_{\eta(\sigma),s+1}(D_{s+1}, p)$. In either case, we end the stage and initialise all nodes weaker than σ. If σ does not act, then the unique immediate successor of σ on the tree of strategies is accessible at stage s.

Verification

Let η be a node that works for P^{γ}. At stage s, we only define a new $\Lambda_{\eta,s+1}(D_s, p)$ computation if $p \notin \operatorname{dom} \Lambda_s(D_s)$. Lemma 1.9 ensures that each $\Lambda_{\eta,s}$ is consistent for D_s, and so that each Λ_{η} is consistent for D.

Lemma 3.26. *Let s be a stage, and let σ be a node working for $P^{e,\gamma}$ which has a follower p at the beginning of stage s.*

(1) If $p \notin \operatorname{dom} \Lambda_{\eta(\sigma),s}(D_s)$, then at the last stage $t<s$ at which $\eta(\sigma)$ was accessible, so was σ, and σ acted at stage t.

(2) If σ' is a node weaker than σ, working for $P^{e',\gamma'}$, and has a follower p' at the beginning of stage s, then $p<p'$. If in addition $p \in \operatorname{dom} \Lambda_{\eta(\sigma),s}(D_s)$ then $\lambda_{\eta(\sigma),s}(p) < p'$. Consequently, if also $p' \in \operatorname{dom} \Lambda_{\eta(\sigma'),s}(D_s)$ then $\lambda_{\eta(\sigma),s}(p) < \lambda_{\eta(\sigma'),s}(p')$.

Proof. Both parts of the lemma are proved simultaneously, by induction on s. Assume both parts hold at all stages before stage s. Let $\eta = \eta(\sigma)$.

For (1), let $t<s$ be the last stage before s at which $\eta(\sigma)$ was accessible, and suppose that σ does not act at stage t. Then p is already a follower for σ at the beginning of stage t, and so σ was not initialised at any stage $r \in [t,s)$. If $p \notin \operatorname{dom} \Lambda_{\eta,t}(D_t)$, then at stage t, η defines a new computation $\Lambda_{\eta,t+1}(D_t, p)$, and ends the stage. This means that $D_{t+1} = D_t$, and so $p \in \operatorname{dom} \Lambda_{\eta,t+1}(D_{t+1})$ with $\lambda_{\eta,t}(p) = \lambda_{\eta,t+1}(p)$. By (2) at all stages $r \in [t+1, s)$, this computation cannot be injured at stage r without initialising σ, so $p \in \operatorname{dom} \Lambda_{\eta,s}(D_s)$. If, on the other hand, $p \in \operatorname{dom} \Lambda_{\eta,t}(D_t)$, then by (2) at all stages $r \in [t,s)$, this computation cannot be injured without initialising σ.

For (2), let σ' and p' be as described. That $p<p'$ follows as usual from the fact that the stage at which p' was chosen is later than the stage at which p was chosen.

For the second part, let $t<s$ be the stage at which the computation $\Lambda_{\eta,s}(D_s, p)$ was defined. To show that $\lambda_{\eta,s}(p) < p'$, we show that the follower p' was chosen after stage t. We know that $\Lambda_\eta(D,p)[t]\uparrow$. Let u be the last stage prior to stage t at which η was accessible. By (1) at stage t, σ acted at stage u, and so σ' was initialised at stage u. Since $\eta \prec \sigma$, η is stronger than σ'. If σ' lies to the right of η, then it is initialised at stage t, and so p' is chosen after stage t. Otherwise, $\sigma' \succ \eta$, and so σ' is not accessible at any stage $r \in (u,t)$ and also not accessible at stage t (as η ends the stage). Thus, again, p' was chosen after stage t. □

As a corollary we can conclude that for σ and p as above, if $p \in \operatorname{dom} \Lambda_{\eta(\sigma),s}(D_s)$, then $\lambda_{\eta(\sigma),s}(p) \notin D_s$. An analogue of Lemma 3.8(d) also holds, with a similar argument.

Lemma 3.27. *Let $t>s$ be stages and let σ be a node which works for some positive subrequirement. Suppose that p is a follower for σ at the beginning of stage s. Suppose that σ is not initialised at any stage $r \in [s,t)$.*

(1) $D_s \restriction_p = D_t \restriction_p$.

(2) If in addition $p \in \operatorname{dom} \Lambda_{\eta(\sigma),s}(D_s)$, then $D_s \restriction_{\lambda_{\eta(\sigma),s}(p)} = D_t \restriction_{\lambda_{\eta(\sigma),s}(p)}$.

(3) If, further, σ does not act at any stage $r \in [s,t)$, then $p \in \operatorname{dom} \Lambda_{\eta(\sigma),t}(D_t)$ and $\lambda_{\eta(\sigma),s}(p) = \lambda_{\eta(\sigma),t}(p)$.

Lemma 3.28. *Suppose that σ works for $P^{e,\gamma}$, and that p is a follower which is appointed for σ at some stage and is never cancelled. Suppose that σ does not act infinitely often. Suppose also that $\eta(\sigma)$ is accessible infinitely often. Then $p \in \operatorname{dom} \Lambda_{\eta(\sigma)}(D)$.*

Proof. Let s be the last stage at which σ acts; since p is not cancelled after stage s, σ is not initialised after stage s. Let t be the least stage after stage s at which $\eta = \eta(\sigma)$ is accessible. If $p \in \operatorname{dom} \Lambda_{\eta,s}(D_s)$ then the action of σ at stage s removes p from $\operatorname{dom} \Lambda_{\eta,s+1}(D_{s+1})$; in any case, $p \notin \operatorname{dom} \Lambda_{\eta,t}(D_t)$. At stage t, η defines a new computation $\Lambda_{\eta,t+1}(D_{t+1}, p)$. By Lemma 3.27, this computation is D-correct. □

Lemma 3.9 holds for the current construction as well: if σ, working for $P^{e,\gamma}$, acts at stages $s < t$ and has the same follower p at the end of stage s and the end of stage t, then $o_t^{e,\gamma}(p) < o_s^{e,\gamma}(p)$. The proof is similar; the computation $\Lambda_{\eta(\sigma),t}(D_t, p)$ must have been defined by $\eta(\sigma)$ at a stage $u > s$, and so its value is $u+1$ which is bigger than s, so $f_s^{e,\gamma}(p) < s < u+1 = f_t^{e,\gamma}(p)$. Now an argument, identical to the argument proving Lemma 3.10, shows that if σ, working for $P^{e,\gamma}$ is on the true path, and the construction is fair to σ, then σ eventually appoints a follower p which is never cancelled, eventually stops acting, and $\Lambda_{\eta(\sigma)}(D, p) \neq f^{e,\gamma}(p)$. It follows that the true path is infinite, that the construction is fair to every node on the true path, and that if η on the true path works for P^γ, then $\Lambda_\eta(D)$ is total, and is not γ-c.a. Since the true path has a node in every level, each P^γ is met, so $\deg_{\mathrm{T}}(D)$ is not totally γ-c.a. for any $\gamma < \alpha$.

To conclude the proof of Theorem 3.25, we need to show that for all $e < \omega$ such that $\Phi_e(D)$ is total, $\Phi_e(D)$ is γ-c.a. for some $\gamma < \alpha$. Fix such e, and let τ be the node on the true path that works for requirement Q_e. At first, we proceed as in the proof of Theorem 3.6. Let s^* be the last stage at which τ is initialised, and let

$$S = \{s > s^* : \tau\hat{\ }\infty \in \delta_s\} = \{s_0, s_1, \dots\}$$

as again, S is infinite. For $x < \omega$ we define $i(x)$ as before, to be the least i such that $x < \operatorname{dom} \Phi_e(D)[s_i]$. And again, for $j \geqslant i(x)$ we let $a_j(x)$ be the set of nodes $\sigma \succcurlyeq \tau\hat{\ }\infty$ which at the beginning of stage s_j have a follower $p = p(\sigma, x)$ which was appointed before stage $s_{i(x)}$. Lemma 3.14 holds for the current construction, with the same proof, except that now we use $p > u$ rather than $\lambda_{\eta(\sigma),t}(p) > u$; so we use part (1) of Lemma 3.27 instead of part (2).

We now find an ordinal bound below α for the complexity of $\Phi_e(D)$. Fix $x < \omega$. For $\sigma \in a(x)$, since $\sigma \succcurlyeq \tau\hat{\ }\infty$, $\eta(\sigma)$ is comparable with $\tau\hat{\ }\infty$.

Lemma 3.29. *Let $\sigma \in a(x)$, and suppose that $\eta(\sigma) \succcurlyeq \tau\hat{\ }\infty$. Then there is at most one $j \geqslant i(x)$ such that σ acts at stage s_j and injures the computation $\Phi_e(D, x)[s_j]$.*

Proof. Let s_j be a stage at which σ acts, where $j \geqslant i(x)$. We show by induction that for all $i > j$ in S, if p is a follower for σ at the beginning of stage s_i, and $p \in \operatorname{dom} \Lambda_{\eta(\sigma)}(D)[s_i]$, then $\lambda_{\eta(\sigma),s_i}(p) > \varphi_{e,s_i}(x)$, so σ cannot injure $\Phi_e(D,x)[s_i]$ at stage s_i. Let $\eta = \eta(\sigma)$.

The base step is vacuous, and this is the main point of the proof. At stage s_j, σ's action extracts its follower p from $\operatorname{dom} \Lambda_{\eta(\sigma)}(D)$. The assumption $\eta \succcurlyeq \tau\hat{}\infty$ means that η is not accessible at any stage $r \in (s_j, s_{j+1})$, and so $p \notin \operatorname{dom} \Lambda_\eta(D)[s_{j+1}]$. Note that p is still the follower for σ at the beginning of stage s_{j+1}.

Let $i > j+1$ and suppose the inductive claim holds for all $i' \in (j,i)$. Let p be a follower for σ at the beginning of stage s_i, and suppose that $p \in \operatorname{dom} \Lambda_\eta(D)[s_i]$. The proof follows the idea for Lemma 3.14. Let t be the stage at which the computation $\Lambda_\eta(D,p)[s_i]$ was defined. Since $\eta \succcurlyeq \tau\hat{}\infty$, $t \in S$; and $t \geqslant s_{j+1}$. Thus $\lambda_{\eta,t}(p)$ is chosen to be larger than $u = \varphi_{e,t}(x)$. Lemma 3.27(2) now shows that $D_{s_i}\restriction_u = D_t\restriction_u$ and so $\varphi_{e,s_i}(x) = u < \lambda_{\eta,s_i}(p)$ as required. □

Since α is a limit of ordinals which are closed under addition, and τ has only finitely many predecessors on the tree of strategies, find some ordinal $\delta < \alpha$, closed under addition, such that for all $\eta \prec \tau$ which work for some P^γ we have $\gamma < \delta$. We give a δ-computable approximation for $\Phi_e(D)$, along the lines of the proof of Theorem 3.6.

Again fixing x, for $j \geqslant i(x)$ and $\sigma \in a_j(x)$ we again let $t_j(\sigma)$ be the greatest stage $t < s_j$ at which σ acts. The main part is defining the ordinal $\beta_j(\sigma)$:

- If $\eta(\sigma) \prec \tau$, then we let $\beta_j(\sigma) = o^{i,\gamma}_{t_j(\sigma)}(p(\sigma,x))$, where σ works for $P^{i,\gamma}$.
- If $\eta(\sigma) \succcurlyeq \tau\hat{}\infty$, then we let $\beta_j(\sigma) = 0$ if there is some $i \in [i(x), j)$ for which σ acts at stage s_i and destroys the computation $\Phi_e(D,x)[s_i]$. If there is no such i, then we let $\beta_j(\sigma) = 1$.

We then mimic the rest of the proof of Theorem 3.6, ordering $a_j(x)$ by descending priority, and defining $m_j(x) = \sum_{\sigma \in a_j(x)} \beta_j(x)$. The proof of Theorem 3.25 is complete once we show that Lemma 3.15 holds for the current construction (with δ replacing α). The proof of this lemma is identical to the previous proof, except for one case: showing that $\beta_{j+1}(\sigma) < \beta_j(\sigma)$ if $g_{j+1}(x) \neq g_j(x)$, where σ is the weakest node in $a_{j+1}(x)$, in the case that $\eta(\sigma) \succcurlyeq \tau\hat{}\infty$. But in this case we appeal to Lemma 3.29.

3.5 DOMINATION PROPERTIES

In [31], Downey, Jockusch, and Stob extend the notion of array computability from the c.e. degrees to all the Turing degrees. This they do by using domination properties of degrees. Such properties have been used early on, to characterise classes such as the hyperimmune-free degrees, the high degrees, and the non-low$_2$ degrees. More recently [53], a combination of domination and measure characterisations have yielded a characterisation of LR-hardness.

Recent work [20, 71] has indicated that the generalisations of array computability defined in this chapter can also be extended to the non-c.e. degrees by considering domination. We give the results for completeness.

Recall that if $\mathcal{C}$ is a class of functions from ω to ω, then a Turing degree $\mathbf{d}$ is *$\mathcal{C}$-dominated* if every function $g \in \mathbf{d}$ (equivalently $g \leqslant_{\mathrm{T}} \mathbf{d}$) is dominated by some function $f \in \mathcal{C}$. For example, the hyperimmune-free degrees are the degrees which are Δ^0_1-dominated, where Δ^0_1 denotes the collection of all computable functions.

Definition 3.30. A Turing degree is *α-c.a. dominated* if it is $\mathcal{C}$-dominated, where $\mathcal{C}$ is the class of all α-c.a. functions. That is, if every $\mathbf{d}$-computable function is dominated by some α-c.a. function.

Theorem 3.31 (Diamondstone, Greenberg, Turetsky [20]). *Let $\alpha \leqslant \varepsilon_0$. A c.e. degree is totally α-c.a. if and only if it is α-c.a. dominated.*

Proof. Let $\mathbf{d}$ be a c.e. degree and let $D \in \mathbf{d}$ be a c.e. set.

In the nontrivial direction, suppose that $\mathbf{d}$ is α-c.a. dominated. Let $g \in \mathbf{d}$, $g = \Gamma(D)$ for a functional Γ. Since D is c.e. it can compute the modulus m for the approximation $\langle g_s \rangle$ for g given by $g_s = \Gamma_s(D_s)$; here the modulus m is defined by $m(k) = s$ if s is the least stage such that for all $t \geqslant s$, $g_t \upharpoonright_{k+1} = g \upharpoonright_{k+1}$.

Let h be an ω-c.a. function which majorises m, and let $\langle h_t, o_t \rangle_{t<\omega}$ be an α-computable approximation for h. Letting $\tilde{g}_t(k) = g_{h_t(k)}(k)$ we get that $\langle \tilde{g}_t, o_t \rangle$ is an α-computable approximation for g.[5] □

The same argument yields an analogous result for the special limit classes.

Theorem 3.32. *Let $\alpha \leqslant \varepsilon_0$ be a limit of powers of ω. A c.e. degree $\mathbf{d}$ is totally $<\alpha$-c.a. if and only if it is $<\alpha$-c.a. dominated, i.e., if for every $\mathbf{d}$-computable function g there is some $\gamma < \alpha$ and some γ-c.a. function which dominates g.*

For the uniform version, for a class of functions $\mathcal{C}$, say that a Turing degree $\mathbf{d}$ is *uniformly $\mathcal{C}$-dominated* if there is some function $f \in \mathcal{C}$ which dominates every function in $\mathbf{d}$. In other words, if $\mathbf{d}$ is $\{f\}$-dominated for some $f \in \mathcal{C}$. For example, a Δ^0_2 degree is low$_2$ if and only if it is uniformly Δ^0_2-dominated. A Turing degree is *uniformly α-c.a. dominated* if, as expected, it is uniformly $\mathcal{C}$-dominated, where $\mathcal{C}$ is the collection of all α-c.a. functions.

[5] Essentially, this argument repeats the proof of Proposition 2.46, after noticing that g is weak truth-table reducible to any function dominating the modulus m.

The following is a generalisation of the aforementioned result by Downey, Jockusch, and Stob: a c.e. degree is array computable if and only if it is uniformly ω-c.a. dominated.

Theorem 3.33 (with McInerney). *Let $\alpha \leqslant \varepsilon_0$ be a power of ω. A c.e. degree $\mathbf{d}$ is uniformly totally α-c.a. if and only if it is uniformly α-c.a. dominated: some α-c.a. function dominates all functions in $\mathbf{d}$.*

Proof. In one direction the argument is similar to the argument for Theorem 3.31, but noticing the uniformity. Assuming that $\mathbf{d}$ is uniformly α-c.a. dominated, let g be an α-c.a. function which dominates every function in $\mathbf{d}$; fix an α-c.a. order function h such that g is h-c.a. Let $f \in \mathbf{d}$, and let μ be the modulus function for f, by an approximation given by a c.e. set in $\mathbf{d}$, so $\mu \leqslant_{\mathrm{T}} \mathbf{d}$. Then g dominates μ, and the argument above shows that f is h-c.a.

In the other direction, we show the slightly stronger fact, that for any α-order function h, there is an α-c.a. function which dominates every h-c.a. function. Fix an α-order function h. Let $\langle f^e \rangle$ be an effective listing of all h-c.a. functions, each with a tidy $(h+1)$-computable approximation $\langle f^e_s, o^e_s \rangle$ (Lemma 3.23). Let $\tilde{f}(n) = \max_{e \leqslant n} f^e(n)$. Certainly $\tilde{f}$ dominates every h-c.a. function. For $n < \omega$ and $s < \omega$, let $\tilde{f}_s(n) = \max_{e \leqslant n} f^e_s(n)$ and let $\tilde{o}_s(n) = \bigoplus_{e \leqslant n} o^e_s(n)$; see the discussion of commutative addition of ordinals in Subsection 2.3.3. Lemmas 2.44 and 2.45 show this is an α-computable approximation for $\tilde{f}$. □

Downey, Jockusch, and Stob also showed that one can pick a single ω-c.a. function dominating all array computable degrees: the modulus of $\varnothing'$. A similar result holds for the higher uniform levels as well.

Proposition 3.34. *Let $\alpha \leqslant \varepsilon_0$ be a power of ω. There is an α-c.a. function q such that any Turing degree $\mathbf{d}$ is uniformly α-c.a. dominated if and only if it is $\{q\}$-dominated.*

Proof. Let $\beta \leqslant \varepsilon_0$ such that $\alpha = \omega^\beta$. Recall (Theorem 2.40) that $I^{\varnothing}_\beta$ is α-c.a., and has greatest weak truth-table degree among all α-c.a. functions. Let $\langle p_s, o_s \rangle$ be an α-computable approximation of $I^{\varnothing}_\beta$; let q be the modulus function of this approximation: $q(n)$ is the least s such that for all $t \geqslant s$, $p_t \restriction_n = I^{\varnothing}_\beta \restriction_n$.

The function q is α-c.a.: for $r < \omega$ let $q_r(n)$ be the least $s \leqslant r$ such that for all $t \in [s, r]$, $p_t \restriction_n = p_r \restriction_n$. Also let $m_r(n) = o_r(0) \oplus o_r(1) \oplus \cdots \oplus o_r(n-1)$. Then $\langle q_s, m_s \rangle$ is an α-computable approximation of q.

We now follow the proof in [31, Thm.1.3]. Suppose that $\mathbf{d}$ is not $\{q\}$-dominated; we show that it is not uniformly α-c.a. dominated. Let $h \in \mathbf{d}$ be a function which is not dominated by q. Let f be an α-c.a. function; we define a function $g \leqslant_{\mathrm{T}} h$ which is not dominated by f.

Let (Φ, φ) we a weak truth-table functional such that $\Phi(I^{\varnothing}_\beta) = f$. We may assume that h and φ are strictly increasing. To define $g(n)$, search for the least $s > h(\varphi(n+1))$ such that $\hat{\Phi}_s(p_s, n)\downarrow$; we let $g(n) = \hat{\Phi}_s(p_s, n) + 1$. Let k be such that $h(k) > q(k)$; since φ is strictly increasing, let n such that $\varphi(n) \leqslant k < \varphi(n+1)$. Then

$$q(\varphi(n)) \leqslant q(k) < h(k) \leqslant h(\varphi(n+1)),$$

and so the stage s witnessing the definition of $g(n)$, which was chosen to be greater than $h(\varphi(n+1))$, is greater than $q(\varphi(n))$; so $p_s\restriction_{\varphi(n)} = I_\beta^\varnothing\restriction_{\varphi(n)}$, whence the computation $\hat{\Phi}_s(p_s, n)$ is correct, so $g(n) = f(n) + 1$. □

Chapter Four

Maximal totally α-c.a. degrees

FOR A COLLECTION $\mathcal{F}$ of c.e. degrees, we say that a degree $\mathbf{a} \in \mathcal{F}$ is *maximal* in $\mathcal{F}$ if it is maximal as an element of the partial ordering induced on $\mathcal{F}$ by the ordering on the Turing degrees. In other words, if there is no degree $\mathbf{b} > \mathbf{a}$ in $\mathcal{F}$.

Classes of c.e. degrees which contain maximal elements are rare; they are mostly prevented by density considerations. For example, no jump classes contain maximal elements, and there are no maximal cappable degrees. A notable exception is the example of the *contiguous* degrees—those degrees all of whose c.e. elements have the same weak truth-table degree. Cholak, Downey, and Walk [14] showed that there are maximal contiguous degrees. Since the contiguous degrees are definable in the c.e. degrees (Downey and Lempp [33]), the maximal contiguous degrees form a definable antichain of c.e. degrees.

The relevance of contiguous degrees to the current study is that contiguous degrees are all array computable, that is, uniformly totally ω-c.a. Like the contiguous degrees, the maximality phenomenon occurs at various levels of the hierarchy discussed in Chapter 3.

4.1 EXISTENCE OF MAXIMAL TOTALLY ω^α-C.A. DEGREES

Theorem 4.1. *If $\alpha \leqslant \varepsilon_0$ is a power of ω, then there is a maximal totally α-c.a. c.e. degree.*

To prove Theorem 4.1, fix an ordinal $\alpha \leqslant \varepsilon_0$ which is a power of ω; we enumerate a c.e. set D whose Turing degree will be maximal totally α-c.a. To ensure that $\deg_{\mathrm{T}}(D)$ is totally α-c.a., we meet, for each $e < \omega$, the requirements

Q_e: If $\Phi_e(D)$ is total, then $\Phi_e(D)$ is α-c.a.

To ensure maximality, for each $e < \omega$, we want to ensure that either $W_e \leqslant_{\mathrm{T}} D$, or that there is some $f \leqslant_{\mathrm{T}} D \oplus W_e$ which is not α-c.a. We enumerate a Turing functional Λ_e, with the aim of showing that either $W_e \leqslant_{\mathrm{T}} D$ or $\Lambda_e(D, W_e)$ is not α-c.a. By Proposition 2.7 let $\left\langle \langle f^i_s, o^i_s \rangle_{s<\omega} \right\rangle_{i<\omega}$ be an effective list of tidy $(\alpha+1)$-computable approximations such that letting $f^i = \lim_s f^i_s$, the sequence

$\langle f^i \rangle$ lists the α-c.a. functions; and as above, every α-c.a. function appears as f^i for some i such that the approximation $\langle f^i_s, o^i_s \rangle$ is eventually α-computable. For $e, i < \omega$, we try to meet the requirement

P^i_e: If $\langle f^i_s, o^i_s \rangle$ is eventually α-computable then either $W_e \leqslant_{\mathrm{T}} D$ or $\Lambda_e(D, W_e) \neq f^i$.

Globally we need to ensure that for all e, $\Lambda_e(D, W_e)$ is total.

Discussion

The construction is not difficult. To meet a requirement Q_e we use the mechanism proving the theorems in Chapter 3: a node τ, working for Q_e, measures an approximation to the question, "Is $\Phi_e(D)$ total?" In the case of an affirmative answer, initialisation of weaker nodes that guess incorrectly allows τ to devise an α-computable approximation for $\Phi_e(D)$.

A node σ working for a requirement P^i_d would like to appoint a follower p and follow the strategy of nodes working for positive requirements in the constructions of Chapter 3: whenever $f^i_s(p) = \Lambda_{d,s}(D_s, W_{e,s}, p)$, enumerate $\lambda_{d,s}(p)$ into D_{s+1}. This action may interfere with the work done by a node τ for some requirement Q_e such that $\tau\hat{\ }\infty \preccurlyeq \sigma$. However, unlike previous constructions, when σ picks p we do not know yet the ordinal bound on the "number of times" σ may need to act for p; the functions o^i_s are in some sense partial, since they allow the value α, which for us is useless.

We isolate three principles which guide the interaction between τ and σ extending $\tau\hat{\ }\infty$. These have been followed in previous constructions as well, but sometimes more easily since the approximations were "total." Let p be a follower for σ, a node working for P^i_d.

(a) Suppose that τ first certifies a computation $\Phi_e(D, x)[s]$ at stage s (in previous notation, $s = s_{i(x)}$). If $o^i_s(p) < \alpha$, then τ can incorporate this ordinal to the bound on its mind-changes for $\Phi_e(D, x)$. It can thus allow every future action for p to injure $\Phi_e(D, x)$.
(b) If the use $\lambda_{d,t}(p)$ is chosen at a stage t at which we see $\Phi_e(D, x)$ converge, then the next action for p will not injure $\Phi_e(D, x)[t]$.
(c) Since Λ_d is global, σ needs to define $\lambda_d(p)$ immediately when it appoints p, that is, before it sees $o^i(p) < \alpha$.

We remark that we could have made the definition of each Λ_d local, tied to a "mother node" η as in the proof of Theorem 3.25. However, in this construction this is not necessary and would not give any benefit. The effect of the finitely many mother nodes $\eta \prec \tau$ would be the same as the effect of having every Λ_d be global, i.e., the root of the tree is the mother node for every Λ_d.

The principles outlined leave one potentially problematic sequence of events. First σ appoints p and defines $\lambda_d(p)$; then τ certifies $\Phi_e(D, x)$; and only later do we see $o^i(p) < \alpha$. In this case, the use is too small, so action for p would injure the certified computation; but τ did not know how many times σ will act for

p when it certified the computation. Note that τ could not wait for this later event, since we may never see $o^i(p) < \alpha$. Of course, this is where we use the additional computational power of W_d. Before we see $o^i(p) < \alpha$, σ does not need to act for p. Once we see $o^i(p) < \alpha$, if $W_d \not\leq_T D$, then W_d will permit σ to lift the use $\lambda_d(p)$ beyond the use of a computation $\Phi_e(D, x)$, in fact beyond the use of a D-correct such computation. Only then is p *cleared* by τ and σ can attack with impunity. We cannot expect that every follower we appoint is permitted, and so σ will need to appoint a sequence of followers $p_0, p_1, \ldots$; one of them will be permitted.

We note two issues. One is that while W_d will permit some follower appointed by σ, the stage at which it gives this permission is not necessarily a stage at which σ is accessible, and this permission cannot "remain open" until σ is next visited: σ may never be visited again, and we need to define $\lambda_d(p)$ to keep $\Lambda_d(D, W_d)$ total. So we act on permissions immediately, even if σ is not accessible; this does no harm to the rest of the construction.

The other issue is that of totality. For each follower p, we note which computations $\Phi_e(D, x)$ it is not allowed to injure, and seek permission from W_d at a stage at which $\Phi_e(D, x)\downarrow$ for all such computations. We are guaranteed eventual permission only if these are D-correct computations. How do we know that such a stage will occur? Of course σ, since it extends $\tau\hat{}\infty$, *guesses* that $\Phi_e(D)$ is total. But there are constructions in which $\tau\hat{}\infty$ lies on the true path but the measured function $\Phi_e(D)$ is in fact not total. This is avoided in this construction because we make D totally α-c.a. and so low$_2$.

The tree of strategies

As usual, to define the tree, we specify recursively the association of nodes to requirements, and specify the outcomes of nodes working for particular requirements. To specify the priority ordering of nodes, we specify the ordering between outcomes of any node.

We order all of the requirements Q_d and P^i_e in order-type ω; all nodes of length k work for the kth requirement on the list. The outcomes of a node working for Q_e are ∞ and `fin`, with $\infty < \texttt{fin}$. A node working for P^i_e has only one outcome.

Clearing followers

A follower p for a node σ working for P^i_e can be in one of three states.

(1) When p is first appointed, it is *unready*.
(2) At a later stage (at which σ is accessible) we may see that $o^i_s(p) < \alpha$; then p becomes *ready*: we have determined which computations $\Phi_d(D, x)$ it is allowed to injure.
(3) At a later stage yet, W_e may give permission to lift the use $\lambda_e(p)$ and begin an attack with p. We say that p is *in the clear*.

Let $\texttt{prec}(\sigma)$ be the collection of nodes τ such that τ works for a requirement Q_d and $\tau\hat{\ }\infty \preccurlyeq \sigma$. This is the collection of nodes that may need to restrain σ's action to protect computations they are monitoring. For each follower p for σ, if p becomes ready (by observing that $o^i(p) < \alpha$) then we define, for each $\tau \in \texttt{prec}(\sigma)$, a value $m^\tau(p)$, which serves as a watermark. If τ works for Q_d, then action by σ for p is allowed to injure computations $\Phi_{d,s}(D_s, x)$ for $x \geqslant m^\tau(p)$, but not for smaller values of x.

Construction

At each stage we will do one of two things. Normally we will build the path of accessible nodes and act accordingly. But at some stages we will observe W_e permissions that will allow us to clear a follower for some σ. In that case no node is accessible at that stage and no other action is taken by any node. In both cases, though, after the main action, we *maintain* functionals (work toward making them total).

OPTION A. At stage s we first ask: is there some node σ working for a positive requirement P^i_e which currently has a ready follower p such that:

- $p \notin \operatorname{dom} \Lambda_{e,s}(D_s, W_{e,s+1})$; and
- for all $\tau \in \texttt{prec}(\sigma)$, working for Q_d, we have $m^\tau(p) \leqslant \operatorname{dom} \Phi_{d,s}(D_s)$.

If so, then we let σ be the strongest such node. We pick such a follower p for σ, and declare it to be *in the clear*. We cancel all other followers for σ. We let $D_{s+1} = D_s$. We define $\Lambda_{e,s+1}(D_{s+1}, W_{e,s+1}, p) = s+1$ with large use (the D-use and the W_e-use will always be equal). We initialise all nodes weaker than σ. For any pair $(d,q) \leqslant s$ distinct from (e,p) we maintain $\lambda_d(q)$ as follows, and then end the stage.

MAINTAINING $\lambda_d(q)$: If $q \notin \operatorname{dom} \Lambda_{d,s}(D_{s+1}, W_{d,s+1})$, then we define a new computation $\Lambda_{d,s+1}(D_{s+1}, W_{d,s+1}, q) = s+1$ with use $\lambda_{d,s+1}(q)$ determined by cases:

- If q is currently a follower for a node σ' working for P^j_d for some j (in particular, q was not just cancelled), then we set $\lambda_{d,s+1}(q) = \lambda_{d,s}(q)$.
- Otherwise, $\lambda_{d,s+1}(q) = -1$.

The instructions will ensure that in the first case, $\lambda_{d,s}(q)$ is indeed defined, that is, $q \in \operatorname{dom} \Lambda_{d,s}(D_s, W_{d,s})$. The point of the first clause is to keep $\Lambda_d(D, W_d)$ total when we have W_d-changes which are not beneficial, i.e., occur when the follower q is unready or $\operatorname{dom} \Phi_c(D)[s] < m^\tau(q)$ for some $\tau \in \texttt{prec}(\sigma')$.

OPTION B. If option A was not taken, then we let, by recursion, the collection of accessible nodes δ_s be an initial segment of the tree of strategies. So the root of the tree is accessible at stage s.

Suppose that a node τ that works for requirement Q_e is accessible at stage s. If s is the least stage at which τ is accessible then we let $\tau\hat{\ }\infty \in \delta_s$. Otherwise we let t be the last stage before s at which $\tau\hat{}\infty$ was accessible. If $t < \operatorname{dom} \Phi_{e,s}(D_s)$ then we let $\tau\hat{}\infty \in \delta_s$. Otherwise we let $\tau\hat{}\mathtt{fin} \in \delta_s$.

Suppose that a node σ, working for requirement P_e^i, is accessible at stage s. There are two cases: either σ has a unique follower which is in the clear; or no follower for σ is in the clear. In the latter case, σ possibly has a number of ready followers, and possibly one unready follower.

1. Suppose that σ has follower p in the clear.
If $\Lambda_{e,s}(D_s, W_{e,s}, p) = f_s^i(p)$ then we enumerate $\lambda_{e,s}(p)$ into D_{s+1} and redefine $\Lambda_{e,s+1}(D_{s+1}, W_{e,s+1}, p) = s+1$ with large use. We initialise all nodes weaker than σ and halt the stage.
If $\Lambda_{e,s}(D_s, W_{e,s}, p) \neq f_s^i(p)$ then the unique immediate successor on the tree of strategies is next accessible.
2. Suppose that σ has no follower in the clear. There are two things we may do.
   (a) If σ has a currently unready follower p and $o_s^i(p) < \alpha$, then we declare p to be *ready*. For each $\tau \in \mathtt{prec}(\sigma)$, working for Q_d, we define $m^\tau(p) = \operatorname{dom} \Phi_{d,s}(D_s)$.
   (b) If either the action in part (a) has just been performed, or σ currently has no followers, then currently all followers for σ are ready. We then appoint a new, large follower p' for σ (which is *unready*) and define $\Lambda_{e,s+1}(D_{s+1}, W_{e,s+1}, p') = s+1$ with large use.

If neither (a) nor (b) are performed then σ already has one unready follower p with $o_s^i(p) = \alpha$, and we do nothing.

If $|\sigma| < s$, then the unique immediate successor on the tree of strategies is next accessible; otherwise we halt the stage. In case 2, we do not initialise weaker nodes even if we appoint a new follower. This is because if $W_e \leqslant_{\mathrm{T}} D$, it is possible that infinitely many followers will be appointed.

At the end of the stage, we maintain $\lambda_d(q)$ for pairs $(d, q) \leqslant s$ (other than pairs for which $\Lambda_d(D, W_d, q)[s+1]$ has just been defined) as above.

Verification

For a while, we follow the verifications for Theorem 3.6. We have an analogue of Lemma 3.8. In the verification, we say that a node σ *acts* at a stage s if either it is accessible at stage s and enumerates a number into D_{s+1} on behalf of a follower in the clear; or if stage s option A is taken and a follower for σ is cleared.

As indicated in the construction, if a follower p for σ is cleared at some stage s, then all other followers for σ are cancelled at that stage. Until possibly a later stage at which σ is initialised, p remains σ's unique follower.

Lemma 4.2. *Let s be a stage.*

(a) Every functional $\Lambda_{e,s}$ is consistent for the pair $D_s, W_{e,s}$.

Suppose that at the beginning of stage s, p is a follower for a node σ which works for P_e^i.

(b) *$\Lambda_{e,s}(D_s, p)\downarrow$ and $\lambda_{e,s}(p) \notin D_s$.*
(c) *Suppose that p' is a follower for a node σ', weaker than σ, working for $P_{e'}^{i'}$. Then $\lambda_{e,s}(p) \neq \lambda_{e',s}(p')$. If p is in the clear at the beginning of stage s, then $\lambda_{e,s}(p) < p'$. As usual $p' < \lambda_{e',s}(p')$.*

Let $t < s$, and suppose that p was already a follower for σ at the beginning of stage t.

(d) *If p was in the clear at stage t, then $D_t \restriction_{\lambda_{e,t}(p)} = D_s \restriction_{\lambda_{e,t}(p)}$; if, in addition, σ did not act at any stage $r \in [t, s)$, then $D_t \restriction_{\lambda_{e,t}(p)+1} = D_s \restriction_{\lambda_{e,t}(p)+1}$.*
(e) *If p is not in the clear at the beginning of stage s then $\lambda_{e,t}(p) = \lambda_{e,s}(p)$.*

Proof. As it is similar to the proof of Lemma 3.8, we note the differences. For (b), that $\Lambda_{e,s}(D_s, p)\downarrow$ is immediate here, from the maintenance round we do at the end of every stage. To show that $\lambda_{e,s}(p) \notin D_s$, the new case is if at stage $s-1$, when performing maintenance, we saw that $\Lambda_{e,s-1}(D_s, W_{e,s}, p)\uparrow$, and defined a new computation with $\lambda_{e,s}(p) = \lambda_{e,s-1}(p)$. However, by induction, $y = \lambda_{e,s-1}(p) \notin D_{s-1}$. The node σ does not act at stage $s-1$, and the first part of (c) (at stage $s-1$) shows that no other node can enumerate y into D_s.

For (c), we note that, as usual, new uses $\lambda_{e,s}(p)$ are chosen to be large, and so distinct from existing uses. The second part follows from the fact that at the stage at which p is cleared, σ' is initialised. The proof of (d) is identical to the previous proof. (e) is new, and follows immediately by induction, since σ never acts for p before p is cleared, and once the use $\lambda_{e,t}(p)$ is picked (at the stage at which p is appointed), the use is never lifted (see maintenance step). □

The proof of Lemma 3.9 gives its analogue, recalling, though, that we say that σ acts for p at stage s only if p is cleared at stage s, or if σ enumerates $\lambda_{e,s}(p)$ into D_{s+1} (when p is already in the clear); not when p is appointed or is declared ready.

Lemma 4.3. *Let σ be a node that works for requirement P_e^i. Let p be a follower for σ at stages $s < t$, and suppose that at both stages, σ acts for p. Then $o_t^i(p) < o_s^i(p)$.*

It follows that for each p, σ enumerates $\lambda_e(p)$ into D at only finitely many stages. If the construction is fair to σ, then it follows that σ halts the stage at most finitely many times after it is last initialised: at most once when a follower p becomes cleared, and then finitely many times when it enumerates $\lambda_{e,s}(p)$ into D.

Lemma 4.4. *The true path δ_ω is infinite, and the construction is fair to every node on the true path.*

Proof. The point is that there are infinitely many stages at which we do not take option A and stop the stage: there are infinitely many stages at which δ_s is nonempty. Suppose for a contradiction that there is a last stage s^* at which we take option B. There are only finitely many nodes σ which have followers at the end of stage s^*. But for each such node σ there is at most one stage $s > s^*$ at which we act for σ. At that stage, a follower for σ is cleared. Either this follower is never cancelled and σ does not act again. Or σ is initialised at some later stage but never has the chance to appoint new followers. This is a contradiction. □

Lemma 4.5. *For all e, $\Lambda_e(D, W_e)$ is total.*

Proof. The difference from the proof of Lemma 3.12 is that W_e-changes may make Λ_e-computations diverge. The maintenance step, and in particular keeping the use fixed unless a follower becomes cleared, addresses this issue. Formally, the convergence of $\Lambda_e(D, W_e, p)$ for a permanent follower p for σ follows from Lemma 4.2(e) if p is never cleared, and from Lemma 4.3 if it is. □

The argument of Lemma 3.10 now shows that if a node σ on the true path, working for requirement P^i_e, has a follower which is eventually cleared but never cancelled, then $\Lambda_e(D, W_e) \neq f^i$.

As mentioned above, perhaps surprisingly, in order to show that each finitary requirement P^i_e is met, we need to investigate the infinitary requirements first. The verification for the finitary requirements will use the fact that $\deg_{\mathrm{T}}(D)$ is low$_2$.

Fix a node τ, working for requirement Q_e, such that $\tau\hat{\ }\infty$ lies on the true path. By Lemma 4.4, let s^* be the last stage at which τ is initialised. Let $S = \{s_0, s_1, \dots\}$ be the collection of stages $s > s^*$ at which $\tau\hat{\ }\infty$ is accessible. For $x < \omega$, let $i(x)$ be the least i such that $x < \operatorname{dom} \Phi_e(D)[s_i]$. For $x < \omega$, we let $a(x)$ be the collection of pairs (σ, p) such that $\sigma \succcurlyeq \tau\hat{\ }\infty$ (in other words $\tau \in \texttt{prec}(\sigma)$), and p is a follower for σ which became ready at some stage *prior* to stage $s_{i(x)}$, but is not cancelled by stage $s_{i(x)}$. For $j \geqslant i(x)$ we let $a_j(x)$ be the collection of pairs $(\sigma, p) \in a(x)$ such that σ is not initialised at any stage $r \in [s_{i(x)}, s_j)$, and p is still a follower for σ at the beginning of stage s_j.

The set $a(x)$ plays the same role as it did in the proof of Theorem 3.6: only action by σ for some p such that $(\sigma, p) \in a_j(x)$ can injure a computation $\Phi_e(D, x)$ at stage s_j. This will show that $\Phi_e(D)$ is α-c.a., as $a(x)$ is finite, effectively obtained from x, and at stage $s_{i(x)}$, we already know an ordinal bound $o^k_t(p)$ on the "number of times" σ can attack with p. Note that for each σ there is at most one p such that $(\sigma, p) \in a(x)$ and σ will attack with p at a later stage s_j. However, the identity of this p—the one follower for σ that will be cleared, if there is one—is not yet known at stage $s_{i(x)}$.

Lemma 4.6. *Let $\sigma \succcurlyeq \tau\hat{\ }\infty$, working for P_d^i, and let p be a follower for σ which is already in the clear at the beginning of stage $s \geqslant s_{i(x)}$. Suppose that $(\sigma, p) \notin a(x)$. Then:*

(1) $m^\tau(p) > x$.
(2) Let t be the stage at which p is cleared. Then $x \in \operatorname{dom} \Phi_e(D)[t]$ and $D_t \upharpoonright_{\varphi_{e,t}(x)} = D_s \upharpoonright_{\varphi_{e,t}(x)}$. It follows of course that $x \in \operatorname{dom} \Phi_e(D)[s]$ and that $\varphi_{e,s}(x) = \varphi_{e,t}(x)$.
(3) $\lambda_{d,s}(p) > \varphi_{e,s}(x)$.

Proof. For (1), let w be the stage at which p is declared ready. If $w < s_{i(x)}$ then $(\sigma, p) \in a(x)$, so $w \geqslant s_{i(x)}$ (and it follows that $t > s_{i(x)}$). At stage w, σ is accessible, and so $w = s_j$ for some $j \geqslant i(x)$, whence $x < \operatorname{dom} \Phi_{e,w}(D_w) = m^\tau(p)$.

At stage t we have $\operatorname{dom} \Phi_e(D)[t] \geqslant m^\tau(p)$—this is one of the conditions for p to be cleared. Hence $x < \operatorname{dom} \Phi_e(D)[t]$, so $\varphi_{e,t}(x)$ is indeed defined. Let $u = \varphi_{e,t}(x)$. At stage t, we define $\lambda_{d,t+1}(p)$ to be large, and so larger than u.

At stage t no node is accessible, so $D_{t+1} = D_t$. Lemma 4.2(d) applied to $t+1 \leqslant s$ says that $D_s \upharpoonright_{\lambda_{d,t+1}(p)} = D_{t+1} \upharpoonright_{\lambda_{d,t+1}(p)}$, and (2) follows.

As $\lambda_{d,r}(p)$ is non-decreasing with r, it follows that $\lambda_{d,s}(p) > u = \varphi_{e,s}(x)$. ☐

We are now ready to prove an analogue of Lemma 3.14.

Lemma 4.7. *Let $j \geqslant i(x)$. Let $u = \varphi_{e,s_j}(x)$. Suppose that $D_{s_{j+1}} \upharpoonright_u \neq D_{s_j} \upharpoonright_u$. Then there is some $(\sigma, p) \in a_j(x)$ such that σ acts for p at stage s_j and enumerates $\lambda_{d,s_j}(p) < u$ into D_{s_j+1}.*

Proof. The argument follows the proof of Lemma 3.14. Suppose that at stage $s \in [s_j, s_{j+1})$, a node σ acts for some follower p and enumerates $\lambda_{d,s}(p) < \varphi_{e,s_j}(x)$ into D_{s+1}. The argument that σ extends $\tau\hat{\ }\infty$, and so $s = s_j$, is the same as above. Note that p is already in the clear at the beginning of stage s_j. Lemma 4.6(3) shows that $(\sigma, p) \in a(x)$, and so $(\sigma, p) \in a_j(x)$. ☐

The next lemma shows that D is low$_2$.

Lemma 4.8. *Let τ be a node on the true path that works for requirement Q_e. Then $\tau\hat{\ }\infty$ lies on the true path if and only if $\Phi_e(D)$ is total.*

Proof. The nontrivial direction is left-to-right. Let $x < \omega$. To show that $x \in \operatorname{dom} \Phi_e(D)$, we observe that there are only finitely many $j \geqslant i(x)$ such that $D_{s_{j+1}} \upharpoonright_{\varphi_{e,s_j}(x)} \neq D_{s_j} \upharpoonright_{\varphi_{e,s_j}(x)}$. This follows from the fact that $a(x)$ is finite, and that for each $(\sigma, p) \in a(x)$, σ acts for p at most finitely many times. ☐

We can now show that the positive requirements are met.

Lemma 4.9. *For all e and i, the requirement P_e^i is met.*

Proof. Let σ be a node on the true path, working for P^i_e. We observed above that if there is a follower p for σ which is at some point cleared and is never cancelled, then P^i_e is met. Let r^* be the last stage at which σ is initialised, and suppose that no follower for σ is cleared after stage r^*. If $\langle f^i_s, o^i_s \rangle$ is not eventually α-computable, then P^i_e is met vacuously, so we assume that it is. Then every follower that σ appoints after stage r^* eventually becomes ready (of course, using the fact that σ is accessible during infinitely many stages). Then σ appoints infinitely many followers. We show that $W_e \leqslant_{\mathrm{T}} D$.

Let p be a follower for σ, appointed after stage r^*; let s_0 be the stage at which p is appointed, and let $u = \lambda_{e,s_0}(p)$. As $u > s_0$, the numbers u are unbounded, as p ranges over the followers for σ. To compute $W_e \restriction u$ from D, we first go to the stage t at which p becomes ready. At that stage we observe the numbers $m^\tau(p)$ for $\tau \in \mathtt{prec}(\sigma)$. For all $\tau \in \mathtt{prec}(\sigma)$, $\tau\hat{\ }\infty$ lies on the true path. By Lemma 4.8, there is a stage s at which for all $\tau \in \mathtt{prec}(\sigma)$, for all $x < m^\tau(p)$, $x \in \operatorname{dom} \Phi_e(D)[s]$ by a D-correct computation. Certainly D can find such a stage s; and $W_{e,s} \restriction u = W_e \restriction u$, for otherwise p would be cleared at some stage $s' > s$. $\square$

We now rejoin the proof of Theorem 3.6, using Lemma 4.7 to show that for every e such that $\Phi_e(D)$ is total, the node τ on the true path working for Q_e is successful in devising an α-computable approximation for $\Phi_e(D)$. Fix such e and τ; we again use the stages s_i, the indices $i(x)$ and the sets $a_j(x)$ discussed above. Fix $x < \omega$. We note, and this is the main point, that for all $(\sigma, p) \in a(x)$, if σ works for P^i_d then $o^i_{s_{i(x)}}(p) < \alpha$.

Let $j \geqslant i(x)$ and let σ be a node, working for P^i_d, which appears in $a_j(x)$ (i.e., $(\sigma, p) \in a_j(x)$ for some p). If no follower for σ is cleared by the beginning of stage s_j, we let

$$\beta_j(\sigma) = \max \left\{ o^i_{s_j}(p) : (\sigma, p) \in a(x) \right\}.$$

Otherwise, let p be the unique follower for σ at stage s_j; $(\sigma, p) \in a_j(x)$. We let $t_j(\sigma)$ be the greatest stage $t < s_j$ at which σ acts (for p); such a stage exists, since p becomes cleared at some stage $t < s_j$. We then let $\beta_j(\sigma) = o^i_{t_j(\sigma)}(p)$. Finally, we order the nodes appearing in $a_j(x)$ in descending priority as $\sigma_0, \sigma_1, \dots, \sigma_{k(j)}$, and let $m_j(x) = \sum_{k \leqslant k(j)} \beta_j(\sigma_k)$. We note that if σ_k acts at stage s_j then $k(j+1) \leqslant k$. Lemma 3.15 holds for the current construction, with much the same proof. This completes the proof of Theorem 4.1.

4.1.1 Maximal uniformly totally ω^α-c.a. degrees

Not only are there maximal uniformly totally ω^α-c.a. degrees, but there are such degrees which are also maximal totally ω^α-c.a.

Theorem 4.10. *If α is a power of ω, then there is a uniformly totally α-c.a. degree which is maximal totally α-c.a.*

Proof. To prove Theorem 4.10, we run the construction for Theorem 4.1 with but one modification: a follower p for a node σ working for P^i_e becomes ready at a stage t_1 if σ is accessible at stage t_1, and at the previous stage $t_0 < t_1$ at which σ was accessible we saw that $o^i_{t_0}(p) < \alpha$. That is, we only let p become ready at the *second* stage at which σ is accessible and at which we see $o^i_t(p) < \alpha$. It is easily verified that this delay in declaring a follower to be ready does not affect the success of the construction, so the degree $\deg_T(D)$ produced under this new definition of readiness is also maximal totally α-c.a.; we show though that the degree produced is also uniformly totally α-c.a.

We follow the argument for proving part (1) of Theorem 3.20. By design of the current construction, a node σ accessible at stage s has length at most s. We fix some τ, working for Q_e, such that $\tau\hat{\ }\infty$ lies on the true path. Now we examine the proof of Lemma 3.22. For $x \geqslant \operatorname{dom} \Phi_e(D)[s_1]$, again let $u_0 < u_1 < s_{i(x)}$ be successive stages at which $\tau\hat{\ }\infty$ is accessible. Let $(\sigma, p) \in a(x)$, with σ working for P^i_d. Then $o^i_{u_0}(p) < \alpha$, and $u_0 < x$. Since $|\sigma| \leqslant u_0$, we may assume that $i < x$. It follows that $m_{i(x)}(x)$ is an ordinal which can be observed at stage x of the construction, and this is independent of τ. This gives an α-order function h such that every $f \leqslant_T D$ is h-c.a. □

In Section 3.3.2 we explained why we could not combine the proofs of the two parts of Theorem 3.25 and obtain a contradiction (a degree which both is and is not uniformly totally α-c.a.). The explanation focussed on the stage $u_1 = s_{i(x)-1}$, the last stage in S before stage $s_{i(x)}$. A follower p appointed at stage u_1 would have bound $o^i_{u_1}(p)$ which can be arbitrarily large with relation to p, but will be able to destroy computations $\Phi_e(D, x)[s_j]$ for $j \geqslant i(x)$. In the previous chapter there is no way around this; we have to allow such a p to destroy the computations, or σ will not be able to meet its requirement. In the current situation, using W_d to lift the use $\lambda_e(p)$ when p is cleared allows us to choose which followers to restrain, and this makes possible the proof of Theorem 4.10.

For the case $\alpha = \omega$, Theorem 4.10 says that there is an array computable c.e. degree which is maximal totally ω-c.a. In fact, we suspect that combining the methods of this chapter together with the construction of a contiguous degree, one can show that there is a contiguous degree which is maximal totally ω-c.a. Since every contiguous degree is array computable, such a degree is also maximal contiguous.

The following theorem, for $\alpha = \omega$, shows that not all maximal totally ω-c.a. degrees are maximal contiguous degrees.

Theorem 4.11. *If α is a power of ω, then there is a maximal totally α-c.a. degree which is not uniformly totally α-c.a.*

Sketch of proof. We combine the construction for Theorem 4.1 with the technique proving Theorem 3.20(2). To the construction for Theorem 4.1 we add the enumeration of a functional Γ, with the aim of making $\Gamma(D)$ witness that

$\deg_T(D)$ is not uniformly totally α-c.a. Again we fix an α-order function h, and enumerate h-c.a. functions $\langle g_i \rangle$ along with tidy $(h+1)$-computable approximations for these functions. We add a third kind of requirement, R^i, namely that $\Gamma(D) \neq g_i$. The action for these requirements is identical to that of the previous chapter. There is no interaction (other than mutual initialisations) between nodes working for R^i and nodes working for P^j_d; and the interaction between nodes working for R^i and nodes working for Q_e is as in the previous chapter. That is, when showing that D is low$_2$, and then devising an α-computable approximation for $\Phi_e(D)$ if it is total, the sets $a(x)$ may contain pairs (σ, p) where σ works for either a requirement R^i or for a requirement P^j_d. In either case, the ordinal bound on the number of times σ will act for p can be observed at stage $s_{i(x)}$, and if (σ, p) is not in $a(x)$, then action by σ for p cannot injure a computation $\Phi_{e,s}(D_s, x)$ observed at a τ-expansionary stage. □

4.2 LIMITS ON FURTHER MAXIMALITY

One might wish for even stronger maximality properties than those provided by Theorem 4.1. Could there be, for example, a totally ω-c.a. degree which is a maximal totally ω^2-c.a. degree? In general, can a degree in one level of our crudest hierarchy be maximal for a higher level? The following theorem says it cannot.

Theorem 4.12. *Let $\beta < \varepsilon_0$. Every totally ω^β-c.a. c.e. degree is bounded by a strictly greater totally $\omega^{\beta+1}$-c.a. c.e. degree.*

To prove Theorem 4.12, fix an ordinal $\beta < \varepsilon_0$, and let $\alpha = \omega^\beta$. Let V be a c.e. set whose Turing degree is totally α-c.a. We enumerate a set D such that $\deg_T(V \oplus D)$ is strictly greater than $\deg_T(V)$ and is totally $\alpha \cdot \omega = \omega^{\beta+1}$-c.a. The requirements to meet are:

P_e: $\Psi_e(V) \neq D$;

and

Q_e: If $\Phi_e(V, D)$ is total then it is $\alpha \cdot \omega$-c.a.

Discussion

The main idea for meeting the requirement Q_e is as follows. We track $\Phi_e(V, D, x)$ for some x. Changes to such a computation can come from two sources: a V-change or a D-change. To keep track of the V-changes—the ones we do not control ourselves—we build what we call a "shadow functional" $\hat{\Phi}_e$, with intended oracle V alone. We pick an input c and define $\hat{\Phi}_e(V, c)$ with the same use as that of $\Phi_e(D, V, x)$ (recall that we assume that the V-use and the D-use are identical).

The input c is called the *tracker* for x. We ensure that if $\Phi_e(D,V)$ is total, then $\hat{\Phi}_e(V)$ is total as well. Since $\deg_{\mathrm{T}}(V)$ is totally α-c.a., $\hat{\Phi}_e(V)$ will equal f^i for some i, where $\langle f^i \rangle$ lists α-c.a. functions. We guess the correct index i; this will be done using the fact that V is low$_2$. This is a Δ^0_3-guessing process, which is very similar to a Π^0_2/Σ^0_2 process, except that infinitely many outcomes are required. The correct guess will observe $o^i(c)$ and bound the V-changes in $\Phi_e(D,V,x)$.

We have to think though what happens when we cause a D-change (for the sake of meeting some P_d). The computation $\Phi_e(D,V,x)$ is gone, but it is possible that the V-part of the computation was correct. In this case $\hat{\Phi}_e(V,c)$ is a correct computation, and we cannot use the tracker c to shadow new $\Phi_e(D,V,x)$ computations. We need to replace c by a new tracker and repeat the process. This is how we get $\alpha \cdot \omega$: when we first certify $\Phi_e(D,V,x)$, we put a bound on the number of D-changes that we allow to destroy such a computation; say it is n. We appoint a tracker c_0 and observe $\beta_0 = o^i_0(c_0)$. We then declare that $\Phi_e(D,V,x)$ will not change more than $\alpha \cdot n + \beta_0$ many "times." While we only see V-changes, the associated ordinal is still $\alpha \cdot n + o^i_s(c_0)$. Once we cause a D-change that destroys a $\Phi_e(D,V,x)$ computation, we appoint a new tracker c_1, observe $\beta_1 = o^i_s(c_1)$, decrease our ordinal to $\alpha \cdot (n-1) + \beta_1$, and repeat the process.

We could be tempted to improve the bound. If we know in advance (i.e., when $\Phi_e(D,V,x)$ is first certified) a bound n on the number of D-injuries to the computation, we could immediately appoint n trackers $c_0, \dots, c_{n-1}$ and start our approximation knowing $\beta_k = o^i_0(c_k)$ for all of these trackers. Then the bound would be $\beta_{n-1} + \beta_{n-1} + \cdots + \beta_0$ which in fact is smaller than α. We would prove that there is no maximal totally α-c.a. degree. The fallacy is easy to see: we do not know whether we will actually see n-many D-injuries to the computation; n is just a bound. While we are using the tracker c_0 we cannot define computations $\hat{\Phi}_e(V,c_k)$ for the other trackers ($k>0$); we need to keep them open, because the use of these computations is the use of $\Phi_e(x)$-computations we have not yet observed. This would make $\hat{\Phi}_e(V)$ partial even if $\Phi_e(D,V)$ is total, and so void the whole plan.

We now discuss how to meet P_e, bearing in mind the severe restriction imposed by the negative requirements: such requirements need to know in advance (relative to the input x) the number of times (in this instance without quotation marks) a D-change could ruin a computation $\Phi_d(D,V,x)$.

We pick a follower p and wait for $\Psi_e(V,p)$ to converge, with the intention of ensuring that $\Psi_e(V,p) \neq D(p)$. Of course the difficulty is that we do not know, when presented with such a computation, whether the presented computation is V-correct. If V were low we could apply R. Robinson's guessing technique. However V need not be low. But it is low$_2$, and again we use this to guess the answer to the question, "Is $\Psi_e(V)$ total?"

Independent of the restrictions imposed by the negative requirements, ensuring that $D \nleq_{\mathrm{T}} V$ would now be easy. Define a D-computable function $\Lambda(D)$. Each outcome of P_e which believes that $\Psi_e(V)$ is total appoints a follower p. If such an outcome is believed and we currently see that $\Psi_e(V,p) = \Lambda(D,p)$

then we diagonalise. If such an outcome lies on the true path then its guess is correct: $\Psi_e(V)$ is indeed total, and so the outcome would act only finitely many times.

Such action causes conflict with stronger negative requirements. To keep $\Lambda(D)$ total, a new value for $\lambda(D,p)$ needs to be picked immediately when an outcome of P_e acts. This means that such an outcome will repeatedly injure a computation $\Phi_d(D,V,x)$. We could try to use the fact that $\deg_{\mathrm{T}}(V)$ is totally α-c.a., rather than the weaker fact that it is low_2. We guess that $\Psi_e(V)=f^i$ for some α-c.a. function f^i on our list; the node following $\Phi_d(D,V,x)$ will observe how many "times" the P_e-child will act, and incorporate it into its bound. The bound though is α rather than ω. In this way we could try to make $D\oplus V$ totally α^2-c.a., but not totally $\alpha\cdot\omega$-c.a. Of course for $\alpha=\omega$ this is sufficient.

To overcome this difficulty we modify the action of P_e as follows. The problem was that even though we have certification that $\Psi_e(V)$ is total, many single computations we see will be incorrect. To respect the main restriction, after a failed attack with a follower we abandon that follower altogether. To ensure that this does not go on indefinitely we build a shadow functional $\hat{\Psi}_e$, with intended oracle V. We need to ensure that if $\Psi_e(V)$ is total then so is $\hat{\Psi}_e(V)$. Each node that guesses totality appoints an *anchor* q which will serve many followers p. We ensure that the uses of $\Psi_e(V,p)$ and $\hat{\Psi}_e(V,q)$ are the same. If the node is correct then the fact that $\hat{\Psi}_e(V,q)$ stabilises ensures that only finitely many followers are ever appointed by that node.

We need to discuss in greater detail how a node τ working for Q_e can tolerate the action of a node σ working for P_d. Assuming that the node σ guesses that $\limsup_s \operatorname{dom}\Phi_e(V,D)[s]=\infty$, it also needs to guess whether $\liminf_s \operatorname{dom}\Phi_e(V,D)[s]=\infty$, that is, if $\Phi_e(V,D)$ is total or not. If σ guesses that $\Phi_e(V,D)$ is total then for each x we allow an enumeration of a follower for σ to injure $\Phi_e(V,D,x)$ at most once. As in the construction of a maximal totally α-c.a. degree, we set a "watermark" $m_s(\sigma)$, differentiating between large inputs whose computations σ is allowed to injure, and smaller inputs which need to be protected. Each time σ attacks, the watermark is updated. It is possible that due to a V-change, a follower p is smaller than the use $\varphi_{e,s}(x)$ for some protected input $x<m_s(\sigma)$. In this case the V-change makes $\hat{\Psi}_d(V,q)\uparrow$, and we can discard the follower and choose a new, large one. Note that when this is done we do not need to update $m_s(\sigma)$: the node τ only cares about the number of followers that will injure a computation $\Phi_e(D,V,x)$, not about the identity of the follower that will inflict the injury.

What at first appears to be a trickier situation is when σ guesses that $\Phi_e(V,D)$ is partial. We still need to protect computations $\Phi_e(V,D,x)$ for small x, because we don't know that σ's guess is correct. This means cancelling a follower p for σ when we see a V-change that causes $\varphi_e(x)$ to increase. But if $\Phi_e(V,D,x)\uparrow$ then this can happen infinitely often. However, σ can guess the exact place at which $\Phi_e(V,D)$ becomes partial, that is, the value of $\liminf_s \operatorname{dom}\Phi_e(V,D)[s]$. Say that value is y. Inputs $x<y$ will eventually settle and stop causing the

cancellation of σ's follower. When we guess the value y we delay the definition of $\hat{\Phi}_\tau(V,c)$ where c is the current tracker for x. Action by σ at such a stage will not cause problems for stronger "totality outcomes" of τ: if $\hat{\Phi}_\tau(V,c)\uparrow [s]$ then enumeration of a number into D at stage s does not mean that we need to abandon the tracker. On the other hand if $\Phi_e(V,D)$ is total then such y will be guessed only finitely often and so $\hat{\Phi}_\tau(V,c)$ will eventually be defined and we can ensure that $\hat{\Phi}_\tau(V)$ is total as well, which is necessary for τ's strategy to work.

The tree of strategies and Δ^0_3 guessing

We define the tree of strategies and assign strategies to nodes on the tree by recursion.

We start with the empty node, to which we assign the requirement Q_0. Suppose that τ is a node on the tree which was assigned the requirement Q_e. The node will have a number of children on the tree which help τ meet its goal. The outcomes of τ are $\infty < \mathtt{fin}$. These outcomes measure $\limsup_s \operatorname{dom} \Phi_e(V,D)[s]$. The node $\tau\hat{}\mathtt{fin}$ is assigned to the requirement P_e.

The outcomes of $\tau\hat{}\infty$ on the tree are ∞_n and $\mathtt{fin}_n$ for $n<\omega$ (ordered by $\infty_0 < \mathtt{fin}_0 < \infty_1 < \mathtt{fin}_1 < \infty_2 < \cdots$). These outcomes participate in the Δ^0_3 guessing process of whether $\hat{\Phi}_\tau(V)$ is total or not. The nodes $\tau\hat{}\infty\hat{}\mathtt{fin}_n$, which guess that $\hat{\Phi}_\tau(V)$ is not total, are assigned to the requirement P_e. The outcomes of nodes of the form $\tau\hat{}\infty\hat{}\infty_n$ are all $i<\omega$ (ordered naturally). A node $\tau\hat{}\infty\hat{}\infty_n$ guesses that $\hat{\Phi}_\tau(V)$ is total. If it is correct then $\hat{\Phi}_\tau(V)$ must equal f^i for some i, where $\langle f^i\rangle$ as usual is a list of the α-c.a. functions equipped uniformly with tidy $(\alpha+1)$-computable approximations $\langle f^i_s, o^i_s\rangle$; this is guessed by the node $\tau\hat{}\infty\hat{}\infty_n\hat{}i$. We assign each such node the requirement P_e.

Suppose that a node π is assigned the requirement P_e. The node π has infinitely many outcomes ∞_n and $\mathtt{fin}_n$, ordered as above. Again this is for guessing the totality of $\hat{\Psi}_\pi(V)$, a shadow functional enumerated by the node π. The children of π—its immediate successors on the tree—combine forces to help π meet its requirement. They each have a single immediate extension on the tree, which is assigned to the requirement Q_{e+1}.

As discussed, nodes τ working for Q_e define a shadow functional $\hat{\Phi}_\tau$ and nodes π working for P_e define a shadow functional $\hat{\Psi}_\pi$. Since V is low$_2$, the set of indices of functionals Θ such that $\Theta(V)$ is total is Σ^0_3. Membership in a Π^0_2 set can be translated to the question whether a given non-decreasing computable sequence is bounded or not. By the recursion theorem we know the indices of the functionals enumerated by the nodes τ and π on the tree. Thus we obtain for each such node μ a computable list $\ell_s(\mu,n)$ of sequences, non-decreasing in s, such that the functional enumerated by μ is total if and only if for some n, the sequence $\langle \ell_s(\mu,n)\rangle$ is unbounded.

As mentioned above, a node τ working for Q_e appoints trackers $\mathtt{tr}_s(\tau,x)$ for inputs $x<\omega$. If σ is a child of a node π working for P_e which believes that $\hat{\Psi}_\pi(V)$ is total (i.e., $\sigma=\pi\hat{}\infty_n$ for some $n<\omega$) then σ may appoint both an anchor $\mathtt{ac}_s(\sigma)$ and a follower $\mathtt{fl}_s(\sigma)$. All followers, anchors, and trackers are cancelled when the node which appointed them is initialised.

Suppose that π is a node which works for P_e. We let $\mathtt{prec}(\pi)$ be the set of nodes τ working for some Q_d such that $\tau\hat{}\infty \prec \pi$. We split this set into two parts: $\mathtt{prec}_\infty(\pi)$ is the set of nodes $\tau \in \mathtt{prec}(\pi)$ such that $\tau\hat{}\infty\hat{}\infty_n \prec \pi$ for some n; $\mathtt{prec}_{\mathtt{fin}}(\pi)$ is the set of nodes $\tau \in \mathtt{prec}(\pi)$ such that $\tau\hat{}\infty\hat{}\mathtt{fin}_n \preccurlyeq \pi$ for some n. If σ is a child of π which believes that $\hat{\Psi}_\pi(V)$ is total then during the construction we may define markers $m_s(\sigma)$. Let $\tau \in \mathtt{prec}(\pi)$ and let $x < \omega$. We say that σ *respects* the input x (for τ) at stage s if:

- $\tau \in \mathtt{prec}_\infty(\pi)$ and $x < m_s(\sigma)$; or
- $\tau \in \mathtt{prec}_{\mathtt{fin}}(\pi)$ and $x < y$, where $\tau\hat{}\infty\hat{}\mathtt{fin}_y \preccurlyeq \pi$.

Construction

Let s be a stage. We let, by recursion, the collection of accessible nodes δ_s be an initial segment of the tree of strategies.

Suppose that a node τ, working for requirement Q_e, is accessible at stage s. Let $t < s$ be the last stage prior to stage s at which $\tau\hat{}\infty$ was accessible, $t = 0$ if there is no such stage. If $\operatorname{dom} \Phi_{e,s}(V_s, D_s) \leqslant t$ then we let $\tau\hat{}\mathtt{fin}$ be next accessible; otherwise we let $\tau\hat{}\infty$ be next accessible.

Suppose that $\tau\hat{}\infty$ is accessible at stage s. For each $n < s$ let t_n be the last stage prior to stage s at which $\tau\hat{}\infty\hat{}\infty_n$ was accessible, $t_n = 0$ if there was no such stage. Also, let y be the least such that either $\Phi_{e,t}(V_t, D_t, y)\uparrow$ or the computation $\Phi_{e,t}(V_t, D_t, y)$ was destroyed since stage t, that is, either $D_t \upharpoonright_u \neq D_s \upharpoonright_u$ or $V_t \upharpoonright_u \neq V_s \upharpoonright_u$, where $u = \varphi_{e,t}(y)$. Note that $y \leqslant t$. If there is some $n \leqslant y$ such that $\ell_s(\tau, n) \geqslant t_n$ then we let $\tau\hat{}\infty\hat{}\infty_n$ be next accessible for the least such n. Otherwise we let $\tau\hat{}\infty\hat{}\mathtt{fin}_y$ be next accessible.

Before we proceed we maintain the functional $\hat{\Phi}_\tau$. Let $x < \omega$ such that $c = \mathtt{tr}_s(\tau, x)$ is already defined. If either

- $\tau\hat{}\infty\hat{}\mathtt{fin}_y$ is next accessible, and $x < y$; or
- $\tau\hat{}\infty\hat{}\infty_n$ is next accessible, and $x < t$

and $\hat{\Phi}_{\tau,s}(V_s, D_s, c)\uparrow$ then we define $\hat{\Phi}_{\tau,s+1}(V_s, D_s, c) = s$ with use $\varphi_{e,s}(V_s, D_s, x)$. Also, if $c < s$ is not currently a tracker for any input for τ and $\hat{\Phi}_{\tau,s}(V_s, D_s, c)\uparrow$ then we define $\hat{\Phi}_{\tau,s+1}(V_s, D_s, c) = 0$ with use 0.[1] Finally for every $x < s$ for which $\mathtt{tr}_s(\tau, x)$ is undefined, we define a new, large tracker $\mathtt{tr}_{s+1}(\tau, x)$.

Suppose that $\tau\hat{}\infty\hat{}\infty_n$ is accessible (for some n). For each $i < s$ let r_i be the last stage at which $\tau\hat{}\infty\hat{}\infty_n\hat{}i$ was last accessible; $r_i = 0$ if there was no such stage. We let $\tau\hat{}\infty\hat{}\infty_n\hat{}i$ be next accessible for the least $i \leqslant s$ such that for all

[1]Recall that since V is not built by us, the use of $\hat{\Phi}$ is not the largest number queried; it is the length of the string appearing in an axiom applying to the oracle.

$x < r_i$, $c = \mathtt{tr}_s(\tau, x)$ is defined, $o_s^i(c) < \alpha$ and $\hat{\Phi}_{e,s}(V_s, c) = f_s^i(c)$. Note that $r_s = 0$ and so such i does exist.

Suppose that a node π, working for P_e, is accessible at stage s. If $|\pi| \geqslant s$ then we end the stage. Suppose that $|\pi| < s$. We first maintain the shadow functional $\hat{\Psi}_\pi$. For every $q < s$ which is not currently an anchor for any child of π, if $\hat{\Psi}_{\pi,s}(V_s, q)\uparrow$ then we define $\hat{\Psi}_{\pi,s+1}(V_s, q) = 0$ with use 0. Now let $q = \mathtt{ac}_s(\sigma)$ be an anchor for a child σ of π, and suppose that $\hat{\Psi}_{\pi,s}(V_s, q)\uparrow$. Let $p = \mathtt{fl}_s(\sigma)$ be the current follower of σ.

- If either $p \in D_s$, or for some $\tau \in \mathtt{prec}(\pi)$ working for Q_d and some x which σ currently respects (for τ) we have $p < \varphi_{d,s}(x)$, then we cancel p and appoint a new, large follower $\mathtt{fl}_{s+1}(\sigma)$. We leave $\hat{\Psi}_{\pi,s+1}(V_s, q)$ undefined. In the first case we have already attacked with p, but now the computation against which we diagonalised has disappeared. In the second case, as described earlier, we need to protect the computation $\Phi_d(D, V, x)$ from the action of σ.
- Otherwise, if $p \in \operatorname{dom} \Psi_{e,s}(V_s)$ then we define $\hat{\Psi}_{\pi,s+1}(V_s, q) = s$ with use $\psi_{e,s}(p)$. If $p \notin \operatorname{dom} \Psi_{e,s}(V_s)$ then we leave $\hat{\Psi}_{\pi,s+1}(V_s, q)$ undefined.

For $n < s$ let t_n be the last stage at which $\pi\hat{\ }\infty_n$ was accessible; $t_n = 0$ if there is no such stage. Also let $y = \operatorname{dom} \hat{\Psi}_{\pi,s}(V_s)$. If there is some $n \leqslant y$ such that $\ell_s(\pi, n) \geqslant t_n$ then we let $\pi\hat{\ }\infty_n$ be next accessible for the least such n. Otherwise we let $\pi\hat{\ }\mathtt{fin}_y$ be next accessible.

Suppose that $\sigma = \pi\hat{\ }\infty_n$ is accessible.

- If σ has no anchor then we appoint a new large anchor $q = \mathtt{ac}_{s+1}(\sigma)$ and a new, large follower $p = \mathtt{fl}_{s+1}(\sigma)$. We let $m_{s+1}(\sigma) = s$.
- If $p = \mathtt{fl}_s(\sigma)$ is defined, $p \notin D_s$, $\Psi_{e,s}(V_s, p) = 0$, and $\mathtt{ac}_s(\sigma) \in \operatorname{dom} \hat{\Psi}_{\pi,s}(V_s)$ then we enumerate p into D_{s+1}. Redefine $m_{s+1}(\sigma) = s$. For all $\tau \in \mathtt{prec}_\infty(\pi)$ and all inputs x which σ does not currently respect (for τ), that is, $x \geqslant m_s(\sigma)$, cancel the tracker $\mathtt{tr}_s(\tau, x)$.

If either of these happen, we stop the stage and initialise all nodes weaker than σ. If the stage was not ended, then the unique child of σ is next accessible.

Verification

First we note that for the functionals Ξ we define, $\hat{\Psi}_\pi$ and $\hat{\Phi}_\tau$, we only define a new axiom $\Xi_{s+1}(V_s, x)$ if $x \notin \operatorname{dom} \Xi_s(V_s)$. This shows that these functionals are consistent for V, indeed at every stage.

We will need to show that these shadow functionals behave properly. The $\hat{\Psi}$-functionals are easy.

Lemma 4.13. *Let π be a node working for a requirement P_e. Let σ be a child of π. Let s be a stage and suppose that $q = \mathtt{ac}_s(\sigma)$ and $p = \mathtt{fl}_s(\sigma)$ are defined. If $\hat{\Psi}_\pi(V, q)\downarrow[s]$ then $\Psi_e(V, p)\downarrow[s]$ and $\hat{\psi}_{\pi,s}(q) = \psi_{e,s}(p)$.*

Proof. Suppose that $\hat{\Psi}_\pi(V,q)\downarrow[s]$; let $u=\hat{\psi}_{\pi,s}(q)$. Let $t<s$ be the stage at which we defined this computation. So $V_t\restriction_u = V_s\restriction_u$. At stage t we have $\Psi_e(V,p)\downarrow[s]$ with use u. Hence this computation persists until stage s. We may assume that while $\Psi_e(V,p)\downarrow$, no new computations (with different use) are enumerated into Ψ_e. Thus $\psi_{e,s}(p)=u$. □

Lemma 4.14. *Suppose that a node π working for P_e is accessible infinitely often and is initialised only finitely often. There is a child σ of π which is accessible infinitely often. Let σ be the strongest such child. Then:*

(1) σ ends the stage only finitely many times;
(2) σ believes that $\hat{\Psi}_\pi(V)$ is total if and only if $\hat{\Psi}_\pi(V)$ is indeed total;
(3) if $\hat{\Psi}_\pi(V)$ is total then the requirement P_e is met.

Proof. Suppose that $\hat{\Psi}_\pi(V)$ is not total. Then for every n, $\pi\hat{\ }\infty_n$ is accessible only finitely often (otherwise $\lim_s \ell_s(\pi,n)=\infty$ and this implies that $\hat{\Psi}_\pi(V)$ is total). On the other hand, because $\hat{\Psi}_\pi$ is defined only at stages at which π is accessible, we know that $y=\liminf_s \operatorname{dom}\hat{\Psi}_\pi(V)[s]$ is finite, and $y=\operatorname{dom}\hat{\Psi}_\pi(V)[s]$ at infinitely many stages s at which π is accessible. Hence $\pi\hat{\ }\mathtt{fin}_y$ is accessible infinitely often, and is the strongest child of π which is accessible infinitely often. This node never ends the stage.

Suppose that $\hat{\Psi}_\pi(V)$ is total. There is some n such that $\lim_s \ell_s(\pi,n)=\infty$; let n be the least such. For almost every stage s, $\operatorname{dom}\hat{\Psi}_\pi(V)[s]>n$. Hence $\pi\hat{\ }\infty_n$ is accessible infinitely often, and is the strongest such child of π.

At the first stage at which $\sigma=\pi\hat{\ }\infty_n$ is accessible after that last stage at which it is initialised we define an anchor $q=\mathtt{ac}(\sigma)$; this anchor is never cancelled. Let t be the stage at which the V-correct computation $\hat{\Psi}_\pi(V,q)$ is defined (note that σ need not be accessible at that stage). The follower $p=\mathtt{fl}_t(\sigma)$ is never cancelled. After stage t, the node σ ends the stage at most once, when p is enumerated into D.

We claim that $\Psi_e(V,p)\neq D(p)$. We have $p\notin D_t$ (for otherwise p would be cancelled at stage t). By Lemma 4.13, $\Psi_e(V,p)\downarrow[t]$ is a V-correct computation. If $\Psi_e(V,p)=0$ then at the next stage $s>t$ at which σ is accessible, p is enumerated into D. If $\Psi_e(V,p)=1$ then at no stage do we enumerate p into D. □

Lemma 4.15. *Let π be a node which works for requirement P_d. Let $\tau\in\mathtt{prec}(\pi)$. Let σ be a child of π which guesses that $\hat{\Psi}_\pi(V)$ is total. Let s be a stage at which π is accessible, and let x be an input for τ which σ respects at stage s. Suppose that $p=\mathtt{fl}_s(\sigma)$ and $q=\mathtt{fl}_s(\sigma)$ are defined. Then $\Phi_e(V,D,x)\downarrow[s]$ and either (i) $p\in D_s$; or (ii) $\hat{\Psi}_\pi(V,q)\uparrow[s]$; or (iii) $\varphi_{e,s}(x)\leqslant p$.*

Proof. Suppose that $\hat{\Psi}_\pi(V,q)\downarrow[s]$ and that $p\notin D_s$. Let $t<s$ be the stage at which the computation $\hat{\Psi}_\pi(V,q)[s]$ is defined. When the anchor is chosen it is large, and it is not large at stage t; hence $q=\mathtt{ac}_t(\sigma)$. The follower $\mathtt{fl}_t(\sigma)$ is not

enumerated into D at stage t since $\hat{\Psi}_\pi(V,q)\uparrow[t]$. The follower is not cancelled at stage t; otherwise $\hat{\Psi}_\pi(V,q)$ is not defined at stage t. The follower is not cancelled at any stage in the interval (r,s) since $\hat{\Psi}_\pi(V,q)\downarrow$ at these stages. Hence $p=\mathtt{fl}_t(\sigma)$. Since $p\notin D_s$, $m_s(\sigma)<t$.

If $\tau\in\mathtt{prec}_\infty(\pi)$ then $x<m_s(\sigma)$. If $\tau\in\mathtt{prec}_{\mathtt{fin}}(\pi)$ then $x<y$ where $\tau\hat{\ }\infty\hat{\ }\mathtt{fin}_y\preccurlyeq\pi$. Since π is accessible at stage $m_s(\sigma)$ we have $y<m_s(\sigma)$ so again $x<m_s(\sigma)$. Hence $\Phi_e(V,D,x)\downarrow[r]$ at every stage $r>m_s(\sigma)$ at which $\tau\hat{\ }\infty$ is accessible. In particular this holds for $r=t$. Since $m_t(\sigma)=m_s(\sigma)$, x is respected by σ at stage t. If $p<\varphi_{e,t}(x)$ then since $\hat{\Psi}_\pi(V,q)\uparrow[t]$, p would be cancelled at stage t. Hence $p\geqslant\varphi_{e,t}(x)$.

For brevity let $u=\varphi_{e,t}(x)$. We may assume that $\psi_{d,t}(p)\geqslant p$, and $\hat{\psi}_{\pi,t}(q)=\psi_{d,t}(p)$. The fact that the computation $\hat{\Psi}_\pi(V,q)[t]$ survives until stage s implies that $V_t\restriction_u=V_s\restriction_u$. The lemma would be proved once we show that $D_t\restriction_u=D_s\restriction_u$; this would imply that the computation $\Phi_e(V,D,x)[t]$ survives until stage s and so $u=\varphi_{e,s}(x)\leqslant p$ as required.

Suppose for a contradiction that at some stage $r\in[t,s)$ a number $p'<u$ is enumerated into D_{r+1}; let r be the least such stage. So the computation $\Phi_e(V,D,x)[t]$ survives until stage r; $\varphi_{e,r}(x)=u$. The number p' is the follower $\mathtt{fl}_r(\sigma')$ for some node σ', a child of a node π' working for $P_{d'}$. The node π' must extend π: it must be weaker than σ, since it does not initialise σ at stage r; and it is not initialised at stage t, because the follower p' is large when it is chosen, and so p' is chosen prior to stage t. The node σ' is initialised at stage $m_s(\sigma)$. Hence $m_r(\sigma')>m_s(\sigma)$. This shows that x is respected (for τ) by σ' at stage r (if $\tau\in\mathtt{prec}_{\mathtt{fin}}(\pi)$ then we use the fact that both π and π' extend the same child of $\tau\hat{\ }\infty$). Applying the lemma at stage r, since $p'\notin D_r$ and $\hat{\Psi}_{\pi'}(V,\mathtt{ac}_r(\sigma'))\downarrow[r]$ (otherwise p' is not enumerated into D_{r+1}), it must be that $p'\geqslant\varphi_{e,r}(x)=u$, a contradiction. □

Lemma 4.16. *Let τ be a node which works for requirement Q_e. Let s be a stage; let x be an input such that $c=\mathtt{tr}_s(\tau,x)$ is defined. Suppose that $\hat{\Phi}_\tau(V,c)\downarrow[s]$. Let $u=\hat{\varphi}_{\tau,s}(c)$. Then:*

(1) $\Phi_e(V,D,x)\downarrow[s]$ and $u=\varphi_{e,s}(x)$;
(2) if $D_s\restriction_u\neq D_{s+1}\restriction_u$ then the tracker c is cancelled at stage s.

Proof. Both parts of the lemma are proved by simultaneous induction on the stage s. Suppose the lemma has been verified for all stages prior to stage s. Assume the hypotheses of the lemma hold at stage s. Let $t<s$ be the stage at which the computation $\hat{\Phi}_\tau(V,c)[s]$ was defined. So $V_t\restriction_u=V_s\restriction_u$. At stage t we have $\Phi_e(V,D,x)\downarrow[t]$ with use $\varphi_{e,t}(x)=u$. Because trackers are chosen large, $c=\mathtt{tr}_t(\tau,x)$.

The conditions of the lemma hold at every stage in the interval $[t,s)$. Since the tracker c is not cancelled at any stage in that interval, by induction on these stages (using (2)) we see that $D_s\restriction_u = D_t\restriction_u$. This shows that the computation $\Phi_e(V, D, x)[t]$ is preserved up to stage s, and so establishes (1) at stage s.

Suppose that a number $p < u$ is enumerated into D at stage s. Then $p = \mathtt{fl}_s(\sigma)$ for some node σ, a child of a node π. The follower p must be chosen prior to stage t. If σ is stronger than τ then τ is initialised at stage s, whence c is cancelled at stage s. Assuming otherwise, it must be the case that $\sigma \succ \tau\hat{\ }\infty$, as σ is not initialised at stage t.

Lemma 4.15 ensures that σ does not respect x (for τ) at stage s. Suppose that $\tau\hat{\ }\infty\hat{\ }\mathtt{fin}_y \preccurlyeq \pi$ for some y. Let r be the last stage prior to stage s at which $\tau\hat{\ }\infty$ was accessible. Then $r \geqslant t$. It follows that $\Phi_e(V, D, x)\downarrow[r]$ and the computation is preserved until stage s. Hence $y > x$. But then σ respects x. So $\tau \in \mathtt{prec}_\infty(\pi)$. Then σ is instructed to cancel c at stage s; so (2) holds. □

Lemma 4.17. *Let τ be a node which works for Q_e. Suppose that τ is initialised only finitely often, and that $\tau\hat{\ }\infty$ is accessible infinitely often.*

(1) For every x we eventually appoint a tracker $\mathtt{tr}(\tau, x)$ which is never cancelled.
(2) There is an outcome $o \in \{\infty_n, \mathtt{fin}_n\}$ such that $\tau\hat{\ }\infty\hat{\ }o$ is accessible infinitely often.

Let $\rho = \tau\hat{\ }\infty\hat{\ }o$ be the strongest child of $\tau\hat{\ }\infty$ which is accessible infinitely often.

(3) If $\Phi_e(V, D)$ is total then so is $\hat{\Phi}_\tau(V)$, and $o = \infty_n$ for some n. Further, for some i, $\tau\hat{\ }\infty\hat{\ }\infty_n\hat{\ }i$ is accessible infinitely often.
(4) Otherwise $o = \mathtt{fin}_y$ where $y = \operatorname{dom}\Phi_e(D, V)$.

Proof. Let $x < \omega$. At any stage $t > x$ at which $\tau\hat{\ }\infty$ is accessible, if $\mathtt{tr}_t(\tau, x)$ is undefined then we appoint a new tracker $\mathtt{tr}_{t+1}(\tau, x)$. Suppose that a tracker $\mathtt{tr}_s(\tau, x)$ is defined and is cancelled at stage s. The stage s is ended by a child σ of a node π working for some P_d; $\tau \in \mathtt{prec}_\infty(\pi)$ and the node σ enumerates its follower $p = \mathtt{fl}_s(\sigma)$ into D_{s+1}. We have $x \geqslant m_s(\sigma)$. The marker $m_s(\sigma)$ is chosen at stage $m_s(\sigma)$, at which σ is accessible. Thus there are only finitely many nodes σ which can ever cancel the tracker $\mathtt{tr}_s(x, \tau)$. Each such node does so at most once, since when it does, it updates $m_{s+1}(\sigma) = s > x$. This gives (1).

Suppose that $\Phi_e(V, D)$ is not total; let $y = \operatorname{dom}\Phi_e(V, D)$. Let c be the eventual tracker for y, which is never cancelled. Then $c \notin \operatorname{dom}\hat{\Phi}_\tau(V)$. This is ensured by part (1) of Lemma 4.16: if $\hat{\Phi}_\tau(V, c)\downarrow$ with use u then at a late stage at which both V and D are correct up to u we would get a V, D-correct computation of $\Phi_e(y)$. Since $\hat{\Phi}_\tau(V)$ is partial, no totality outcome ∞_n is guessed infinitely often. Since $\Phi_e(V, D, x)$ is eventually fixed for all $x < y$, eventually, no outcome stronger than $\mathtt{fin}_y$ is ever guessed; but $\mathtt{fin}_y$ is guessed infinitely often. This gives (4).

Suppose that $\Phi_e(V, D)$ is total. For every y, $\mathtt{fin}_y$ is guessed only finitely many times. We show that $\hat{\Phi}_\tau(V)$ is total. This will imply that some ∞_n is guessed infinitely often. Let $c < \omega$. As usual, if c is never chosen as a follower or is chosen and later cancelled, then $\hat{\Phi}_\tau(V,c)\downarrow$. Suppose that c is chosen as a tracker for x at stage r, and is never cancelled. Eventually no $\mathtt{fin}_y$ for $y \leqslant x$ is ever guessed; so eventually, at every stage s at which $\tau\hat{\ }\infty$ is accessible, if $\hat{\Phi}_\tau(V,c)\uparrow[s]$ then a new computation $\hat{\Phi}_{\tau,s+1}(V_s,c)$ is defined. The use is $\varphi_{e,s}(x)$. This use stabilizes, and eventually V stabilizes below that use, and so eventually a V-correct computation must be made.

Since $\deg_{\mathrm{T}}(V)$ is totally α-c.a., there is some $i<\omega$ such that $\hat{\Phi}_\tau(V)=f^i$ and $\langle f^i_s, o^i_s\rangle$ is eventually α-computable. Since every input eventually receives a permanent tracker, the outcome i is guessed infinitely often for the least such i. ☐

Lemmas 4.14 and 4.17 together show that the true path is infinite and that the construction is fair to every node on the true path.

Lemma 4.18. *Every positive requirement P_e is met.*

Proof. Let π be the node on the true path which works for P_e. Suppose that $\Psi_e(V)$ is total. We show that $\hat{\Psi}_\pi(V)$ is total (and then appeal to Lemma 4.14).

Let $q<\omega$. To show that $\hat{\Psi}_\pi(V,q)\downarrow$ we may, as usual, assume that q is chosen as an anchor of a child σ of π at some stage r, and is never cancelled. We show that followers for σ are cancelled only finitely many times. This suffices: if p is a follower for σ which is never cancelled, then eventually we see the V-correct computation $\Psi_e(V,x)$. At any stage s at which π is accessible, if $\hat{\Psi}_\pi(V,q)\uparrow[s]$ then a new computation is defined with use $\psi_{e,s}(p)$, which eventually stabilizes.

The node σ believes that $\hat{\Psi}_\pi(V)$ is total. Hence if σ is accessible infinitely often then $\hat{\Psi}_\pi(V)$ is total and we are done. We assume that σ is accessible only finitely many times. The marker $m_s(\sigma)$ is updated only when σ is accessible, so reaches a final value $m(\sigma)$ at stage $t \geqslant r$.

Suppose that the follower $p=\mathtt{fl}_s(\sigma)$ is cancelled at a stage after stage t. This is done on behalf of a node $\tau \in \mathtt{prec}(\pi)$ (working for some Q_d) and an input x. There are two cases. If $\tau \in \mathtt{prec}_\infty(\pi)$ then a totality outcome for $\tau\hat{\ }\infty$ lies on the true path. This implies that $\Phi_d(V,D)$ is total. Also, $x<m(\sigma)$. If the follower for σ is cancelled after the correct computation $\Phi_d(V,D,x)$ appears then a new follower is chosen to be large, and so is greater than $\psi_{d,s}(x)$ for all later s. This implies that this τ can cause only finitely many cancellations of $\mathtt{fl}_s(\sigma)$.

The other case is $\tau \in \mathtt{prec}_{\mathtt{fin}}(\pi)$; say $\tau\hat{\ }\infty\hat{\ }\mathtt{fin}_y \preccurlyeq \pi$; so $x<y$. By Lemma 4.17, $y=\operatorname{dom}\Phi_d(V,D)$, so again $\Phi_d(V,D,x)$ eventually converges by a correct computation. The argument is now the same as in the first case. ☐

To finish the verification we show that every requirement Q_e is met. Let τ be the node on the true path which works for Q_e, and suppose that $\Phi_e(V,D)$ is total. Then $\tau\hat{\ }\infty$ lies on the true path; and Lemma 4.17 says that for some n and i, $\rho=\tau\hat{\ }\infty\hat{\ }\infty_n\hat{\ }i$ lies on the true path. Then $\langle f^i_s, o^i_s\rangle$ is eventually α-computable and $\hat{\Phi}_\tau(V)=f^i$. As in previous proofs let s^* be the last stage at which ρ is initialised, and let $s_0 < s_1 < s_2 < \cdots$ be the stages after stage s^* at which ρ is accessible.

Fix $x < \omega$. We let $j(x)$ be the least j such that $x < s_{j-1}$. For all $j \geqslant j(x)$, $\Phi_e(V, D, x)\downarrow [s_j]$, $c_j = c_j(x) = \mathtt{tr}_{s_j}(\tau, x)$ is defined, $o^i_{s_j}(c_j) < \alpha$ and $\hat{\Phi}_\tau(V, c_j)\downarrow = f^i(c_j)\ [s_j]$. For $j \geqslant j(x)$ let $a_j = a_j(x)$ be the set of nodes σ, children of nodes π working for some P_d such that $\rho \preccurlyeq \pi$, such that $m_{s_j}(\sigma) \leqslant x$. Since $m_s(\sigma)$ is non-decreasing, if $j < j'$ then $a_{j'} \subseteq a_j$.

The following lemma is an analogue of Lemmas 3.14 and 4.7.

Lemma 4.19. *Let $x < \omega$ and $j \geqslant j(x)$. Let $u = \varphi_{e,s_j}(x)$.*

(1) If $a_{j+1} = a_j$ then $c_{j+1} = c_j$.
(2) If $D_{s_{j+1}} \restriction u \neq D_{s_j} \restriction u$ then $c_{j+1} \neq c_j$.
(3) If $D_{s_{j+1}} \restriction u = D_{s_j} \restriction u$ but $V_{s_{j+1}} \restriction u \neq V_{s_j} \restriction u$ then $o^i_{s_{j+1}}(c_j) < o^i_{s_j}(c_j)$.

Proof. The instructions ensure that only a node σ (with parent π) such that $\tau \in \mathtt{prec}_\infty(\pi)$ and $m_s(\sigma) \leqslant x$ can cancel $\mathtt{tr}_s(\tau, x)$. Say that a node π with $\tau \in \mathtt{prec}_\infty(\pi)$ is accessible at a stage $r \in (s_j, s_{j+1})$; then π is initialised at stage s_j and so $m_s(\sigma) > s_j > x$. So if c_j is cancelled by stage s_{j+1}, then it is cancelled by a node $\sigma \in a_j$. But then we define $m_{s_j+1}(\sigma) = s_j > x$ and so $\sigma \notin a_{j+1}$. This gives (1).

The same argument shows that if $D_{s_{j+1}} \restriction u \neq D_{s_j} \restriction u$ then $D_{s_j+1} \restriction u \neq D_{s_j} \restriction u$. (2) is given by Lemma 4.16(2).

Suppose that $a_{j+1} = a_j$ but $V_{s_{j+1}} \restriction u \neq V_{s_j} \restriction u$. Let $s \geqslant s_j$ be the least stage such that $V_{s+1} \restriction u \neq V_s \restriction u$.

By Lemma 4.16(1), $u = \hat{\varphi}_{\tau,s_j}(c_j)$, and so $\hat{\Phi}_\tau(V, c_j)\uparrow [s+1]$. When we redefine a value for $\hat{\Phi}_\tau(V, c_j)$, it is the stage number, and so $\hat{\Phi}_\tau(V, c_j)[s_{j+1}] > s_j$. In particular $\hat{\Phi}_\tau(V, c_j)[s_{j+1}] \neq \hat{\Phi}_\tau(V, c_j)[s_j]$. But then $f^i_{s_{j+1}}(c_j) \neq f^i_{s_j}(c_j)$, and (3) follows. $\square$

Now let for all $j \geqslant j(x)$

$$\gamma_j = \gamma_j(x) = \alpha \cdot |a_j| + o^i_{s_j}(c_j).$$

Since α is closed under addition, for all n and all $\beta < \alpha$ we have $\alpha \cdot n + \beta < \alpha \cdot (n+1)$. Thus if $a_{j+1} \neq a_j$ then (as $a_{j+1} \subsetneq a_j$) $\gamma_{j+1} < \gamma_j$. Suppose that $a_{j+1} = a_j$. Then $c_{j+1} = c_j$ and so $o^i_{s_{j+1}}(c_{j+1}) = o^i_{s_{j+1}}(c_j) = o^i_{s_j}(c_j)$; so $\gamma_{j+1} \leqslant \gamma_j$. Suppose further that $\Phi_e(V, D, x)[s_j] \neq \Phi_e(V, D)[s_{j+1}]$. Since $c_{j+1} = c_j$, $D_{s_{j+1}} \restriction u = D_{s_j} \restriction u$. Then Lemma 4.19(3) ensures that $\gamma_{j+1} < \gamma_j$. This concludes the verification.

4.2.1 Uniformity again

Inspecting the construction we see that $|a_{j(x)}(x)| < x$. This is because $m_{s_{j(x)}}(\sigma)$ is distinct for distinct $\sigma \in a_{j(x)}(x)$ (when $m_s(\sigma)$ is set the stage ends). This shows that in fact $\deg_{\mathrm{T}}(V \oplus D)$ is uniformly totally $\alpha \cdot \omega$-c.a., as every $\Phi_e(V, D)$ is h-c.a. for $h(n) = \alpha \cdot (n+1)$.

4.2.2 Maximal $< \alpha$-c.a. degrees

Suppose that α is a limit of powers of ω, and that $\deg_{\mathrm{T}}(V)$ is totally $<\alpha$-c.a. We can modify the construction above by letting the sequence $\langle f^i \rangle$ range over all functions which are β-c.a. for some $\beta < \alpha$. Examining the proof above, we see that the ordinal bound on the number of changes of $\Phi_e(V, D, x)$ is given by a finite multiple of $o^i(c)$ for a variety of c but for fixed i. Thus, if f^i is β-c.a., then $\Phi_e(V, D)$ is $\beta \cdot \omega$-c.a. We thus obtain:

Theorem 4.20. *If $\alpha \leqslant \varepsilon_0$ is a limit of powers of ω, then no c.e. degree is maximal totally $<\alpha$-c.a.*

Chapter Five

Presentations of left-c.e. reals

IN THIS CHAPTER we prove Theorem 1.4:

(1) If a c.e. degree $\mathbf{d}$ is not totally ω-c.a. then there is a left-c.e. real $\varrho \leqslant_{\mathrm{T}} \mathbf{d}$ and a c.e. set $B <_{\mathrm{T}} \varrho$ such that every presentation of ϱ is B-computable.
(2) If a left-c.e. real ϱ has a totally ω-c.a. degree then there is a presentation of ϱ which is Turing equivalent to ϱ.

5.1 BACKGROUND

One of the main ideas of this book is unifying the combinatorics of constructions in various subareas of computability theory. In this chapter we will look at one such subarea: *algorithmic randomness* [27, 69, 74]. Algorithmic randomness seeks to give meaning to our intuition that a sequence like 010111010101111... is not random, whereas ones obtained from, e.g., tosses of an unbiased coin would be. The idea is that we should not be able to give algorithmic tests for predictability and if a sequence *fails* such a test, then it cannot be random. The test above would be that every even bit is a 1.

By way of motivation, we now give a brief account of the basics of algorithmic randomness, and include the basic definitions required in this chapter.

The "playing ground" of basic algorithmic randomness is Cantor space, 2^ω, the space of all infinite binary sequences (later these concepts can be extended to other spaces such as the unit interval). The topology on Cantor space is the product topology starting with the discrete topology on $\{0, 1\}$. This topology is generated by sets of the form $[\sigma] = \{X \in 2^\omega : \sigma \prec X\}$, where $\sigma \in 2^{<\omega}$ is a finite binary string. In general, the open subsets of Cantor space are of the form $[W]^\prec = \bigcup\{[\sigma] : \sigma \in W\}$ for subsets W of $2^{<\omega}$. By coding finite binary strings by natural numbers in a reasonable fashion, we can consider notions such as computable and c.e. sets of finite strings. This turns Cantor space into an effective topological space. The *effectively open* subsets of Cantor space are those of the form $[W]^\prec$ for c.e. sets $W \subseteq 2^{<\omega}$. These are also called *c.e. open* sets. These sets are important in their own right: the study of the effective topology of Cantor space is essentially the study of restricted notions of Cohen genericity. For example, an element of Cantor space is 1-generic if and only if it is not an element of the boundary of any effectively open set.

A notion of restricted genericity is determined by considering a countable collection of meagre sets. Those sets are considered "small" and so their elements are considered atypical, at least topologically. The main idea of algorithmic randomness is to replace category with measure, and so replace meagre sets by null sets. To do that we need to work with a measure on our space, and we choose the "fair-coin" measure, which we denote by λ. This is the product measure starting by giving 0 and 1 equal probability, namely 1/2. For every finite binary string σ, $\lambda([\sigma]) = 2^{-|\sigma|}$. If we think of 0 as tails and 1 as heads, then this measure represents the probability that an infinite sequence of independent tosses begins with the coin tosses represented by the string σ. The fair-coin measure λ is also often referred to as Lebesgue measure, because of the measure-preserving almost isomorphism between Cantor space and the unit interval, which we will mention below.

Whilst algorithmic randomness has a history going back to the early work of Borel [8] on normal numbers, von Mises [100], and even Turing [99] (see Downey-Hirschfeldt [26]), the key concept in the modern incarnation of algorithmic information theory is *Martin-Löf randomness.* A notion of randomness is determined by a countable collection of null sets, with each null set considered a statistical test. Elements of the null sets are those which have *failed* the test; they are atypical, in the sense of measure. For Martin-Löf (ML) randomness we use the collection of *effectively* G_δ*, effectively null* sets: intersections $\bigcap \mathcal{U}_n$ of uniformly effectively open sets whose measure goes to 0 computably (we can require for example $\lambda(\mathcal{U}_n) \leqslant 2^{-n}$). Such intersections are called *ML-null*, and an element of Cantor space is *ML-random* if it is an element of no ML-null set.

One of the reasons the notion of ML-randomness is central is that it is robust. It has many equivalent characterisations, and one of them is in terms of Kolmogorov complexity [56, 68]. The motivating idea here is that *finite* strings should be considered random if they cannot be effectively compressed: if the only way to convey the information they store is in writing them down. In other words, they do not have short descriptions. To formalise this, we consider any partial computable function $M \colon 2^{<\omega} \to 2^{<\omega}$ as a "description system"; if $M(\sigma) = \tau$ then we say that σ is an M*-description* of τ. We call M a "machine." The idea is that all information in τ is already stored in σ, as M can effectively produce τ given σ. The M*-complexity* of τ is the length of the shortest M-description of τ. Kolmogorov's intuition is that τ is M*-random* if its M-complexity is no smaller than its length. That is, M thinks τ is so random it cannot be compressed at all.

There is a slight problem with this definition in that in decompressing, M actually uses not only the information stored in σ but also its length $|\sigma|$, since we know that M halts on σ. In some sense this means that M has actually "used" $|\sigma| + \log_2 |\sigma|$ many bits of information. In many applications of Kolmogorov complexity this slight difference does not matter, but in the definition of random infinite sequences it does. This consideration leads to a central concept in algorithmic randomness, namely *prefix-free machines.*

Definition 5.1. A set of strings $C \subset 2^{<\omega}$ is *prefix-free* if no two distinct strings in C are comparable.

A machine M is prefix-free if its domain is prefix-free. Prefix-free machines are those that have the "telephone number" property (in most countries, no two telephone numbers are prefixes of each other). For a prefix-free machine M we write $K_M(\tau)$ for the M-complexity of τ. Schnorr's theorem links Kolmogorov complexity with ML-randomness:

Theorem 5.2 (Schnorr, see [11]). *$X \in 2^\omega$ is ML-random if and only if for all prefix-free machines M, $K_M(X\upharpoonright_n) \geqslant^+ n$.*

That is, if for some constant d, for all n, $K_M(X\upharpoonright_n) \geqslant n - d$.

Prefix-free machines occupy a central role in the theory of algorithmic randomness. This connection is evidenced by a number of further results. The easiest way to exhibit an ML-random sequence was observed by Chaitin. For a set of strings $C \subseteq 2^{<\omega}$ we write $\lambda(C)$ for the measure of the open set generated by C; if C is prefix-free then $\lambda(C) = \sum_{\sigma \in C} 2^{-|\sigma|}$. For a prefix-free machine M, $\lambda(\operatorname{dom} M)$ is also known as the *a priori* probability that a string is in the domain of M; it is referred to as the *halting probability of M*.

Levin and others observed that there are *universal* prefix-free machines: machines U which simulate any other machine, in that for any prefix-free machine M there is some $\rho \in 2^{<\omega}$ such that for all σ, $M(\sigma) = U(\rho\hat{\ }\sigma)$. For such machines U we have $K_U \leqslant^+ K_M$ for any prefix-free machine M. Thus in Schnorr's theorem above, we can replace all prefix-free machines M by a single universal machine U. Chaitin observed that if U is a universal prefix-free machine, then $\lambda(\operatorname{dom} U)$ (written in binary) is ML-random. For a universal machine U, the quantity $\lambda(\operatorname{dom} U)$ is now called Chaitin's Ω.[1]

While the halting problem is c.e., the binary expansion of the halting probability Ω is not c.e. To characterise halting probabilities we find an analogue of computable enumerability on the real line.

Definition 5.3. A real number $\varrho \in \mathbb{R}$ is *left-c.e.* if its left cut, the set of rational numbers $q < \varrho$, is c.e.

A real number is left-c.e. if and only if it is the limit of an increasing, computable sequence of rational numbers. Left-c.e. reals are also known as *lower semicomputable* reals. If A is a c.e. set then $0.A$ (the real whose binary expansion is A) is a left-c.e. real. However, not all left-c.e. reals are of this form. To see this we use that the following are equivalent for a real ϱ in the unit interval $[0, 1]$:

---

[1]More precisely, Ω_U since it depends on the universal machine in the same way that $\varnothing' = \{e : \varphi_e(e)\downarrow\}$ depends on the enumeration of the partial computable functions in classical computability theory.

(1) ϱ is left-c.e.;
(2) ϱ is the measure of an effectively open set;
(3) $\varrho = \lambda(C)$ for a c.e., prefix-free set of strings C;
(4) ϱ is the halting probability of a prefix-free machine

(see Soare [90], Calude et al. [10]). Thus Ω is left-c.e.; since no c.e. set can be ML-random, it is not of the form $0.A$ for a c.e. set A. Kučera and Slaman [58] gave much more information about left-c.e. random reals. Using an analogue of many-one reducibility $\leqslant_m$ introduced by Solovay [92], they showed that a left-c.e. real is ML-random if and only if it is the halting probability of a universal prefix-free machine.

Theorem 1.4, which we will prove in this chapter, characterises the dynamic properties of a construction of unusual left-c.e. reals. The equivalence above motivates the following definition.

Definition 5.4. A *presentation* of a left-c.e. real $\varrho \in [0,1]$ is a c.e. prefix-free set $C \subset 2^{<\omega}$ such that $\lambda(C) = \varrho$.

As this chapter is concerned with presentations, all real numbers from now on are in the unit interval $[0,1]$.

As we saw, every left-c.e. real has presentations. Indeed, every left-c.e. real has computable presentations: if C is a presentation of ϱ then, fixing an effective enumeration $\langle C_s \rangle$ of the c.e. set C, we replace C by a computable prefix-free set D of the same measure ϱ as follows: if a string σ is enumerated into C at stage s, we enumerate into D all extensions of σ of length s.

Every presentation of a left-c.e. real ϱ is computable from ϱ. To see this, suppose that C is a presentation of ϱ. Again let $\langle C_s \rangle$ be a computable enumeration of C; let $\varrho_s = \lambda(C_s)$; so $\langle \varrho_s \rangle$ is a computable increasing sequence of rational numbers converging to ϱ. To determine whether a string σ is in C, search for some s such that $\varrho - \varrho_s < 2^{-|\sigma|}$. This can be done if ϱ is given as an oracle. Then $\sigma \in C \Leftrightarrow \sigma \in C_s$.

In light of this fact, it is natural to ask if the complexity of ϱ (as measured by its Turing degree) is reflected in the complexity of some of its presentations. Namely, is every left-c.e. real computable from one of its presentations? In [32], Downey and LaForte gave a strong negative answer to this question: they constructed a noncomputable left-c.e. real ϱ, *all* of whose c.e. presentations are computable. On the other hand they showed that any left-c.e. real with promptly simple degree (for example, Ω) has a noncomputable presentation. Stephan and Wu [94] showed that the same holds for all noncomputable K-trivial left-c.e. reals. See also [36, 104].

Theorem 1.4 extends these results. It characterises the computational power required to compute one of the "unusual" left-c.e. reals, those with no presentation computing them, precisely as non-total ω-c.a.-ness. Indeed it gives a stronger dichotomy, with the unusual examples ϱ having a single bound B strictly below ϱ bounding the complexity of all presentations of ϱ.

Computing with real numbers

In this chapter we view real numbers as elements both of the computable metric space $[0,1]$ and as infinite binary sequences in $2^{\mathbb{N}^+}$ (where $\mathbb{N}^+ = \{1,2,3,\dots\}$) by using their binary expansion.

As the former, an oracle determining a real number ϱ is a sequence $\langle I_k \rangle$ of closed intervals satisfying:

- the endpoints of each I_k are binary rational numbers, indeed integer multiples of 2^{-k};
- the length of I_k is 2^{-k};
- the sequence is nested: $I_{k+1} \subset I_k$; and
- $\{\varrho\} = \bigcap_k I_k$.

The sequence $\langle I_k \rangle$ is coded by an element of Baire space; we ignore this detail. In computable analysis, the sequence $\langle I_k \rangle$ is called a *name* of ϱ.

On the other hand, given $X \subseteq \mathbb{N}^+$ we let $0.X = \sum_{k\in X} 2^{-k}$. Thinking of X as an element of $2^{\mathbb{N}^+}$, it is a binary expansion of $\varrho = 0.X$. We abuse notation by referring to X as ϱ: we write $\varrho(k)$ for the kth bit of ϱ's binary expansion.

If ϱ is not a binary rational, then it has both a unique binary expansion, and a unique name $\langle I_k \rangle$, and these are Turing equivalent; their degree is also called the Turing degree of ϱ.

Passing between names and binary expansions is uniform, *provided that we are guaranteed that ϱ is not a binary rational number.* Of course binary rational numbers are computable; each has two binary expansions, one ending with zeros and one with ones. They also have two names. The reals ϱ we will construct will not be computable. However their stage s approximations will be binary rationals, and so during constructions, to be definite, we always choose the binary expansion which ends with zeros.

5.2 PRESENTATIONS OF C.E. REALS AND NON-TOTAL ω-C.A. PERMITTING

In this section we prove part (1) of Theorem 1.4.

5.2.1 A simplified construction

Before adding permitting we construct a left-c.e. real ϱ and a c.e. set B such that $B <_{\mathrm{T}} \varrho$ but every presentation of ϱ is B-computable. As mentioned above, this has been done in [32] with $B = \varnothing$. We present the construction proving the weaker statement because it is simpler than the original one. The simplification is compatible with non-total ω-c.a. permitting. The original construction is in some sense compatible with non-total $<\omega^\omega$-c.a. permitting. We discuss this later, in Subsection 5.2.3.

We enumerate a c.e. set B, and give an increasing computable approximation $\langle \varrho_s \rangle$ of a left-c.e. real ϱ.

Let $\langle \Psi_e \rangle$ be an enumeration of functionals which output names of reals in the interval $[0,1]$. So for each k (and oracle X), $\Psi_e(X,k)$ (if it converges) is a closed interval I_k, of length 2^{-k}, with endpoints which are integer multiples of 2^{-k}. We also agree that if $k>0$ and $\Psi_e(X,k)\downarrow = I_k$ then $\Psi_e(X,k-1)\downarrow = I_{k-1}$ and $I_k \subset I_{k-1}$. Thus, if $\Psi_e(X)$ is total then $\langle \Psi_e(X,k) \rangle = \langle I_k \rangle$ is a name of a real number ϱ in the unit interval; we abuse notation by writing $\Psi_e(X)$ for ϱ.

We need to meet the requirements:

P_e: $\Psi_e(B) \neq \varrho$.

Let $\langle C_e \rangle$ be an enumeration of all prefix-free c.e. sets of binary strings. We need to meet the requirements

N_e: If $\lambda(C_e) = \varrho$ then $C_e \leqslant_{\mathrm{T}} B$.

Globally we also need to ensure that $B \leqslant_{\mathrm{T}} \varrho$.

Discussion

Recall the argument above that shows that if C is a presentation of ϱ then $C \leqslant_{\mathrm{T}} \varrho$: if $\varrho - \lambda(C_s) < 2^{-k}$ then no string of length k or shorter can be added to C after stage s. In the other direction, if we know C, then ϱ may still elude us: it may be that no strings of length $\leqslant k$ are added to C after stage s, but a large increase to ϱ after stage s can be made by adding to C many long strings.

To meet a requirement P_e, we can wait for $\Psi_e(B,k)$ to converge and give us an interval $I = I_k$ for some fairly large k, so that I is fairly short. We then aim to ensure that $\varrho \notin I$. If at a current stage s we have $\varrho_s \in I$, then adding a quantity of no more than 2^{-k} to ϱ will suffice to escape I. Of course, we also need to ensure that B will not change below the use $b = \psi_e(k)$ of the Ψ_e computation giving us the interval I.

To meet a requirement N_e, for each length t (or at least for infinitely many t), we need to let B know if strings of length $<t$ will enter C_e after some stage. If indeed $\lambda(C_e) = \varrho$ then $\lambda(C_{e,s})$ and ϱ_s will get closer and closer. One way to ensure that strings of length $\leqslant t$ will not enter C_e after a stage s will be to wait for $|\varrho_s - \lambda(C_{e,s})| < 2^{-t}$ (which we will eventually see); and then ensure that $\varrho - \varrho_s \leqslant 2^{-t}$.

The conflict between requirements is clear: N_e would like to keep $\varrho - \varrho_s < 2^{-t}$, while P_d would like to increase ϱ by at least 2^{-k}, and we may have $2^{-k} > 2^{-t}$.

However, N_e need not be so greedy. It can *temporarily* require that ϱ should not increase by more than 2^{-t}. The positive requirement P_d (positive because it causes increases to ϱ, while the negative requirements N_e want to keep increases small) can increase ϱ by the allotted 2^{-t}. Then, we wait for a later stage s' at which we again see $|\varrho_{s'} - \lambda(C_{e,s'})| < 2^{-t}$. Again, if this does not happen then $\lambda(C_e) \neq \varrho$ and we do not need to worry about N_e. When we see such a stage s', we allow P_d to increase ϱ by *another* 2^{-t}. This cycle can repeat 2^{t-k} times. At the end, P_d gets to increase ϱ by as much as it needs (2^{-k}), while at no stage can strings of length $<t$ enter C_e.[2]

We remark that of course P_d needs to guess whether we will keep seeing "e-expansionary" stages: stages s at which $\lambda(C_{e,s})$ is close to ϱ_s. Thus, as usual, the construction is performed on a tree of strategies.

So far we have not really mentioned B, and it seems that we can arrange for every presentation of ϱ to be computable. And indeed, as we discussed above, this is possible, and this argument is the basic module for the more elaborate construction of [32]. The more difficult issues show up when we consider more than one negative requirement N_e. Actually, the module as described above is imprecise. The point is that the requirement N_e needs to let t go to infinity, as it needs to compute more and more of C_e. On the other hand it needs to wait for P_d to finish 2^{t-k} cycles before it moves to greater t; otherwise the cycle could be infinite. When P_d is accessible, and declares its intention to increase ϱ by 2^{-k}, it sends a message to that effect to the stronger requirement N_e. Because the node working for N_e does not know if the P_d-node will be accessible again, it takes upon itself the task of repeatedly increasing ϱ. It waits for 2^{t-k} many expansionary stages, increases ϱ at each one, while keeping t fixed, and only then allows t to increase, and nodes extending the infinite outcome to be accessible.

When we consider though $N_0, N_1, \ldots, N_e$, all stronger than P_d, the various cycles relating P_d with each N_i need to be nested, which is incompatible with permitting at the level of ω-c.a. We discuss this in greater detail below in Subsection 5.2.3. For the construction we are doing now, B simplifies things. Instead of trying to compute C_e, for each t we set up markers $\eta_e(t)$, intended for B; this is the B-use for determining C_e on strings of length $<t$. If we enumerate $\eta_e(t)$ into B then we are allowed to violate the restriction limiting each increase to ϱ. Now the various restraints of the requirements N_i stronger than P_d can be uniformised as follows. At some stage r the requirement P_d sees that $\Psi_d(B,k){\downarrow}=I$ with some B-use $b=\psi_d(k)$ that it wants to protect. It immediately enumerates any markers $\eta_i(t)$ which are greater than b into B. Its "quota" for each increment of ϱ is then given by the greatest number t with $\eta_i(t)<b$ for any $i\leqslant e$, which is really the stage number r at which we saw the Ψ_d computation. Thus it is allowed to increase ϱ by 2^{-r} at a time, and so needs 2^{r-k} many iterations of such increases; note that this is independent of $i\leqslant e$. We can therefore forget about delegating the task of increasing ϱ to the N_i nodes; we can wait until the P_d node is accessible again, and at each time, enumerate the markers $\eta_i(t)$ for $t\geqslant r$ into B.

These were the main ideas of the construction; we discuss a couple more minor points. First, note that N_e may assume that for all $s<\omega$, $\lambda(C_{e,s})<\varrho_s$, for we will ensure that ϱ is not a dyadic rational. When we see that enumerating

[2]This strategy has been likened to a cautious investor, slowly realising gains by repeatedly selling small amounts of stock, ensuring that the market does not notice their actions: they only sell a further amount once the stock price recovers to the original value.

a string σ into $C_{e,s}$ will make $\lambda(C_{e,s}) \geqslant \varrho_s$, we hold back the enumeration and wait until ϱ_t grows beyond $\lambda(C_{e,s-1} \cup \{\sigma\})$, and only then enumerate σ into C_e. If $\lambda(C_e) = \varrho$ then such a stage will occur.

Next, we note that above we said that P_e only needs to worry if it sees $\varrho_s \in I$. This is not quite true, because if the noncomputability requirements are not met then it is actually possible that ϱ_s would converge to the left endpoint of I. Thus we need to spring into action when we see ϱ_s getting close to I. This will mean that we need a couple more rounds of increases (more than 2^{r-k} detailed above) to ensure that ϱ will eventually lie to the right of I.

The tree of strategies

The requirements P_e and N_e are ordered in order-type ω; the kth level of the tree is devoted to meeting the kth requirement. If σ is a node which works for P_e, then σ has only one outcome. If τ is a node which works for N_e, then the outcomes of τ are $\infty < \texttt{fin}$.

A node σ working for P_e may define first a follower $k_{\sigma,s}$ and then an interval $I_{\sigma,s}$ which it would like ϱ to avoid. It also defines $r_{\sigma,s}$, the amount by which it is allowed to increase ϱ at a single step. When σ is initialised, the follower k_σ, the interval I_σ, and restraint bound r_σ are cancelled. They will be cancelled only when σ is initialised.

Nodes τ working for N_e define markers $\eta_\tau(t)$. We note that it is not necessarily the case that the set of t for which $\eta_\tau(t)$ is defined is an initial segment of ω. In fact $\eta_\tau(t)$ may be defined at most once (at a stage greater than t), and t will be a stage at which τ is accessible. For this reason $\eta_\tau(t)$ is not indexed by the stage number s.

Construction

At stage s we define the path of accessible nodes δ_s to be an initial segment of the tree of strategies, and at the end of the stage define ϱ_{s+1}.

We start with $\varrho_0 = 0$.

Suppose that a node τ, working for N_e, is accessible at stage s. Let $t < s$ be the previous stage at which $\tau\hat{\ }\infty$ was accessible; $t = 0$ if there was no such stage. If $\varrho_s - \lambda(C_{e,s}) < 2^{-t}$ we let $\tau\hat{\ }\infty$ be next accessible and choose $\eta_\tau(t)$ to be large. Otherwise we let $\tau\hat{\ }\texttt{fin}$ be next accessible.

Suppose that a node σ, working for P_e, is accessible at stage s. The node may either let its only immediate successor on the tree of strategies be next accessible or decide to end the stage. In the latter case all nodes weaker than σ are initialised.

First, suppose that a follower $k_{\sigma,s}$ is not defined. Define $k_{\sigma,s+1}$ to be large; let $\varrho_{s+1} = \varrho_s$ and end the stage.

Next, suppose that $k_{\sigma,s}$ is defined but an interval $I_{\sigma,s}$ is not defined. If $\Psi_e(B, k_\sigma)\downarrow[s] = I$ (recall that I is a dyadic rational interval of length $2^{-k_{\sigma,s}}$) then we let $I_{\sigma,s+1} = I$ and $r_{\sigma,s+1} = s$. Let $\varrho_{s+1} = \varrho_s$ and end the stage.

If $\Psi_e(B, k_\sigma)\uparrow[s]$ then σ does not end the stage (and as we said, the unique immediate successor of σ is next accessible).

Suppose that $I_{\sigma,s}$ is defined. If $d(\varrho_s, I_{\sigma,s}) < 2^{-r_{\sigma,s}}$ then for all τ working for some $N_{e'}$ such that $\tau\hat{\ }\infty \preccurlyeq \sigma$, for all $t \geqslant r_{\sigma,s}$ such that $\eta_\tau(t)$ is defined, enumerate $\eta_\tau(t)$ into B_{s+1}. Let $\varrho_{s+1} = \varrho_s + 2^{-r_{\sigma,s}}$ and end the stage.

If the distance $d(\varrho_s, I_{\sigma,s})$ is at least $2^{-r_{\sigma,s}}$ we do not end the stage.

Verification

The global requirement is satisfied:

Lemma 5.5. $B \leqslant_{\mathrm{T}} \varrho$.

Proof. Suppose that x enters B at stage s. Then $x = \eta_\tau(t)$ for some t and τ, and $\varrho_{s+1} = \varrho_s + 2^{-r}$ where $t \geqslant r$. Since $\eta_\tau(t) > t$, we see that once $\varrho - \varrho_s < 2^{-r}$, no numbers below r can enter B. □

We observe that the construction is fair and that the true path δ_ω is infinite. This follows by induction on the length of nodes, using the following lemma.

Lemma 5.6. *Suppose that a node σ, working for a positive requirement P_e, is accessible infinitely often and is initialised only finitely often. Then σ ends the stage only finitely often.*

Proof. Let t be the last stage at which σ is initialised. At the next stage after t at which σ is accessible we appoint a new follower k_σ which is never cancelled. If there is no later stage at which an interval I_σ is defined then σ never stops the stage again.

Otherwise, an interval I_σ is defined at some stage r_σ; the interval (and the bound r_σ) are never cancelled again. If σ is accessible at stage $s > r_\sigma$ then σ ends stage s only if $d(\varrho_s, I_\sigma) < 2^{-r_\sigma}$, in which case it adds 2^{-r_σ} to ϱ_s. Since the length of the interval I_σ is 2^{-k_σ}, this happens at most $2^{r_\sigma - k_\sigma} + 2$ many times. □

To bound the value of ϱ, for a positive node σ (one working for some P_e) and a stage t let

$$\beta(\sigma, t) = \sum (\varrho_{s+1} - \varrho_s) \quad [\![s \geqslant t \ \ \& \ \ \sigma \text{ ends stage } s]\!].$$

So $\varrho - \varrho_t$ is the sum of $\beta(\sigma, t)$ for all positive nodes σ.

Lemma 5.7. *Suppose that a positive node σ is initialised at stage t. Then $\beta(\sigma, t) < 2^{-(3t+1)}$.*

Proof. Suppose that σ is initialised at stage t, that $u > t$ and σ is not initialised at any stage in the interval $(t, u]$. Let k_σ be the value of the follower for σ in the interval $[t, u]$ (if appointed). Since k_σ is chosen large relative to t we assume that $k_\sigma > 3t+3$; and $r_\sigma > k_\sigma$. The proof of Lemma 5.6 shows that the sum

$$\sum(\varrho_{s+1} - \varrho_s) \quad [\![s \in [t, u] \ \ \& \ \ \sigma \text{ ends stage } s]\!]$$

is bounded by $2^{-k_\sigma} + 2 \cdot 2^{-r_\sigma}$ which is bounded by $2^{-(3t+2)}$. We now sum over all the stages $t' \geqslant t$ at which σ is initialised. □

We conclude that $\varrho = \lim_s \varrho_s$ exists and lies in the unit interval.

Lemma 5.8. $\varrho < 1$.

Proof. Every node of length s is initialised at every stage $s' \leqslant s$. Thus for such a node σ we have $\beta(\sigma, 0) = \beta(\sigma, s) < 2^{-(3s+1)}$. There are at most 2^s many nodes of length s as the tree of strategies is at most binary branching. Hence level s contributes at most $2^{-(s+1)}$ to ϱ. Some levels consist of negative nodes and so contribute nothing to ϱ. □

We turn to showing that all requirements are met.

Lemma 5.9. *Each positive requirement P_e is met.*

Proof. Let σ be a node on the true path which works for P_e. Let k_σ be the value of the last follower chosen by σ, the one which is never cancelled. We suppose that $\Psi_e(B)$ is total; so I_σ is eventually defined at a stage $r_\sigma > k_\sigma$. Since σ acts only finitely often, for almost all stages s, $d(\varrho_s, I_\sigma) \geqslant 2^{-r_\sigma}$. Hence $d(\varrho, I_\sigma) \geqslant 2^{-r_\sigma}$ and so $\varrho \notin I_\sigma$.

It remains to show that $\Psi_e(B) \in I_\sigma$, which would follow once we show that the computation $\Psi_e(B, k_\sigma)[r_\sigma]$ is B-correct. Let $b = \psi_e(B, k_\sigma)[r_\sigma]$ be the use of this computation.

Suppose that a number x enters B at stage $s \geqslant r_\sigma$, enumerated by a node ρ. We show that $x > b$. The number x equals $\eta_\tau(t)$ for some $\tau\hat{\ }\infty \preccurlyeq \rho$ and some t. We know that $x = \eta_\tau(t) > t \geqslant r_{\rho,s}$. The node ρ cannot be stronger than σ, for otherwise σ is initialised at stage $s \geqslant r_\sigma$, contradicting the permanence of k_σ and I_σ. Hence $r_{\rho,s} \geqslant r_\sigma$: this is clear if $\rho = \sigma$; otherwise, ρ is initialised at stage r_σ, $s > r_\sigma$ and $r_{\rho,s}$ must be greater than r_σ. Finally the use $b = \psi_e(B, k)[r_\sigma]$ is bounded by r_σ. □

Toward showing that negative requirements are met, let τ be a node, working for N_e, and suppose that $\tau\hat{\ }\infty$ lies on the true path. Let t^* be the last stage at which τ is initialised. We let S be the set of stages $t > t^*$ at which $\tau\hat{\ }\infty$ is accessible. For $t \in S$ let t^+ be the next stage in S.

The markers defined by τ are $\eta_\tau(t)$ for $t \in S$. The marker $\eta_\tau(t)$ is defined at stage t^+.

Lemma 5.10. *Let $u < t$ be two stages in S. Assume that $\eta_\tau(u) \notin B_{t+1}$. Then $\varrho_{t^+} - \varrho_t \leqslant 2^{-u}$. It follows that no strings of length less than u lie in $C_{e,t^+} \setminus C_{e,t}$.*

Proof. We consider various contributions. All nodes that lie to the right of $\tau\hat{}\infty$ are initialised at stage t. The calculation in the proof of Lemma 5.8 shows that $\varrho_{t^+} - \varrho_{t+1} \leqslant 2^{-t} \leqslant 2^{-(u+1)}$.

Next consider nodes $\sigma \succcurlyeq \tau\hat{}\infty$. In the interval of stages $[t, t^+)$, such nodes are only accessible at stage t. At stage t at most one such node σ increases ϱ; the amount of increase $\varrho_{t+1} - \varrho_t$ equals $2^{-r_{\sigma,t}}$. Since $\eta_\tau(u)$ is not enumerated into B at stage t we have $r_{\sigma,t} > u$, and so $\varrho_{t+1} - \varrho_t \leqslant 2^{-(u+1)}$.

As discussed above, the last sentence follows: $\varrho_t - \lambda(C_{e,t}) < 2^{-t}$ and $\lambda(C_{e,t^+}) \leqslant \varrho_{t^+}$ and so $\lambda(C_{e,t^+}) - \lambda(C_{e,t}) < 2^{-u} + 2^{-t} < 2^{-(u-1)}$. □

The verification ends with:

Lemma 5.11. *Each negative requirement N_e is met.*

Proof. We assume that $\lambda(C_e) = \varrho$; we need to show that $C_e \leqslant_{\mathrm{T}} B$. Let τ be a node on the true path working for N_e.

We claim that infinitely many markers $\eta_\tau(u)$ are not enumerated into B. Let $w > t^*$ be a stage. Let σ be the strongest extension of $\tau\hat{}\infty$ which acts (ends the stage) after stage w. Since infinitely many nodes on the true path act, σ cannot lie to the right of the true path. It follows that σ acts only finitely often. Let t be the last stage at which σ acts. The marker $\eta_\tau(t)$ is appointed at stage t^+. Let $\rho \succcurlyeq \tau\hat{}\infty$ be a node which enumerates a marker $\eta_\tau(v)$ into B at some stage $s \geqslant t^+$. The node ρ is initialised at stage t; after stage t it is first accessible not before stage t^+, and so $v \geqslant r_{\rho,s} \geqslant t^{++}$. Hence $\eta_\tau(t)$ (and in fact $\eta_\tau(t^+)$ as well) are never enumerated into B.

Now Lemma 5.10 shows that the following algorithm with oracle B correctly computes C_e: Given $k < \omega$, find a stage $t > k$ in S such that $\eta_\tau(t) \notin B$. Announce that $C_e \restriction_{2^{<t}} = C_{e,t^+} \restriction_{2^{<t}}$. □

5.2.2 Non-totally ω-c.a. permitting

We now add non-totally ω-c.a. permitting to prove part (1) of Theorem 1.4: if $\mathbf{d}$ is not totally ω-c.a. then there is a left-c.e. real $\varrho \leqslant_{\mathrm{T}} \mathbf{d}$ and a c.e. set $B <_{\mathrm{T}} \varrho$ such that every presentation of ϱ is B-computable.

Fix some function $g \in \mathbf{d}$ which is not ω-c.a. Since $\mathbf{d}$ is c.e., we can replace g by its modulus (see the proof of Theorem 3.31). So we have a computable approximation $\langle g_s \rangle$ of g such that:

- if $s<t$ then $g_s(n) \leqslant g_t(n)$ for all n;
- if $g_{s+1}(n) \neq g_s(n)$ then $g_{s+1}(m) \neq g_s(m)$ for all $m > n$.

At first approximation, the idea for reducing ϱ to g (and hence to $\mathbf{d}$) is to declare that if $g_s(k) = g(k)$ then $\varrho - \varrho_s \leqslant 2^{-k}$. Using the notation of the construction above, when a node σ is visited and wants to increase ϱ we must first wait for a change in $g(k_\sigma)$. The number of permissions needed to meet σ's requirement is bounded by 2^{r_σ}. We note that it is the follower k_σ that needs to be permitted, even though at each step we increase ϱ by 2^{-r_σ}, not 2^{-k_σ}. It is the eventual increase in ϱ which counts, because the promise is that if k is not permitted then $\varrho - \varrho_s \leqslant 2^{-k}$.

Of course it is possible that the number of permissions will be insufficient. While waiting for permissions the node σ must appoint more followers k, with the expectation that at least one of them will receive the necessary number of permissions. If the follower k does not receive enough permissions then we can approximate $g(k)$ with fewer than 2^{r_σ} many mind-changes. If no follower receives enough permissions then infinitely many of them will be appointed. This will give an ω-computable approximation of g.

The remaining issues are the timing of permissions and necessary cancellation of followers. The follower k could be permitted at a stage s at which σ is not accessible. We cannot "leave the permission open" and wait to increase ϱ at the next stage at which σ is accessible, since we do not know whether such a stage will occur. We need to act on permissions immediately.

When a follower k receives a permission we increase ϱ by the associated amount $2^{-r_\sigma(k)}$ (determined by the stage $r_\sigma(k)$ at which we see the computation $\Psi_e(B,k)$ converge) and we need to enumerate markers $\eta_\tau(t)$ for $t \geqslant r_\sigma(k)$ into B. This means that the computations $\Psi_e(B,k')$ for followers $k' > k$ for the same node are destroyed. We cannot keep these followers: overall we want action for some follower k to not increase ϱ by more than 2^{-k+1} say. So the larger followers k' are cancelled, and later, larger followers may be appointed.

But this creates a problem when arguing that eventually some follower will be permitted. Suppose that a follower k is eventually cancelled. When approximating $g(k)$ we do not know in advance that it will be cancelled, so we promise that our guesses for $g(k)$ will not change more than $2^{r_\sigma(k)}$ many times. We observe many changes, and then k is cancelled. Henceforth changes in $g(k)$ do not seem to help us to meet σ's requirement, which means that there is no mechanism which will bound these changes. We need to ensure that every change in $g(k)$ is useful.

The solution (as in [25]) is to allow stronger followers to take over the responsibility for approximating greater portions of g. When a follower k is permitted, larger followers $k' > k$ are cancelled. We declare that from now on what would have been permissions for k' must count as permissions for k. Technically we define moveable markers $a_{k,s}$, and we declare that k is permitted if $g(a_k)$ changes (rather than $g(k)$). When k is permitted then we raise $a_{k,s}$ to be greater than the previous values of $a_{k'}$ for the followers k' which were cancelled.

Construction

The tree of strategies is the same as in the construction above. Positive nodes σ appoint followers. All followers are cancelled when σ is initialised or when smaller followers for σ receive attention; otherwise they are retained. For all followers k of σ (except possibly for the largest one) we also define associated intervals $I_\sigma(k)$ (of length 2^{-k}) and bounds $r_\sigma(k)$ as above. Any number can be chosen at most once as a follower for any requirement.

Negative nodes τ define markers $\eta_\tau(t)$ as in the previous construction. Globally we define location markers $a_{k,s}$ for all $k<s$, useful for reducing ϱ to g.

We start with setting $\varrho_0=0$. At stage s we either act on permissions or define the path of accessible nodes δ_s and act for nodes on that path.

We say that a node σ is *already met* by stage s if at stage s there is some follower k for σ such that $I_\sigma(k)$ is defined and ϱ_s lies strictly to the right of $I_\sigma(k)$.

OPTION A: ACTING ON PERMISSIONS. We say that a follower k (for a positive node σ) *requires attention* at stage s if:

- the node σ is not already met at stage s;
- the interval $I_\sigma(k)$ is defined;
- $d(\varrho_s, I_\sigma(k)) < 2^{-r_\sigma(k)}$;
- the follower k did not receive attention since the last stage at which σ was accessible; and
- $g_{s+1}(a_{k,s}) \neq g_s(a_{k,s})$.

If no follower requires attention then we take option B. Otherwise let k be the strongest follower which requires attention: the node σ is the strongest, any of whose followers requires attention at stage s; and k is the strongest (smallest) follower for σ that requires attention at stage s. We say that the follower k *receives attention.*

We execute the following instructions. Let $\varrho_{s+1}=\varrho_s+2^{-r_\sigma(k)}$. For all negative nodes τ such that $\tau\hat{\ }\infty \preccurlyeq \sigma$, for all $t \geqslant r_\sigma(k)$ such that $\eta_\tau(t)$ is defined, enumerate $\eta_\tau(t)$ into B_{s+1}. Initialise all nodes weaker than σ; cancel all followers for σ greater than k and their associated intervals. Redefine $a_{m,s+1}$ to be large for all $m \geqslant k$, and define a new marker $a_{s,s+1}$ to be large as well. End the stage.

OPTION B: BUILDING THE PATH OF ACCESSIBLE NODES. If option A was not taken then we define the path δ_s of accessible nodes. Since no permissions were used, we set $\varrho_{s+1}=\varrho_s$ and $a_{m,s+1}=a_{m,s}$ for all $m<s$; we define $a_{s,s+1}$ to be large.

Suppose that a node τ working for N_e is accessible at stage s. Let $t<s$ be the previous stage at which $\tau\hat{\ }\infty$ was accessible; $t=0$ if there was no such stage. If $\varrho_s-\lambda(C_{e,s})<2^{-t}$ we let $\tau\hat{\ }\infty$ be next accessible and choose $\eta_\tau(t)$ to be large. Otherwise we let $\tau\hat{\ }\texttt{fin}$ be next accessible.

Suppose that a node σ working for P_e is accessible at stage s. The node may either let its only immediate successor on the tree of strategies be next accessible or decide to end the stage. In the latter case all nodes weaker than σ are initialised. If the node σ is already met by stage s then σ takes no action and does not end the stage.

Suppose that σ is not already met. If σ has no followers then a new, large one is appointed, and the stage is ended. Otherwise, let k be the largest follower for σ.

If $I_\sigma(k)$ is defined and $d(\varrho_s, I_\sigma(k)) < 2^{-r_\sigma(k)}$ then appoint a new, large follower for σ and end the stage. If $d(\varrho_s, I_\sigma(k)) \geqslant 2^{-r_\sigma(k)}$ then the stage is not ended.

Suppose that $I_\sigma(k)$ is not defined. If $\Psi_e(B,k)\downarrow[s]$ then set $I_\sigma(k) = \Psi_e(B,k)[s]$ and $r_\sigma(k) = s$; end the stage. If $\Psi_e(B,k)\uparrow[s]$ then no action is taken and the stage is not ended.

Verification

Suppose that a positive node σ is initialised only finitely many times. Every follower for σ is either eventually cancelled, or receives attention only finitely many times. As above the point is that the follower k cannot receive attention more than $2^{r_\sigma(k)-k}+1$ many times, as each time ϱ is increased by $2^{-r_\sigma(k)}$. Indeed if a follower k receives attention the full number of times then the requirement is declared met and no follower for σ receives attention, at least until a later stage at which σ is cancelled.

Since new followers are always chosen large we see that, as promised, each k is chosen at most once to be a follower (for any node). A location marker $a_{m,s}$ is moved only when some follower $k \leqslant m$ receives attention. We conclude that the location markers $a_{m,s}$ reach limits a_m. Thus, for all $m < \omega$ there is some stage s at which $g_s(a_{m,s}) = g(a_{m,s})$. The following lemma then shows that ϱ is computable from g, and so from $\mathbf{d}$.

Lemma 5.12. *Suppose that $g_s(a_{m,s}) = g(a_{m,s})$. Then $\varrho - \varrho_s \leqslant 2^{-(m-1)}$.*

Proof. Note that $a_{m,t+1} \neq a_{m,t}$ only if $g_{t+1}(a_{m,t}) \neq g_t(a_{m,t})$. Hence $a_{m,s} = a_m$ is the final value of this marker. Let $\beta(k)$ be the sum of $\varrho_{t+1} - \varrho_t$, as t ranges over the stages at which the follower k receives attention. As discussed above, $\beta(k)$ is bounded by $2^{-k} + 2^{-r_\sigma(k)} \leqslant 2 \cdot 2^{-k}$ (where σ is the node for which k is a follower), since $r_\sigma(k) > k$. Since no follower of size less than or equal to m receives attention after stage s we know that

$$\varrho - \varrho_s \leqslant \sum_{k>m} \beta(k) \leqslant 2 \cdot 2^{-m}. \qquad \square$$

The proof that $B \leqslant_{\mathrm{T}} \varrho$ is identical to the one given earlier. The proof that $\varrho < 1$ requires minor modifications but is essentially the same. If σ is a positive node which is initialised at stage t then the total contribution to $\varrho - \varrho_t$

due to stages at which followers for σ receive attention is bounded by $2\sum 2^{-k}$ where the sum ranges over follower k for σ appointed after stage t. Since all of these followers are chosen to be large we may assume that this sum is bounded by 2^{-3t-1} as above.

The following lemma ensures that the true path is infinite and that the construction is fair to nodes on the true path. First note that there are infinitely many stages at which option B is taken: if s is the last stage at which option B is taken, then only finitely many followers are ever appointed and each one receives attention at most once after stage s.

Lemma 5.13. *Suppose that σ is a node which works for requirement P_e, is only initialised finitely many times and is accessible infinitely often. Then the unique immediate successor of σ on the tree of strategies is initialised only finitely often and so is accessible infinitely often. Further, the requirement P_e is met.*

Proof. Let t^* be the last stage at which σ is initialised.

Let $s^* > t^*$ and let k be a follower for σ at stage s^* which is never cancelled. No follower stronger than k receives attention after stage s^*.

If the interval $I_\sigma(k)$ is never defined then no larger followers for σ are ever appointed and σ never later ends a stage at which it is accessible. Since all followers receive attention only finitely many times we see that the successor of σ is initialised only finitely many times. Further, in this case $\Psi_e(B,k)\uparrow$ and so the requirement P_e is met.

Suppose then that at some stage r_σ the interval $I_\sigma(k)$ is defined. The argument in the previous construction shows that the computation $\Psi_e(B,k)[r_\sigma(k)]$ is B-correct and so if total, $\Psi_e(B) \in I_\sigma(k)$.

If at all stages $s \geqslant r_\sigma(k)$, ϱ_s lies to the left of $I_\sigma(k)$ and $d(\varrho_s, I_\sigma(k)) \geqslant 2^{-r_\sigma(k)}$ then no follower greater than k is ever appointed for σ, so again the successor of σ is on the true path and the construction is fair to that successor. As before, in this case $d(\varrho, I_\sigma(k)) \geqslant 2^{-r_\sigma(k)}$ so $\varrho \neq \Psi_e(B)$.

Similarly, if at some stage $s \geqslant r_\sigma(k)$ we see that ϱ_s lies strictly to the right of $I_\sigma(k)$ then σ is declared met and no action is taken for σ after stage s. Since $\varrho \geqslant \varrho_s$ again we see that $\varrho \notin I_\sigma(k)$ and so P_e is met.

Further, in this last case we do not need to assume in advance that k is never cancelled: once we see ϱ_s lying to the right of $I_\sigma(k)$, all action for σ ceases and no follower is cancelled.

We claim that there is some follower k for σ which is never cancelled and for which one of the cases described above holds. Assume, for a contradiction, that this is not the case. We show that g is ω-c.a.

The assumption means that:

- for every follower k for σ appointed after stage s, either k is cancelled or $I_\sigma(k)$ is eventually defined and for all but finitely many stages $s \geqslant r_\sigma(k)$, $d(\varrho_s, I_\sigma(k)) < 2^{-r_\sigma(k)}$
- the node σ is never declared met after stage t^*.

For a follower k of σ, if there is such a stage, we let $s_\sigma(k)$ be the least stage $s \geqslant r_\sigma(k)$ such that $d(\varrho_s, I_\sigma(k)) < 2^{-r_\sigma(k)}$ and σ is accessible at stage s. As observed above, if k is a follower for σ at a stage s and is not the largest follower for σ at that stage, then $s > s_\sigma(k)$.

Let $x < \omega$. Let $S(x)$ be the set of stages $s > t^*$ satisfying:

- σ is accessible at stage s; and
- there is some follower k of σ at stage s such that $s \geqslant s_\sigma(k)$ and $x \leqslant a_{k,s}$.

For $s \in S(x)$ let $k_s(x)$ be the smallest follower for σ witnessing that $s \in S(x)$. We first claim that if $s \in S(x)$, $t > s$ and σ is accessible at stage t then $t \in S(x)$ and $k_t(x) \leqslant k_s(x)$. Let $k = k_s(x)$. If k is still a follower for σ at stage t then k witnesses that $t \in S(x)$, because $a_{k,s} \leqslant a_{k,t}$. Otherwise a follower stronger than k receives attention at a stage between stages s and t. Let m be the strongest such follower. Then m is still a follower for σ at stage t. If m receives attention at stage $u \in (s,t)$ then we define $a_{m,u+1}$ to be large, in particular greater than x, and so $x < a_{m,t}$ and m witnesses that $t \in S(x)$.

Suppose that $s < t$ are successive stages in $S(x)$ and that $g_t(x) \neq g_s(x)$. Let $k = k_s(x)$. The fact that $x \leqslant a_{k,s}$ implies that $g_t(a_{k,s}) \neq g_s(a_{k,s})$. Let m be the smallest follower for σ such that $g_t(a_{m,s}) \neq g_s(a_{m,s})$; so $m \leqslant k$. Let u be the least stage $u \in (s,t)$ at which $g_{u+1}(a_{m,s}) \neq g_u(a_{m,s})$. Then m is not cancelled by stage u, and as it did not receive attention at stages between s and u, it requires attention at stage u, and receives it.

Above we calculated for any follower k for which $I_\sigma(k)$ is ever appointed a bound $h(k) = 2^{r_\sigma(k)-k} + 1$ for the number of times k receives attention. It follows that the number of stages $s \in S(x)$ such that $g_t(x) \neq g_s(x)$ (where t is the next stage in $S(x)$) is bounded by $\sum h(m)$, where m is a follower for σ at stage $s = \min S(x)$ and $s_\sigma(m) \leqslant s$. From this we can construct an ω-computable approximation for g. □

It remains to show that every negative requirement is met. Let $e < \omega$ and let τ on the true path work for N_e; in the interesting case $\tau\hat{\ }\infty$ also lies on the true path. The proof of Lemma 5.11, that infinitely many markers $\eta_\tau(t)$ are not enumerated into B, goes through as above: say w is a late stage; let σ be the strongest node which ever acts (ends the stage) or a follower of whose receives attention after stage w. Then σ extends $\tau\hat{\ }\infty$ and does not lie to the right of the true path. Either σ lies to the left of the true path, in which case σ appoints only finitely many followers; each one receives attention infinitely often. If σ lies on the true path then Lemma 5.13 shows that σ acts only finitely often. Hence there is a last stage t at which σ is accessible and ends the stage, or a follower for σ receives attention. Any node ρ which acts after stage t is initialised at stage t. If t' is the least stage $t' \geqslant t$ at which $\tau\hat{\ }\infty$ is accessible then $\eta_\tau(t')$ is not enumerated into B.

Thus we need to prove an analogue of Lemma 5.10. Again let $u < t$ be two late stages at which $\tau\hat{\ }\infty$ is accessible and suppose that $\eta_\tau(u) \notin B_{t+}$, where again t^+ is the next stage after t at which $\tau\hat{}\infty$ is accessible. As above, the total contribution to $\varrho_{t^+} - \varrho_t$ made by nodes that lie to the right of $\tau\hat{}\infty$ is bounded by 2^{-t}, as all such nodes are initialised at stage t. It is no longer true however that nodes extending $\tau\hat{}\infty$ do not act at stages strictly between t and t^+, nor that only one such node acts between these stages. Nonetheless, every follower k for a node $\sigma \succcurlyeq \tau\hat{}\infty$ receives attention at most once between stages t and t^+, and so the total increase in ϱ attributed to such nodes is bounded by $\sum 2^{-r_\sigma(k)}$ where $\sigma \succcurlyeq \tau\hat{}\infty$, k is a follower for σ at stage t and $r_\sigma(k) > u$ (again as $\eta_\tau(u) \notin B_{t^+}$). Since the numbers $r_\sigma(k)$ are distinct for distinct followers k we see that this sum is bounded by 2^{-u}. We conclude that $\varrho_{t^+} - \varrho_t$ is bounded by $2^{-(u-1)}$ and so that strings of length smaller than $u-1$ do not enter C_e between stages t and t^+, completing the proof.

5.2.3 The complexity of the original construction

As mentioned above, the original construction in [32] gives a noncomputable left-c.e. real ϱ, all of whose presentations are computable. That is, $B = \varnothing$. This construction is more complicated than the one presented above. Since we are not allowed to enumerate markers into B, promises that a node τ makes at an expansionary stage are binding to all. Considering one such node τ and one positive node σ extending $\tau\hat{}\infty$, ensuring that σ acts only finitely many times requires τ to delay making stricter restraints. Suppose that an interval I_σ is defined at some stage r_σ. Ignoring subtleties we assume that at that stage the node τ declares that from now on, any increase in ϱ between two successive τ-expansionary stages must be bounded by 2^{-r_σ}.

The node σ issues a request from τ: until σ's mission is accomplished, τ should refrain from imposing stronger bounds on the increase of ϱ between τ-expansionary stages. In turn, since τ does not know if σ will be accessible sufficiently many times to complete its task, it cannot abide by σ's request indefinitely. Hence τ takes upon itself to act on σ's behalf: at the next few τ-expansionary stages, the stage ends when τ is accessible and an amount of 2^{-r_σ} is added to ϱ. This happens finitely many times, until ϱ_s lies to the right of I_σ; after that, σ never acts again and τ is free to make stricter promises about increases of ϱ.

So far the number of actions required is similar to the previous construction, but the story gets more complicated when more than one node τ is considered. Suppose now that τ_1 and τ_2 are two negative nodes with $\tau_1\hat{}\infty \preccurlyeq \tau_2$ and $\tau_2\hat{}\infty \preccurlyeq \sigma$. At stage r_σ both negative nodes promise that between τ_i-expansionary stages, ϱ increases by no more than 2^{-r_σ}. So we cannot increase ϱ by the desired 2^{-r_σ} until the next τ_2-expansionary stage. Now τ_1 is in a bind. It cannot act on its own to help σ, it seems; but it does not know if there are infinitely many τ_2-expansionary stages, so it cannot wait for one while not making its own promises about ϱ stricter.

The solution is to follow a nested loop. Suppose that $t \geqslant r_\sigma$ is τ_2-expansionary. Unlike τ_1, the node τ_2 can afford to wait until σ is done, and so keeps the bound between τ_2-expansionary stages to be 2^{-r_σ}. Until the next τ_2-expansionary stage the entire construction is restricted to the interval $[\varrho_s, \varrho_s + 2^{-r_\sigma})$. At stage s the node τ_1 announces a strict bound, roughly 2^{-t}. At subsequent τ_1-expansionary stages we increase ϱ on σ's behalf, say up to $\varrho_s + 2^{-r_\sigma}/2$. This means that at the next $2^{t-r_\sigma-1}$ many τ_1-expansionary stages, the path of accessible nodes ends at τ_1. After this action, the construction continues without special action on σ's behalf but with sufficient initialisations to the right of σ so that the promise that $\varrho < \varrho_s + 2^{-r_\sigma}$ is honoured. At the next τ_2-expansionary stage we repeat the cycle again: a new, stricter bound $2^{-t'}$ is announced by τ_1; for the next $2^{t'-r_\sigma-1}$ many τ_1-expansionary stages we act on behalf of σ, and then again wait for a new τ_2-expansionary stage. After no more than 2^{r_σ} many such iterations we meet σ's requirement.

Consider now how this argument would translate to a permitting argument. We know in advance that to meet σ on the follower k we will need something like $2^{r_\sigma(k)}$ many τ_2-expansionary stages. If t is one of these stages, we will need roughly 2^t many permissions for τ_1 to act on σ's behalf. We will not know the next value until we actually observe the next τ_2-expansionary stage. So the number of total permissions required is given by an ω^2-c.a. function: the number of times we change our mind about how many permissions we need for follower k is bounded by the computable number $2^{r_\sigma(k)}$. If we know that $\mathbf{d}$ is not totally ω^2-c.a. then we can meet σ's requirement. If there are three negative nodes τ_1, τ_2 and τ_3 below σ, then we have three layers of nesting of loops, and so the number of permissions is now given by an ω^3-c.a. function, and so on. Overall we see that this kind of permission is related to non-total $<\omega^\omega$-c.a. permission.

5.3 TOTAL ω-C.A. ANTI-PERMITTING

We prove part (2) of Theorem 1.4. Let ϱ be a left-c.e. real such that $\deg_{\mathrm{T}}(\varrho)$ is totally ω-c.a. We enumerate a presentation C of ϱ which is Turing equivalent to ϱ.

The technique we use is the so-called "anti-permitting" technique described in [7, 25]. In some sense it is a mirror image of the previous construction. As discussed earlier in this chapter, we view ϱ as an infinite binary sequence via binary expansion. This is unique as we may assume that ϱ is noncomputable. In fact we will later make significant use of the assumption that ϱ is noncomputable; it will help us lift array computable anti-permitting to total-ω-c.a. anti-permitting.

5.3.1 Basic algorithm and plan

Before we describe the construction we discuss one of the algorithms that will be used in the construction and the high-level plan for the construction.

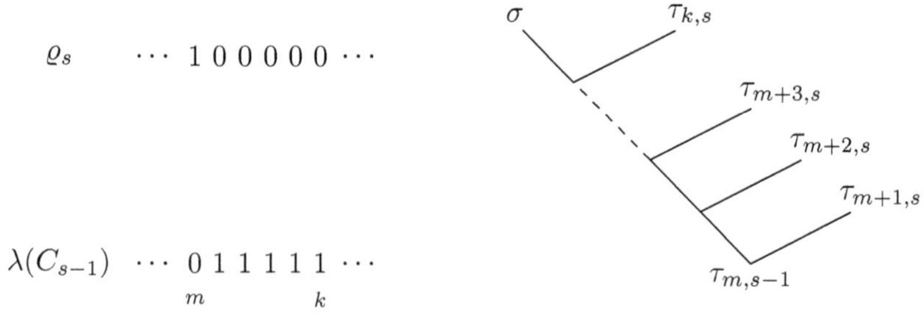

Figure 5.1. $C_s = C_{s-1} \cup \{\sigma\}$.

Building presentations

We want to enumerate a presentation C of ϱ. We follow a proof by J. Miller of the Kraft-Chaitin theorem of algorithmic randomness theory (see [27]). We fix an increasing approximation $\langle \varrho_s \rangle$ of ϱ, where each $\varrho_s \in [0,1)$ is a dyadic rational number. We will not require that $\lambda(C_s) = \varrho_s$ for all stages s. We will only add strings to C_s to bring its measure up to ϱ_s at stages s at which we receive some "certification" that various initial segments of ϱ_s are correct. This process of certification is the heart of the construction. Ignoring the mechanics of certification for the moment, let s be a stage at which we want to add strings to C_{s-1} to ensure that $\lambda(C_s) = \varrho_s$. The instruction will be:

Adding strings to C.

Let $\beta = \varrho_s - \lambda(C_{s-1})$. For each k such that $\beta(k) = 1$, add a single string of length k to C_s.[3]

Since $\beta = \sum \beta(k) 2^{-k}$, it is clear that $\lambda(C_s) = \varrho_s$. The pertinent point is:

if $\beta \geqslant 2^{-k}$ then a string of length at most k enters C_s.

We need to argue though that the instruction can be carried out while keeping C_s prefix-free. This is done by using an auxiliary sequence of strings. At each stage t we will have reserved strings $\tau_{k,t}$ for each k such that $\lambda(C_t)(k) = 0$, with $|\tau_{k,t}| = k$, such that $C_t \cup \{\tau_{k,t} : \lambda(C_t)(k) = 0\}$ is prefix-free. We work with each length at a time, so we may assume that $\beta = 2^{-k}$, i.e., we want to add a single string of length k to C_{s-1}. Since $\lambda(C_{s-1}) < \varrho_s < 1$ there is some $m \leqslant k$ such that $\lambda(C_{s-1})(m) = 0$. Let m be the greatest such. So the change in ϱ_s compared to ϱ_{s-1} is that the mth bit changes from 0 to 1, and the nth bit, for all $n \in (m,k]$ (if $k > m$) changes from 1 to 0. So $\tau_{n,s} = \tau_{n,s-1}$ for $n \notin [m,k]$. We then add $\tau_{m,s-1}\hat{\ }0^{k-m}$ to C_s and for $n \in (m,k]$ we let $\tau_{n,s} = \tau_{m,s-1}\hat{\ }0^{n-m-1}1$. See Figure 5.1.

[3] Recall that we consider β as a string via its binary expansion.

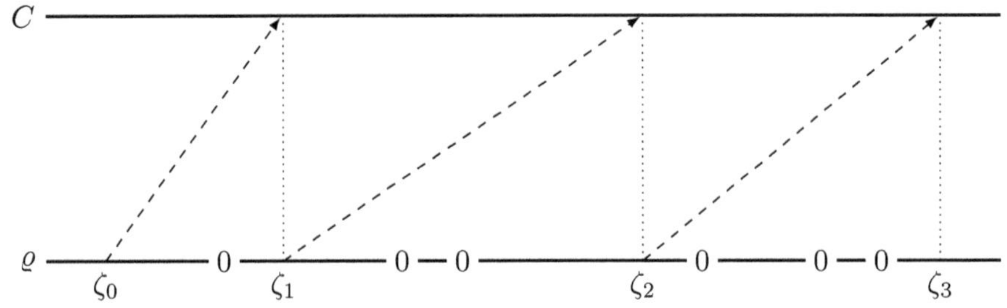

Figure 5.2. Layers. The dashed lines represent the reduction of ϱ to C.

Henceforth the details of the auxiliary strings are assumed, and we only invoke the algorithm above.

Layers

Suppose that we enumerate a presentation C as described above. Why is it the case that C might not compute ϱ? We have arranged for that in the previous construction: we gradually add small amounts to ϱ. If we update C each time, this means that only long strings enter C. However the cumulative effect on ϱ may be big, which is a change that C does not comprehend.

In terms of binary expansions, the problematic case is when ϱ_s contains a long block of 1's. Suppose that $\varrho_s \upharpoonright_{(m,k]}$ is a string of ones. Then adding 2^{-k} to ϱ_s results in adding a string of length k to C but changes the bit $\varrho(m)$.

We can try to prevent this by setting up *layers* which contain sufficiently many zeros, and appropriately set uses for computing ϱ from C. We set markers $\zeta_0 < \zeta_1 < \zeta_2 < \cdots$ such that the block $\varrho\upharpoonright_{[\zeta_n,\zeta_{n+1})}$ contains many zeros (and the idea is that the markers may increase with time, but hopefully settle down eventually). We let ζ_{n+1} be the use for reducing $\varrho\upharpoonright_{\zeta_n}$ to C. See Figure 5.2. Here since C is a set of strings, by use u we mean querying the oracle on strings of length less than u.

Now the point is that if between stages s and $s+1$, ϱ changes on the interval $[\zeta_{n-1}, \zeta_n)$, then since the interval $\varrho_s\upharpoonright_{[\zeta_n,\zeta_{n+1})}$ contains zeros, the increase $\varrho_{s+1} - \varrho_s$ is greater than $2^{-\zeta_{n+1}}$; and so if we update C then some string of length smaller than ζ_{n+1} will enter C and allow us to fix the reduction of $\varrho\upharpoonright_{\zeta_n}$ to C.

After this increase, we may have $\varrho_{s+1}\upharpoonright_{[\zeta_n,\zeta_{n+1})}$ be all ones, but we can increase ζ_{n+1} so that the new interval contains many zeros. However, it is possible that no string of length smaller than ζ_n entered C_{s+1}; so we cannot increase ζ_n, as this is the use of computing $\varrho\upharpoonright_{\zeta_{n-1}}$. This is a problem, since we lost a zero on the interval $[\zeta_{n-1}, \zeta_n)$. Note though that we lose at most one zero, or the increase is beyond $2^{-\zeta_n}$ and strings of length smaller than ζ_n will in fact enter C_{s+1}. *So if we can ensure that the number of times this happens is at most the number of zeros we originally set up in the interval* $[\zeta_{n-1}, \zeta_n)$, the construction will succeed. This is precisely what the certification process gives us.

Certification

The certification process relies on the computational weakness of $\deg_T(\varrho)$. We enumerate a Turing functional Γ with intended oracle ϱ, and ensure that $\Gamma(\varrho)$ is total. We know that the function $\Gamma(\varrho)$ is ω-c.a. Suppose that $\langle g_s, o_s\rangle$ is an ω-computable approximation for $\Gamma(\varrho)$. When a computation $\Gamma_s(\varrho_s, n)$ is destroyed, we redefine it with a new value. It follows that there are fewer than $o_0(n)$ many stages s at which $\Gamma_s(\varrho_s, n) = g_s(n)$ and the computation $\Gamma_s(\varrho_s, n)$ is ϱ-incorrect.

The plan for setting up the layers is then as follows. Given ζ_{n-1}, calculate $o_0(n)$ and let ζ_n be sufficiently large so that the current version of ϱ contains at least $o_0(n)$ many zeros in the interval $[\zeta_{n-1}, \zeta_n)$. Define $\Gamma(\varrho, n)$ with use ζ_n. Recall that since the oracle ϱ is given, our convention is that by use u we mean that $\varrho\upharpoonright_u$ computes $\Gamma(n)$, not $\varrho\upharpoonright_{u+1}$.

We can then carry out our original plan. Suppose that for a while everything is stable, but that at some stage t we see an increase in ϱ_{t+1}, say a quantity $q \in (2^{-\zeta_{n+1}}, 2^{-\zeta_n}]$. As discussed above, this may change the bits of ϱ on the interval $[\zeta_{n-1}, \zeta_n)$. This means that now $\Gamma(\varrho, n)\uparrow$. We define a new value for the computation (say t) with the same use. Before we act, we wait for certification: for a later stage s at which we see that $g_s(n)$ equals that new value t. Only once we've seen this certification do we add strings to C_{s+1} (of lengths between ζ_n and ζ_{n+1}). Compared to $\varrho_t\upharpoonright_{[\zeta_{n-1}, \zeta_n)}$, the interval $\varrho_s\upharpoonright_{[\zeta_{n-1}, \zeta_n)}$ contains one zero fewer. But this is compensated by the change in g, which ensures that $o_s(n) < o_t(n)$. Note though that while waiting, further increases can occur. If the amount increases beyond $2^{-\zeta_n}$ then we can abandon ζ_n and repeat the work on the interval $[\zeta_{n-2}, \zeta_{n-1})$.

Uniformity and simple permitting

All is well, except that even if we ensure that $\Gamma(\varrho)$ is total, we cannot effectively find an ω-computable approximation for $\Gamma(\varrho)$. We need to guess one. Let $\langle g^e\rangle$ be an enumeration of the ω-c.a. functions, equipped with tidy $(\omega+1)$-computable approximations $\langle g^e_s, o^e_s\rangle$ (Proposition 2.7). We perform countably many constructions which are almost independent of each other. The eth construction guesses that $\Gamma(\varrho) = g^e$, and based on this guess enumerates a prefix-free set C^e and a reduction of ϱ to C^e. If the guess is correct then the construction will succeed.

Since they enumerate distinct sets and reductions, there is very little interaction between the different constructions. However they do combine forces in defining $\Gamma(\varrho)$. To keep things simple, the eth construction defines $\Gamma(\varrho, n)$ for inputs $n \in \omega^{[e]}$ (the eth column of ω). The catch is that even if the guess that $\Gamma(\varrho) = g^e$ is incorrect, an eventual ϱ-correct definition of $\Gamma(\varrho, n)$ must be made by the eth construction, for all $n \in \omega^{[e]}$.

Even while waiting for an agreement between $\Gamma(\varrho, n)$ and $g^e(n)$, the eth construction can keep defining new values of markers ζ_m for $m > n$ in $\omega^{[e]}$, and with them computations $\Gamma(\varrho, m)$. If $\deg_T(\varrho)$ were array computable this would not be a problem. Recall that we need to ensure that the block ending with ζ_m must contain at least $o_s^e(m)$ many zeros (where s is the stage at which we make the definition). If we know that $\Gamma(\varrho)$ is say id-c.a., then we can work with a list of tidy (id+1)-computable approximations, and so $o_0^e(m) = m$ for all e and m, and we can find how large ζ_m must be. However under the weaker assumption that $\deg_T(\varrho)$ is totally ω-c.a., we need to work with what are essentially (if not formally) partial approximations. So the conflict is that we need to define ζ_m even if $o_s^e(m) = \omega$ for all s; so we cannot wait for a value $o_s^e(m) < \omega$ to show up. But if we define ζ_m before seeing $o_s^e(m) < \omega$ then we will not have enough zeros and will not be able to carry out the construction outlined above, *even if the eth guess is correct.*

The solution (as in [25]) is to make use of the fact that ϱ is noncomputable. We actually use simple permitting. This is perhaps paradoxical in an anti-permitting argument. But of course the point is that noncomputable (simple) permitting is weaker than non-total ω-c.a. permitting, and so the former can co-exist with the negation of the latter.

What we do is go ahead and define a computation $\Gamma(\varrho, m)$ without waiting for $o_s^e(m)$ to give us a natural number. But we wait with the definition of the reduction of ϱ to C^e (which is fine, as it is local to the eth construction). Once we see the value $o_s^e(m)$ we wait for a voluntary change in ϱ below the use $\gamma(m)$. Simple permitting will ensure that for infinitely many m we will see such changes (provided of course that the approximation is eventually ω-computable). If we see such a change then we can now define a new large value for $\gamma(m)$, bounding sufficiently many zeros, and declare it to be one of our markers ζ_m. Note again that to move ζ_m we need not only a ϱ-change below ζ_m, but also a change in C^e on strings of length below ζ_m, *if the reduction of $\varrho\restriction_{\zeta_{m-1}}$ to C^e has already been defined.* This is why it is important to keep this reduction undefined until we see $o_s^e(m) < \omega$.

This discussion contained all the ideas needed for the proof, and so we turn to giving the formal details.

5.3.2 Total ω-c.a. anti-permitting: the details

As discussed, we are given a noncomputable left-c.e. real $\varrho \in [0, 1)$ with an increasing approximation $\langle \varrho_s \rangle$. We use a list $\langle g^e \rangle$ of all ω-c.a. functions, with tidy $(\omega + 1)$-computable approximations $\langle g_s^e, o_s^e \rangle$.

We enumerate a Turing functional Γ, with intended oracle ϱ, viewed as an element of Cantor space.

For every $e < \omega$ we perform the eth construction. These constructions are independent of each other. Fix some $e < \omega$. In the eth construction we enumerate a prefix-free c.e. set C^e and define $\Gamma(\varrho, m)$ for all $m \in \omega^{[e]}$. Also, we define an increasing sequence of numbers $k^e(0) < k^e(1) < \ldots$ (the list may eventually be finite or infinite). All of the numbers $k^e(n)$ are elements of $\omega^{[e]}$. These will be the numbers that are permitted (simply) and so they will be the ones that will

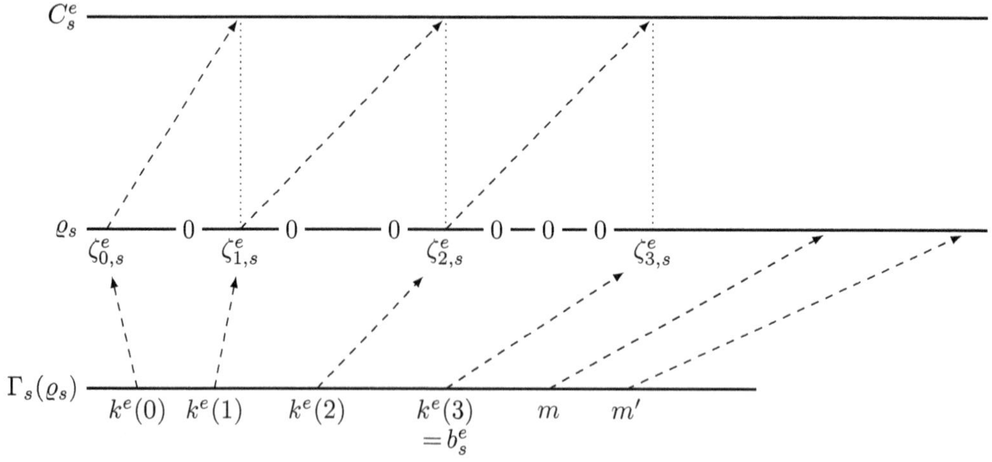

Figure 5.3. The eth construction at stage s. In this example $v_s^e = 3$, and $\Gamma_s(\varrho_s)$ is also defined on $m' > m > b_s^e = k^e(3)$, the two next elements in $\omega^{[e]}$.

be used as inputs for defining the layers. We renumber our markers by letting $\zeta_n^e = \gamma(k^e(n))$.

The beginning of stage s

By the beginning of a stage s we will have already:

(1) enumerated the set C_s^e;
(2) defined the sequence $k^e(0), k^e(1), \ldots, k^e(v)$ for some $v = v_s^e$, such that each $k^e(n) \in \omega^{[e]} \cap s$; for brevity we let $b_s^e = k^e(v_s^e)$ be the last element of this sequence;
(3) defined computations $\Gamma_s(\varrho_s, m)$ for all $m \in \omega^{[e]} \cap s$, with uses $\gamma_s(m)$; for $n \leqslant v_s^e$ we let $\zeta_{n,s}^e = \gamma_s(k^e(n))$.

The uses $\gamma_s(m)$ are not quite monotone:

- If $k^e(n-1) < m < k^e(n)$ for some $n \leqslant v_s^e$ then $\gamma_s(m) = 0$. That is, the computation $\Gamma_s(\varrho_s, m)$ does not look at the oracle and so is never destroyed. These inputs m were discarded when we got permission to use $k^e(n)$ to define the next layer ending with $\zeta_{n,s}^e$.
- Otherwise, the uses are monotone:
  - $\zeta_{n,s}^e = \gamma_s(k^e(n)) < \gamma_s(k^e(n+1)) = \zeta_{n+1,s}^e$ for all $n < v_s^e$;
  - if $n \leqslant v_s^e$ and $m > b_s^e$ then $\zeta_{n,s}^e < \gamma_s(m)$;
  - if $b_s^e < m < m'$ then $\gamma_s(m) < \gamma_s(m')$.

See Figure 5.3.

The eth *construction*

The construction begins at stage $s = \min \omega^{[e]}$. At that stage we define $k^e(0) = s$ and $C^e_{s+1} = \varnothing$. We define a new computation $\Gamma_{s+1}(\varrho_{s+1}, s) = 0$ with use 1. So $\zeta^e_{0,s+1} = 1$. Recall our convention that $\varrho = 0.\varrho(1)\varrho(2)\cdots$. This means that $\varrho\restriction_k$ is the bit-sequence $\varrho(1)\varrho(2)\cdots\varrho(k-1)$. If we define a computation with use 1 this means that the oracle is not consulted and so this computation is never destroyed.

Now suppose that $s > \min \omega^{[e]}$. We give the instructions for the eth construction at stage s.

STEP 1: REDEFINING DESTROYED COMPUTATIONS $\Gamma(\varrho, m)$.

We may see that some of the computations $\Gamma_s(\varrho_s, m)$ are destroyed by the change from ϱ_s to ϱ_{s+1}. If none of these computations are destroyed then we skip to step 2 below.

Otherwise we need to define new computations $\Gamma_{s+1}(\varrho_{s+1}, m)$ for m for which the computations were destroyed. In all but one case the value of the new computations will be $s+1$, and so to define these computations we only need to specify their use $\gamma_{s+1}(m)$.

Let p be the smallest element of $\omega^{[e]}$ such that $p < s$ and $\Gamma_s(\varrho_{s+1}, p)\uparrow$. There are three cases.

FIRST CASE. USELESS CHANGE: $p > b^e_s$ BUT $o^e_s(p) = \omega$.

For all $m \in [p, s) \cap \omega^{[e]}$ set $\gamma_{s+1}(m) = \gamma_s(m)$. We don't increase the uses, to ensure that they do not go to infinity.

SECOND CASE. MAKING USE OF SIMPLE PERMISSION: $p > b^e_s$ AND $o^e_s(p) < \omega$.

In this case we add p as the new last element of the list of useful inputs. That is, we define $k^e(n) = p$ where $n = v^e_s + 1 = v^e_{s+1}$; so $p = b^e_{s+1}$.

- For $m \in (b^e_s, b^e_{s+1}) \cap \omega^{[e]}$ define $\Gamma_{s+1}(\varrho_{s+1}, m) = \Gamma_s(\varrho_s, m)$ with use 1. We use the previous value to keep the functional consistent.
- Set $\gamma_{s+1}(p)$ (which of course equals $\zeta^e_{n,s+1}$) to be the least $u > \zeta^e_{n,s} + 1$ such that the block $\varrho_{s+1}\restriction_{[\zeta^e_{n-1,s}, u)}$ contains at least $o^e_s(p) + 2$ many zeros.
- For $m \in (p, s) \cap \omega^{[e]}$ set $\gamma_{s+1}(m)$ to be large.

THIRD AND MAIN CASE: $p = k^e(q)$ FOR SOME $q \leqslant v^e_s$.

Let $n \in [q, v^e_s]$. Let $\beta = \varrho_{s+1} - \lambda(C^e_s)$.

- If $\beta \leqslant 2^{-\zeta^e_{n,s}}$ then set $\zeta^e_{n,s+1} = \zeta^e_{n,s}$ (in other words, set $\gamma_{s+1}(k^e(n)) = \gamma_s(k^e(n))$).
- If $\beta > 2^{-\zeta^e_{n,s}}$ then set $\zeta^e_{n,s+1}$ to be the least possible value greater than $\zeta^e_{n-1,s} + 1$ so that the block $\varrho\restriction_{[\zeta^e_{n-1}, \zeta^e_n)}[s+1]$ contains at least $o^e_s(n) + 2$ many zeros.

As in the second case, for all $m \in (b^e_s, s) \cap \omega^{[e]}$ set $\gamma_{s+1}(m)$ to be large.

This defines the computations $\Gamma_{s+1}(\varrho_{s+1}, m)$ for all $m < s$ in $\omega^{[e]}$ and concludes the first step.

STEP 2: UPDATING C^e.

Let $\beta = \varrho_{s+1} - \lambda(C^e_s)$. Suppose that $\beta > 0$ and that for all $n \leqslant v^e_s$, $n > 0$ such that $\beta \leqslant 2^{-\zeta^e_{n-1,s}}$ we have $\Gamma(\varrho, k^e(n)) = g^e(n)\,[s+1]$. Then enumerate strings into C^e_{s+1} following the algorithm above to ensure that $\lambda(C^e_{s+1}) = \varrho_{s+1}$.

STEP 3: A NEW COMPUTATION.

At the very end of the stage, if $s \in \omega^{[e]}$ we define a new computation Γ_{s+1} (ϱ_{s+1}, s) with new, large use.

This concludes the instructions for stage $s > \min \omega^{[e]}$.

Verification

Each functional Γ_s is consistent for ϱ_s. This uses the fact that $\langle \varrho_s \rangle$ is an increasing approximation, and that at every stage we define a new computation $\Gamma_{s+1}(\varrho_{s+1}, m)$ only if $\Gamma_s(\varrho_{s+1}, m)\uparrow$, or otherwise we let $\Gamma_{s+1}(\varrho_{s+1}, m) = \Gamma_s(\varrho_s, m)$.

Lemma 5.14. $\Gamma(\varrho)$ *is total.*

Proof. Let $m < \omega$; let e be such that $m \in \omega^{[e]}$. Let $v^e = \sup_s v^e_s$.

If there is some $n < v^e$ such that $k^e(n) < m < k^e(n+1)$ then at the stage at which $k^e(n+1)$ is defined we define a new computation $\Gamma(m)$ with use 1. Recall that this means that the oracle is not consulted. So certainly $\Gamma(\varrho, m)\downarrow$.

Suppose that $m = k^e(n)$ for some $n \leqslant v^e$ or that $v^e < \omega$ and $m > b^e = k^e(v^e)$. By induction on such m we show that $\Gamma(\varrho, m)\downarrow$. For every $s > m$ the computation $\Gamma_s(\varrho_s, m)$ converges. To show that $\Gamma(\varrho, m)\downarrow$ it is sufficient to show that the sequence $\langle \gamma_s(m) \rangle$ is bounded. For if it is bounded by some value u and $\varrho_s\restriction_u = \varrho\restriction_u$, then $\Gamma_s(\varrho_s, m)$ is an ϱ-correct computation.

First suppose that $m = k^e(n)$ for some n. If $n = 0$ then $\gamma_{m+1}(m) = 0$ which implies that the computation $\Gamma_{m+1}(\varrho_{m+1}, m)$ is ϱ-correct and so is never destroyed. Suppose that $n > 0$. By induction we assume that $\zeta^e_{n-1,s}$ reaches a limit ζ^e_{n-1}. Let $r > m$ be a stage sufficiently late so that $n \leqslant v^e_r$ and $\zeta^e_{n-1,s} = \zeta^e_{n-1}$ for all $s \geqslant r$. We note that the fact that $n \leqslant v^e_r$ implies that $o^e_r(m) < \omega$. Let u be the least number greater than ζ^e_{n-1} such that the block $\varrho\restriction_{[\zeta^e_{n-1}, u)}$ contains at least $o^e_r(m)$ many zeros; such a number exists since ϱ is not a dyadic rational. By increasing r we may assume that $\varrho_r\restriction_u = \varrho\restriction_u$ (and so $\varrho_s\restriction_u = \varrho\restriction_u$ for all $s \geqslant r$). If $s \geqslant r$ is a stage at which $\gamma_{s+1}(m)$ is redefined then we choose $\gamma_{s+1}(m) \leqslant u$.

Now suppose that $v^e < \omega$ and $m > b^e$. Let m' be m's predecessor in $\omega^{[e]}$. By induction find a stage $r > m$ sufficiently late so that the computation $\Gamma_r(\varrho_r, m')$

is ϱ-correct. At every stage $s \geqslant r$ at which we redefine $\gamma_{s+1}(m)$ we let $\gamma_{s+1}(m) = \gamma_s(m)$. □

Since we assume that $\deg_{\mathrm{T}}(\varrho)$ is totally ω-c.a. there is some e such that $\Gamma(\varrho) = g^e$ and the approximation $\langle g_s^e, o_s^e \rangle$ is eventually ω-computable. We fix such e. From now we only concern ourselves with the eth construction. For clarity of notation we omit the superscript e from all the associated objects (we write g_s for g_s^e, C for C^e, $\zeta_{n,s}$ for $\zeta_{n,s}^e$, and so on).

Lemma 5.15. $\lim_s v_s = \omega$.

Proof. Assume for a contradiction that $v = \lim_s v_s$ is finite. Let r be a stage sufficiently late so that by stage r, $\zeta_{v,s}$ has reached a limit ζ_v and $\varrho_r \restriction_{\zeta_v} = \varrho \restriction_{\zeta_v}$. The assumption for contradiction means that at all stages $s > r$, for all $m \in \omega^{[e]} \cap s$ such that $o_s(m) < \omega$, the computation $\Gamma_s(\varrho_s, m)$ is ϱ-correct. This implies that ϱ is computable. Given $u < \omega$, to compute $\varrho \restriction_u$ we pick $m > u, r$ in $\omega^{[e]}$ and wait for a stage $s > m$ at which $o_s(m) < \omega$; so $\varrho_s \restriction_{\gamma_s(m)} = \varrho \restriction_{\gamma_s(m)}$. But $\gamma_s(m) > m > u$. □

We can show that C is a presentation of ϱ.

Lemma 5.16. $\lambda(C) = \varrho$.

Proof. Suppose not. Let n be sufficiently large so that $2^{-n} < \varrho - \lambda(C)$. But if s is a very late stage then all markers $\zeta_{m,s}$ for all $m \leqslant n$ have stabilised to their final values and are all certified: $g_s(k(m)) = \Gamma(\varrho, k(m))$ for all $m \leqslant n$. Also assume that $\varrho - \varrho_s < 2^{-n}$ and $\varrho_s - \lambda(C_s) > 2^{-n}$. Then at stage s we would increase C to have measure ϱ_s, which is a contradiction. □

The next lemma (really an observation) is trivial but useful. Both parts rely on the fact that for all $\beta \in [0, 1)$ and $k \geqslant 1$, $\beta \restriction_k$ (as a number in binary) is the integral part of $2^{k-1}\beta$.

Lemma 5.17. *Let $t < s$ and $k \geqslant 1$.*

(1) If $\varrho_s - \varrho_t \geqslant 2^{-(k-1)}$ then $\varrho_t \restriction_k \neq \varrho_s \restriction_k$.
(2) If $\varrho_s - \varrho_t \leqslant 2^{-k}$ and further $\varrho_t(k) = 0$ then $\varrho_t \restriction_k = \varrho_s \restriction_k$.

The following is the main combinatorial lemma.

Lemma 5.18. *Let s be a stage, and let $n > 0$, $n \leqslant v_s$. The block $\varrho \restriction_{[\zeta_{n-1}, \zeta_n)} [s]$ contains a zero.*

Proof. Fix n. For brevity let $m = k(n)$. Suppose that s is a stage and $n \leqslant v_s$.

As above, say that the marker $\zeta_{n,s}$ is *certified* at stage s if $\Gamma_s(\varrho_s, k(n)) = g_s(k(n))$. Let $S_{\texttt{cert}}$ be the set of such stages. This set contains a final segment of ω.

We say that the marker $\zeta_{n,s}$ is *redefined* if $\Gamma_s(\varrho_{s-1}, m)\uparrow$ and either

- $v_{s-1} = n-1$, i.e., $\zeta_{n,s}$ is the very first value of this marker; or
- $\beta = \varrho_s - \lambda(C_{s-1}) > 2^{-\zeta_{n,s-1}}$.

Let $S_{\texttt{redef}}$ be the set of such stages s.

Let $S = S_{\texttt{cert}} \cup S_{\texttt{redef}}$. We show by induction on the stages s for which $n \leqslant v_s$ that:

(a) if $s \in S$ then the block $\varrho\restriction_{[\zeta_{n-1}, \zeta_n)}[s]$ contains at least $o_s(m)+2$ many zeros;
(b) if $s \notin S$ then the block $\varrho\restriction_{[\zeta_{n-1}, \zeta_n)}[s]$ contains at least $o_s(m)+1$ many zeros.

In either case the number is positive, and so the lemma follows.

The induction starts with $s = \min S_{\texttt{redef}}$. The instructions ensure that (a) holds at every stage $s \in S_{\texttt{redef}}$.

Let $t \in S$ and suppose that (a) has already been verified for stage t. Let r be the next stage in S after stage t. We verify that (a) holds at stage r and that (b) holds at all stages $s \in (t, r)$.

The marker $\zeta_{n,s}$ is constant for $s \in [t, r)$; we denote this fixed value by ζ_n (note that this is not necessarily the final value of this marker). Similarly define ζ_{n-1}.

Now for brevity let:

- A be the set of stages $u \in (t, r)$ such that $C_u \neq C_{u-1}$;
- if $r \in S_{\texttt{redef}}$ let B be the set of stages $s \in (t, r)$ such that $\varrho_s\restriction_{\zeta_n} \neq \varrho_{s-1}\restriction_{\zeta_n}$; if $r \notin S_{\texttt{redef}}$ let B be the set of such stages in the interval $(t, r]$.

We make two observations.

(1) Let $u \in A$. Then $\varrho_u - \lambda(C_{u-1})$ is strictly greater than $2^{-\zeta_{n-1}}$. This is because $u \notin S_{\texttt{cert}}$.
(2) Let $s \in B$. Then $\varrho_s - \lambda(C_{s-1}) \leqslant 2^{-\zeta_n}$—for otherwise $s \in S_{\texttt{redef}}$.

In particular, A and B are disjoint.

Suppose that B is empty. Then $\varrho_{r-1}\restriction_{\zeta_n} = \varrho_t\restriction_{\zeta_n}$; and if $r \notin S_{\texttt{redef}}$ then $\varrho_r\restriction_{\zeta_n} = \varrho_t\restriction_{\zeta_n}$. Since $o_s(m) \leqslant o_t(m)$ for all $s \in (t, r]$, we see that (b) holds for all $s \in (t, r)$. If $r \in S_{\texttt{redef}}$ then we already know that (a) holds at r. If $r \notin S_{\texttt{redef}}$ then the latter equality ensures that (b) holds at stage r.

We assume therefore that B is nonempty.

Suppose that A is nonempty. We claim that $A < B$. That is, there are no $s \in B$ and $u \in A$ with $s < u$. For a contradiction, suppose there are. By choosing a maximal s and then minimal u we can find $s \in B$, $u \in A$ such that $s < u$ but the interval (s, u) is disjoint from both A and B. Since $A \cap [s, u)$ is empty we see that $C_{s-1} = C_{u-1}$. Let $q = \lambda(C_{s-1})$; then $\varrho_s - q \leqslant 2^{-\zeta_n}$ and $\varrho_u - q > 2^{-\zeta_{n-1}}$. Since

$\zeta_n > \zeta_{n-1} + 1$, this means that $\varrho_u - \varrho_s > 2 \cdot 2^{-\zeta_n}$. By Lemma 5.17, $\varrho_u \restriction_{\zeta_n} \neq \varrho_s \restriction_{\zeta_n}$. This contradicts the assumption that $B \cap (s, u]$ is empty.

Thus, we let $t' = \max A$ if A is nonempty, and $t' = t$ otherwise. Then $\varrho_{t'} \restriction_{\zeta_n} = \varrho_t \restriction_{\zeta_n}$.

Let $r' = \max B$. Then $\varrho_{r'} \restriction_{\zeta_n} = \varrho_{r-1} \restriction_{\zeta_n}$; and if $r \notin S_{\texttt{redef}}$ then $\varrho_{r'} \restriction_{\zeta_n} = \varrho_r \restriction_{\zeta_n}$. Also we note that $C_{r'} = C_{t'}$ and so $\varrho_{r'} - \lambda(C_{t'}) \leqslant 2^{-\zeta_n}$.

Let k be the rightmost zero in the block $\varrho \restriction_{[\zeta_{n-1}, \zeta_n)} [t]$ – the greatest $k < \zeta_n$ such that $\varrho_t(k) = 0$. Such k exists by induction.

Since $\varrho_{r'} - \varrho_{t'} \leqslant 2^{-\zeta_n}$ and $\varrho_{t'}(k) = \varrho_t(k) = 0$, Lemma 5.17 says that $\varrho_{r'} \restriction_k = \varrho_{t'} \restriction_k$. Overall, we see that $\varrho_{r-1} \restriction_k = \varrho_t \restriction_k$; and if $r \notin S_{\texttt{redef}}$ then $\varrho_r \restriction_k = \varrho_t \restriction_k$.

The block $\varrho_t \restriction_{[\zeta_{n-1}, k)}$ contains at least $o_t(m) + 1$ many zeros. Since $o_s(m) \leqslant o_t(m)$ for all $s > t$, we see that (b) holds for all stages $s \in (t, r)$.

Now consider r. We may assume that $r \notin S_{\texttt{redef}}$. Then the argument above shows that the block $\varrho_r \restriction_{[\zeta_{n-1}, \zeta_n]}$ contains at least $o_t(m) + 1$ many zeros. Further, $\zeta_{n,r} = \zeta_n$ and $\zeta_{n-1,r} = \zeta_{n-1}$.

We assumed that $B \neq \varnothing$. Indeed, a new computation $\Gamma_{r'}(\varrho_{r'}, m)$ is defined and $\Gamma_r(\varrho_r, m) = \Gamma_{r'}(\varrho_{r'}, m) = r'$. Since $r \in S$ it must be that $r \in S_{\texttt{cert}}$. Thus $g_r(m) = r' > t > g_t(m)$. It follows that $o_r(m) < o_t(m)$, and so $o_r(m) + 2 \leqslant o_t(m) + 1$. This establishes (a) for stage r. □

Finally we show that C computes ϱ.

Lemma 5.19. *Let s be a stage and let $n < v_s$. Suppose that for all strings σ of length at most $\zeta_{n+1,s}$, $\sigma \in C$ if and only if $\sigma \in C_s$. Then $\varrho \restriction_{\zeta_{n,s}} = \varrho_s \restriction_{\zeta_{n,s}}$.*

Proof. Let s be a stage as described. The assumption means that for all $u > s$, if $C_u \neq C_{u-1}$ then $\lambda(C_u) - \lambda(C_{u-1}) < 2^{-\zeta_{n+1,s}}$.

For brevity let $\zeta_n = \zeta_{n,s}$ and $\zeta_{n+1} = \zeta_{n+1,s}$. By induction on $t \geqslant s$ we show that $\zeta_{n,t} = \zeta_n$, $\zeta_{n+1,t} = \zeta_{n+1}$ and $\varrho_t \restriction_{\zeta_n} = \varrho_s \restriction_{\zeta_n}$. Suppose this is known for $t - 1 \geqslant s$.

We claim that $\beta = \varrho_t - \lambda(C_{t-1}) \leqslant 2^{-\zeta_{n+1}}$. Suppose otherwise; let u be the least stage $u \geqslant t$ such that $C_u \neq C_{u-1}$. Then $\lambda(C_{u-1}) = \lambda(C_{t-1})$ and $\varrho_u \geqslant \varrho_t$ and so $\lambda(C_u) - \lambda(C_{u-1}) \geqslant \beta$, contradicting our assumption on C_s.

The instructions (third case) now show that at stage $t - 1$ we set $\zeta_{n,t} = \zeta_{n,t-1}$ and $\zeta_{n+1,t} = \zeta_{n+1,t-1}$.

Further, $\varrho_t \restriction_{\zeta_n} = \varrho_s \restriction_{\zeta_n}$. Since $\zeta_{n+1,t-1} = \zeta_{n+1}$ and $\zeta_{n,t-1} = \zeta_n$, Lemma 5.18 implies that the block $\varrho_{t-1} \restriction_{[\zeta_n, \zeta_{n+1})}$ contains a zero. If $\varrho_t \restriction_{\zeta_n} \neq \varrho_{t-1} \restriction_{\zeta_n}$ then by Lemma 5.17, $\varrho_t - \varrho_{t-1} > 2^{-\zeta_{n+1}}$, and of course $\varrho_{t-1} \geqslant \lambda(C_{t-1})$. □

Chapter Six

m-topped degrees

IT WAS POST [78] who first pointed out that many reducibilities occurring in practice were stronger than Turing reducibility; indeed most codings of the halting problem into a concrete undecidable problem like the word problem for groups were m-reducibilities. For example, for the word problem, for each instance e, we could compute a word $w(e)$ such that $e \in \varnothing' \Leftrightarrow w(e) = 1_G$. The thrust of Post's problem was whether all instances of undecidable c.e. problems were simply the halting problem in disguise. Myhill characterised the c.e. sets as those that are m-reducible to $\varnothing'$. Thus interactions of Turing and m-reducibilities would seem a natural thing to study.

Downey and Jockusch [29] answered a longstanding question of Odifreddi and Degtev (see Odifreddi [76]) by proving the existence of incomplete c.e. sets which resembled the Halting problem in the sense of these interactions. That is, they constructed what are now called *m-topped* degrees: degrees containing c.e. sets A such that for every c.e. set $B \leqslant_{\mathrm{T}} A$ we have $B \leqslant_m A$. In other words, a c.e. Turing degree $\mathbf{a}$ is m-topped if among the m-degrees of c.e. sets inside $\mathbf{a}$ there is a greatest one. Thus, *locally* they resemble $\varnothing'$. Such sets seem strange, and have some remarkable properties. For example, they were one of the first "natural classes" all of whose members are low_2. Moreover, Downey and Jockusch showed that m-topped degrees cannot be low. Finally, Downey and Shore [34] showed that every low_2 c.e. degree is bounded by an m-topped degree. Thus, the m-topped degrees c.e. degrees can be used to define the low_2 c.e. degrees in the degree structure with both reducibilities.

In [23] we investigated the dynamics required for the Downey-Jockusch construction. We showed that the cascading effect that happened in the construction led to an ω^ω-type behaviour. Specifically, we showed that there is an m-topped degree which is totally ω^ω-c.a. We also hinted at a proof that this is the best possible:

Theorem 6.1. *No m-topped degree is totally $<\omega^\omega$-c.a.*

In this chapter we flesh out the details of this construction. Apart from the intrinsic interest in this result, this argument will serve as a preparation for the next chapter.

The dynamics of the cascading phenomenon occurring in the construction of m-topped degrees strongly resemble the dynamics of the embedding of the

1-3-1 lattice in the c.e. degrees, which we discussed in the introduction. These dynamics are captured by the class of totally $<\omega^\omega$-c.a. degrees, in that, as we show in the next chapter, the 1-3-1 lattice can be embedded precisely below the not totally $<\omega^\omega$-c.a. degrees. Similar dynamics occurred in the original construction of a noncomputable left-c.e. real with only computable presentations, which we discussed in the previous chapter; this will be made more formal when we discuss prompt not total $<\omega^\omega$-c.a. permitting in the last chapter of this monograph. However, the similarity has some limits. Unlike the 1-3-1 embedding, the m-topped construction cannot be captured precisely by the hierarchy of totally $<\alpha$-c.a. degrees: it is *not* the case that every c.e. degree which is not totally $<\omega^\omega$-c.a. bounds an m-topped degree. This is because, as mentioned above, m-topped degrees cannot be low, and every level of our hierarchy contains low (as well as non-low) degrees. It would be interesting to see if there is a permitting argument combining non-total $<\omega^\omega$-c.a.-ness and non-lowness that would yield bounding of m-topped degrees.

Before we give the full argument we start with easier, weaker results. We show that no totally ω-c.a. degree is m-topped; then that no totally ω^2-c.a. degree is m-topped; and then give the full proof.

6.1 TOTALLY ω-C.A. DEGREES ARE NOT m-TOPPED

Let $\mathbf{d}$ be a totally ω-c.a. c.e. degree. To show that $\mathbf{d}$ is not m-topped we need, given a c.e. $D \in \mathbf{d}$, to enumerate some c.e. set $V \leqslant_T D$ which is not many-one reducible to D.

The basic module is as follows. Suppose that we want to show that the dth computable function φ_d is not a many-one reduction of V to D. We set up a finite set X of followers and wait for them to be realised, which means that $\varphi_d(x)\downarrow$ for all $x \in X$. While we wait we prevent the enumeration of the followers into V. When they get realised we may assume that $\varphi_d(x) \notin D_s$ for all $x \in X$; otherwise we get an easy win. We then attack by enumerating some $x \in X$ into V. The opponent can respond by enumerating $\varphi_d(x)$ into D, in which case we will attack with another follower in X. We need to ensure two things:

- V is Turing reducible to D; and
- X is sufficiently large so that the opponent cannot always respond.

For the first we will define a functional Ψ with the intention of having $\Psi(D) = V$. To be able to attack without violating this reduction we will ensure that the use $\psi_s(x)$ of any follower is greater than $\varphi_d(y)$ for any other follower. Thus a response by our opponent to our attack with y will be the D-change which allows us to attack next with x.

For the second we use the "anti-permitting" method used in Chapter 5. We tie the set of followers X with some input n for a function $\Gamma(D)$ we build which will serve as an "anchor" (or "anti-permitting number"). Since $\Gamma(D)$ is ω-c.a. we find a bound m on the number of times an approximation for $\Gamma(D, n)$

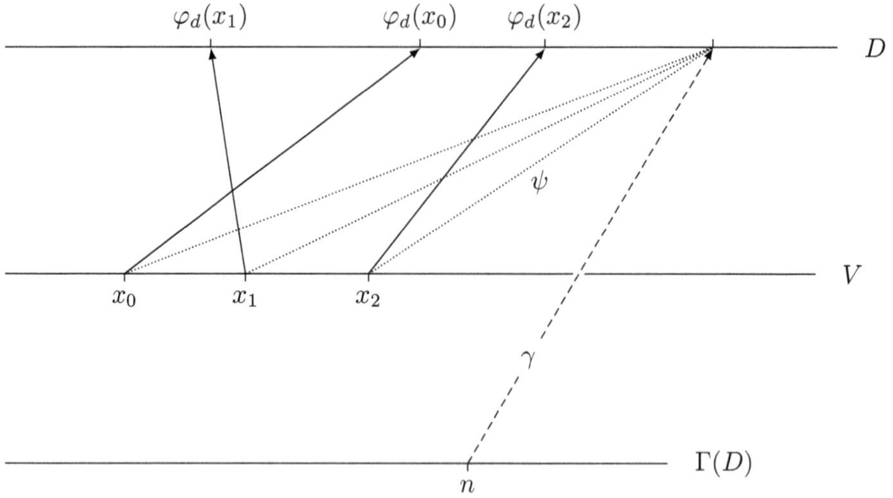

Figure 6.1. ω-c.a. degrees are not m-topped.

changes. We ensure that the use $\gamma(n)$ of $\Gamma(D,n)$ is the same as the use $\psi_s(x)$ for followers $x \in X$. So the opponent's D-change that allows us to attack with another follower also allows us to redefine $\Gamma(D,n)$ to have a new value and so reduce the number of changes left to the opponent. If $|X| > m$ then the opponent will not be able to always respond. See Figure 6.1.

As in the previous chapter we need to add a simple permitting step. Previously this was only necessary because we were working with a degree which is totally ω-c.a. and not necessarily an array computable one: the number m is revealed to us eventually but is not fixed in advance; if we guess incorrectly about our approximation for $\Gamma(D)$ it may never be given. We nonetheless must make sure that $\Gamma(D,n)\downarrow$ (so that $\Gamma(D)$ is total) even if the guess using n is wrong. In the current construction there is another reason to use simple permitting. We do not know whether φ_d is total or not. This means that we need to set the uses $\psi_s(x)$ for $x \in X$ immediately when we appoint these followers. Before we attack we need to lift these uses beyond $\varphi_d(y)$ for $y \in X$, and these values are revealed to us after we already appoint the followers and define the Ψ-computations. So we wait for a "free pass" to raise these markers, and this will be given as usual by assuming that D is noncomputable.

6.1.1 Construction

We are given a c.e. set D whose Turing degree is totally ω-c.a. We use a list $\langle g^e \rangle$ of all ω-c.a. functions, with tidy $(\omega+1)$-computable approximations $\langle g^e_s, o^e_s \rangle$. We enumerate a Turing functional Γ with intended oracle D.

For every $e < \omega$ we perform an eth construction. These constructions are independent of each other, except that as usual they together define the functional Γ.

The eth construction will guess that $\Gamma(D)=g^e$. For every $d<\omega$ the eth construction will employ an *agent* d, which performs a "sub-construction" of the eth construction. The action of distinct agents is independent of each other; we only need to ensure that they don't share followers.[1]

The eth construction will enumerate a c.e. set V^e. It also defines a Turing functional Ψ^e with the aim of having $\Psi^e(D)=V^e$.

An agent d for construction e aims to define a finite set X of followers. The sets of followers for distinct agents are pairwise disjoint. The agent will choose an *anchor* n (distinct from the numbers chosen by any other agent for any construction). The agent will be responsible for defining $\Gamma(D,n)$ and for defining $\Psi^e_s(D_s,x)$ for $x\in X$. The use $\psi^e_s(x)$ for all $x\in X$ will be the same, namely $\gamma_s(n)$.

We note that the agent must ensure that $n\in\operatorname{dom}\Gamma_s(D_s)$ at every stage s (and that the uses $\gamma_s(n)$ are bounded). However $\Psi^e(D)=V^e$ is required only if the hypothesis that $\Gamma(D)=g^e$ is correct. The agent is thus allowed to leave computations $\Psi^e_s(D_s,x)$ undefined until it gets further evidence that the hypothesis holds.

In this chapter we simplify our notation as follows.

Notation 6.2. The intended oracle for the functionals Γ and Ψ^e is D; at stage s we only define computations $\Gamma_s(D_s,n)$ and $\Psi^e_s(D_s,x)$. Further, the value of these computations is also fixed: at stage s, the value of a new $\Gamma_s(D_s,n)$ computation is always s, and the value of a new $\Psi^e_s(D_s,x)$ computation is $V^e_s(x)$. Thus to specify a computation all we need to provide is the use $\gamma_s(n)$ or $\psi^e_s(x)$. Instead of mentioning the functionals we only mention the uses (which can be thought of as moving markers). So for example we write $\psi^e_s(x)\downarrow$ if $\Psi^e_s(D_s,x)\downarrow$, and when a new computation is defined, we simply say that we define $\psi^e_s(x)$.

As mentioned above, before we can use any followers to diagonalise against many-one reductions we need them to be simply permitted by D. Thus before commencing the attacks, the agent will define distinct sets of followers $X_0,X_1,\dots$ associated with anchors $n_0,n_1,\dots$, one of which we hope will become the X and n we eventually use.

To carry out the construction we need the following, which we will verify after we specify the construction. It says that an agent does not run out of followers to attack with.

Lemma 6.3. *Suppose that at some stage s, an agent d for the eth construction is attacking with a set of followers X. Then $X\not\subseteq V^e_s$.*

[1] We use the term "agent" to refer to entities working independently of each other in parallel constructions; "strategies" or "nodes" lie on a tree of strategies and interact with each other.

The action of agent d for the eth *construction*

We now describe two cycles (subroutines) detailing the action of an agent d for the eth construction. The agent starts with set-up cycles; if some set of followers is set up and permitted then the agent moves to attack cycles. During either cycle the agent is instructed to wait for some event. It is possible that the event does not happen, in which case the agent will wait forever and not act again, other than maintaining the convergence of some functionals. In fact we will show that either we get an easy win, or the agent will get stuck waiting indefinitely from some point onwards, either because g^e is not the correct guess, φ_d is not total, or because some attack succeeds.

The agent starts with setting up the first set of followers.

SETTING UP THE kth SET OF FOLLOWERS.

1. Let s_0 be the stage at which this set-up cycle begins. Choose a large anchor n_k. Define $\gamma_{s_0}(n_k) = n_k$.

2. We wait for a stage s_1 at which $o^e_{s_1}(n_k) < \omega$. At that stage we choose a set X_k of $(o^e_{s_1}(n_k)+2)$-many large followers. For each $x \in X_k$ we define $\psi^e_{s_1}(x) = n_k$.

3. We wait for a stage $s_2 > s_1$ at which $\varphi_{d,s_2}(x)\downarrow$ for all $x \in X_k$.

4. We then wait for a stage $s_3 > s_2$ at which $D_{s_3}\restriction_{n_k} \neq D_{s_3-1}\restriction_{n_k}$. While waiting we (recursively) set up the $(k+1)^{\text{th}}$ set of followers.

When such a stage s_3 is found, we interrupt all set-up cycles. We discard all anchors $n_{k'}$ and sets of followers $X_{k'}$ for $k' \neq k$. We let $X = X_k$ and $n = n_k$. We let $u = 1 + \max\{\varphi_d(x) : x \in X\}$. We start an attack with some $x \in X$.

Throughout the set-up phase, if some anchor n_k is already chosen and $D_s\restriction_{n_k} \neq D_{s-1}\restriction_{n_k}$ then unless we start an attack at stage s, we redefine $\gamma_s(n_k) = n_k$ and if also X_k is defined, $\psi^e_s(z) = n_k$ for all $z \in X_k$.

If we start an attack at some stage t then we will ensure that $\Gamma_{t-1}(D_t, n)\uparrow$ and that $\Psi^e_{t-1}(D_t, z)\uparrow$ for all $z \in X$.

ATTACKING WITH A FOLLOWER x.

1. Let t_0 be the stage at which the attack begins. We define a new Γ computation by setting $\gamma_{t_0}(n) = u$.

2. We wait for a stage $t_1 > t_0$ at which $g^e_{t_1}(n) = \Gamma_{t_1}(D_{t_1}, n)$. While waiting, the markers $\psi^e_s(z)$ for all $z \in X$ remain undefined.

If $\varphi_d(x) \in D_{t_1}$ then we interrupt the attack cycle and discard both n and X; all action for the agent ceases. In this case we get an easy win by keeping x out of V^e.

Otherwise, we enumerate x into $V^e_{t_1}$; we define $\psi^e_{t_1}(z) = u$ for all $z \in X$.

3. We wait for a stage $t_2 > t_1$ at which $\varphi_d(x) \in D_{t_2}$. At that stage we end the current attack and commence a new attack with some $x' \in X \setminus V^e_{t_2}$.

Throughout the attack phase, if $D_s \restriction_u \neq D_{s-1} \restriction_u$ and we do not start a new attack at stage s then we define $\gamma_s(n) = u$, and if further $s > t_1$, then we define $\psi^e_s(z) = u$ for $z \in X$.

Globally, if $n < s$ and n is at stage s not used as anchor by any agent for any construction (either it was never chosen, or was chosen and later discarded) then we define $\gamma(n) = 0$. For all $e < s$, if $x < s$ and x is not at stage s used as a follower by any agent for the eth construction then we define $\psi^e(x) = 0$.

6.1.2 Verification

We first need to show that the construction can be performed as described. Fix an agent d for the eth construction.

Let t be a stage at which an attack cycle begins. We need to show that $\Gamma_{t-1}(D_t, n)\uparrow$ and that $\Psi^e_{t-1}(D_t, z)\uparrow$ for all $z \in X$. Suppose that the set-up phase ended at stage t. Then $D_t \restriction_n \neq D_{t-1} \restriction_n$ and n equals both $\gamma_{t-1}(n)$ and $\psi^e_{t-1}(z)$ for $z \in X$. If on the other hand an attack cycle (with some follower x) ends at stage t then $\varphi_d(x) \in D_t \setminus D_{t-1}$ and $\varphi_d(x) < u$, and u equals both $\gamma_{t-1}(n)$ and $\psi^e_{t-1}(z)$ for $z \in X$.

Next, we prove Lemma 6.3, which stated that an agent never runs out of followers: if an agent for the eth construction is attacking at some stage s with a set of followers X, then $X \nsubseteq V^e_s$.

Proof of Lemma 6.3. During each attack cycle at most one follower is enumerated into D. Let $t < s$ be two stages at which an attack cycle begins. Since $g^e_r(n) = \Gamma_t(D_r, n) \geqslant t$ at some stage $r \in (t, s)$ and by convention $g^e_t(n) < t$ we see that $o^e_s(n) < o^e_t(n)$. It follows that at most $o_{s_0}(n) + 1$ many attack cycles are started, where s_0 is the stage at which X is appointed. Thus at most $o_{s_0}(n) + 1$ many elements of X are enumerated into V^e. The lemma follows from the choice $|X| = o_{s_0}(n) + 2$. ☐

We also observe that $\Gamma(D)$ is total. Let $n < \omega$. If n is not chosen as an anchor by any agent for any construction, or is chosen but is later discarded, then we arranged that $n \in \operatorname{dom} \Gamma(D)$ (with use 0). Otherwise $n = n_k$ for some unique agent for a unique construction. If the agent never enters the attack phase then $\gamma_s(n)$ is defined at every stage after n is chosen, always with use n, and so eventually a correct computation is defined. If the agent enters the attack phase with n then at every stage s during this phase the computation $\gamma_s(n)$ is defined, with use u; so again a correct computation is eventually defined.

We fix some e such that $\Gamma(D) = g^e$ and $\langle f^e_s, o^e_s \rangle$ is eventually ω-computable. We will show that the eth construction succeeds. We drop all superscripts e from now on.

Lemma 6.4. $\Psi(D) = V$.

Proof. Let $x < \omega$. If x is enumerated into V at some stage t then $\Psi_{t-1}(D_t, x)\uparrow$ and a computation with a correct value is defined at stage s. So it suffices to show that $x \in \operatorname{dom} \Psi(D)$.

If x is never chosen as a follower by any agent for the eth construction, or if it is chosen and later discarded, then we arrange that $\Psi(D, x)\downarrow$ with use 0. Suppose that x is chosen by some agent d and is never discarded.

During the set-up phase we ensure that $\psi_s(x)\downarrow$ at every stage after the stage at which x was appointed, with use n_k (if $x \in X_k$). As with $\Gamma(D)$, if the attack phase never begins then this ensures that $n \in \operatorname{dom} \Psi(D)$.

Suppose that the attack phase eventually begins and that $x \in X$. Suppose that s is a stage during the attack phase and that $\psi_s(x)\uparrow$. Let $t \leqslant s$ be the stage at which the attack cycle began which is running at stage s. At stage s we are still waiting to see $g_r(n) = \Gamma_r(D_r, n)$. Since we assume that x is never discarded, the attack phase is never interrupted. Since $g = \Gamma(D)$ we see that a stage r as required will occur, and at that stage we will define $\psi_r(x) = u$. Again we see that eventually a correct computation will be defined. □

Lemma 6.5. $V \not\leqslant_m D$.

Proof. Suppose that φ_d is total; we show that there is some x such that $x \in V \Leftrightarrow \varphi_d(x) \notin D$.

We claim that agent d will enter the attack phase. Otherwise, the fact that φ_d is total and that $\langle g_s^e, o_s^e \rangle$ is eventually ω-computable ensures that anchors n_k are defined for every $k < \omega$. But then we compute D: if X_k is appointed and $X_k \subseteq \operatorname{dom} \varphi_d$ at stage s, then $D_s \restriction_{n_k}$ is correct.

We have argued that only finitely many attack cycles are started by the agent. Let x be the last follower with which we start an attack. If the attack is interrupted then $\varphi_d(x) \in D$ but we keep $x \notin V$. Otherwise, as argued above, we eventually enumerate x into D. Since no new attack is ever started, $\varphi_d(x) \notin D$. □

6.2 TOTALLY ω^2-C.A. DEGREES ARE NOT m-TOPPED

6.2.1 An easy proof

Consider how the construction in the previous section needs to change if $\deg_{\mathrm{T}}(D)$ is totally ω^2-c.a. In this case the ordinal $o_{s_0}(n)$ that we discover is not a natural number m but an ordinal of the form $\omega \cdot m_0 + m_1$, where m_0 and m_1 are natural numbers.

The most natural adaptation is the following. When the ordinal $\omega \cdot m + k$ is revealed, we appoint a set X of followers of size $k + 1$. We wait for φ_d to converge on the followers in X and then for permission to lift the uses $\gamma_s(n) = \psi_s^e(z)$ (for $z \in X$) above the values of $\varphi_d(z)$ for $z \in X$. When permission is granted we attack as above; but it is possible that eventually we exhaust all the followers in X. But when that happens, since $|X| > k$, the ordinal we see when X is exhausted is $\omega \cdot m' + k'$, with $m' < m$: we dropped below the limit ordinal $\omega \cdot m$. We then would like to repeat the process: appoint a new set X' of followers of size $k' + 1$; wait for φ_d to converge on X', and then for permission to lift $\gamma_s(n) = \psi_s^e(z)$ above the values of φ_d; and then attack again. We can go through at most m many cycles of attacks, and so eventually the opponent will not be able to respond.

The only question is why we would get enough permissions. Simple permitting is insufficient here; we need multiple permitting for each attempt to meet the requirement. But it is hopefully clear that the kind of permitting which we need to carry this plan out is non-total ω-c.a. permitting. That is, if we assume that $\deg_{\mathrm{T}}(D)$ is totally ω^2-c.a. but not totally ω-c.a. then this argument will actually work. If $\deg_{\mathrm{T}}(D)$ does happen to be totally ω-c.a. then we just refer to the construction in the previous section.

We can also generalise this argument to $n > 2$; it shows that every c.e. degree which is totally ω^n-c.a. is not m-topped. This approach however does not seem to work when we consider degrees which are totally $< \omega^\omega$-c.a. but not totally ω^n-c.a. for any n (see Theorem 3.25). In that argument we define a single function $\Gamma(D)$ and guess some n such that $\Gamma(D)$ is ω^n-c.a., and guess an appropriate approximation. However, for the permitting part of the argument we cannot just guess some function $\Theta(D)$ which is not ω^{n-1}-c.a.: the point is that to set $\gamma(m)$ in the first place we need $\theta(k)$ where k is the associated permitting number; if $\Theta(D, k)$ never converges then we will fail to make $\Gamma(D)$ total.

We thus give even for the case $n = 2$ a more complicated argument which we will be able to generalise to give the full result.

Nonuniformity

Rather than hope for a voluntary D-change, we manufacture it by using more than one set. Returning to the ω^2 case, suppose that we enumerate two c.e. sets V and W. It suffices to ensure for every pair (c, d) of indices that either φ_d is not a many-one reduction of V to D or φ_c is not a many-one reduction of W to D. The rough idea is to use two sets of followers Y and X. We associate an anchor n with the requirement; if we guess that $\Gamma(D, n)$ will not change more than $\omega \cdot m + k$ many times then we set $|Y| > m$ and $|X| > k$. We attack with the followers $x \in X$ against φ_d (and so enumerate them into V). When X runs out, as discussed above, the new ordinal is smaller than $\omega \cdot m$; we then attack with one follower $y \in Y$ against φ_c (and so aim to enumerate it into W). Before the attack with y commences we appoint a new set of followers to take the role of the new X, sufficiently large to last until we drop below the next limit ordinal. We wait for realisation of the new followers and then attack with y. The failure of this attack will give us the D-change that allows us to lift the new $\Gamma(D)$-use (and $\Psi^e(D)$-use for computing V from D) beyond $\varphi_d(x)$ for all x in the new X.

While we wait for the realisation of the new followers we must leave open the reduction of W to D (in the same way that in the ω-construction, while we wait

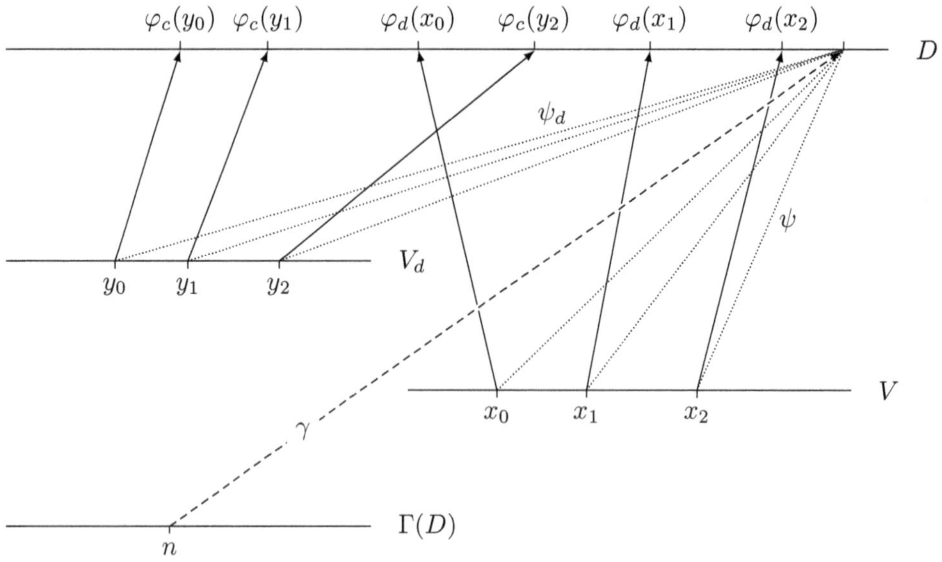

Figure 6.2. ω^2-c.a. degrees are not m-topped.

for a new agreement between g^e and $\Gamma(D)$ to appear, we leave the reduction of V to D open). This means that the totality of the reduction of W to D must rely on the totality of φ_d. We thus enumerate not a single set W but infinitely many, one for each φ_d, and we rename the sets V_d. Assume that the guess g^e is correct. Then we will in any case ensure that $V \leqslant_T D$; and if φ_d is a many-one reduction of V to D, then we will ensure that $V_d \leqslant_T D$ and that it is not many-one reducible to D. See Figure 6.2.

6.2.2 Construction

We are given a c.e. set D whose Turing degree is totally ω^2-c.a. We use a list $\langle g^e \rangle$ of all ω^2-c.a. functions, with tidy (ω^2+1)-computable approximations $\langle g^e_s, o^e_s \rangle$. We enumerate a Turing functional Γ with intended oracle D.

For every $e<\omega$ we perform an eth construction. As above, these constructions are independent of each other. For every pair (d,c) of natural numbers, the eth construction will employ an agent (d,c). The action of distinct agents is independent of each other; we only need to ensure that they don't share followers.

The eth construction will enumerate a c.e. set V^e, and for all $d<\omega$, a c.e. set V^e_d. It also defines a Turing functional Ψ^e with the aim of having $\Psi^e(D)=V^e$, and Turing functionals Ψ^e_d with the aim of having $\Psi^e_d(D)=V^e_d$. We continue to follow Notation 6.2 and mostly refer to the uses of these computations.

As discussed, an agent (d,c) for the eth construction plans to set up sets of followers Y and X. Once it enters the attack phase, the set Y is fixed, but the set X is not: once the followers in X are exhausted, we attack with another follower from Y and appoint a new set of followers to play the role of X. While it is not precise, during the construction we refer to the current version of X simply by "X" rather than give it an index. During the verification we may refer to the version of X at stage s by X_s.

During the set-up phase we appoint a sequence $Y_1, Y_2, \ldots$ of sets, one of which may be chosen to be the set Y we use for attack.

The action of agent (d,c) for the eth construction

The agent starts with setting up the first set Y_1.

SETTING UP Y_k.

1. Let s_0 be the stage at which this set-up cycle begins. We choose a large anchor n_k. Define $\gamma_{s_0}(n_k) = n_k$.

2. We wait for a stage s_1 at which $o^e_{s_1}(n_k) < \omega^2$. Suppose that $o^e_{s_1}(n_k) = \omega \cdot m + p$. At stage s_1 we choose a set Y_k of $(m+2)$-many large followers. For each $y \in Y_k$ we define $\psi^e_{d,s_1}(y) = n_k$.

3. We wait for a stage $s_2 > s_1$ at which $\varphi_{c,s_2}(y)\downarrow$ for all $y \in Y_k$.

4. We then wait for a stage $s_3 > s_2$ at which $D_{s_3} \restriction_{n_k} \neq D_{s_3-1} \restriction_{n_k}$. While waiting we (recursively) set up the set Y_{k+1}.

When such a stage s_3 is found, we interrupt all set-up cycles. We discard all anchors $n_{k'}$ and sets of followers $Y_{k'}$ for $k' \neq k$. We let $Y = Y_k$ and $n = n_k$. We let $u = 1 + \max\{\varphi_c(y) : y \in Y\}$. We start an attack with some $y \in Y$.

Throughout the set-up phase, if some anchor n_k is already chosen and $D_s \restriction_{n_k} \neq D_{s-1} \restriction_{n_k}$ then unless we start an attack at stage s we redefine $\gamma_s(n_k) = n_k$ and if also Y_k is defined, $\psi^e_{d,s}(y) = n_k$.

ATTACKING WITH A FOLLOWER $y \in Y$.

1. Let r_0 be the stage at which the attack begins. We define $\gamma_{r_0}(n) = u$. We appoint a set X of $(p+2)$-many large followers, where $o^e_{r_0}(n) = \omega \cdot m + p$. For each $x \in X$ we define $\psi^e_{r_0}(x) = u$. For now, we leave $\psi^e_{d,s}(y')$ for $y' \in Y$ undefined.

2. We wait for a stage $r_1 > r_0$ at which $\varphi_{d,r_1}(x)\downarrow$ for all $x \in X$. If $\varphi_c(y) \in D_{r_1}$ then we interrupt the attack cycle, discard all associated followers and anchor, and cease all action for the agent.

Otherwise we enumerate y into V^e_{d,r_1}. For all $y' \in Y$ we define $\psi^e_{d,r_1}(y') = u$.

3. We wait for a stage $r_2 > r_1$ at which $\varphi_c(y) \in D_{r_2}$. At that stage we end the current attack and commence an attack with some $x \in X$; we let $v = 1 + \max\{\varphi_d(x) : x \in X\}$.

Throughout this attack phase, if $D_s\restriction_u \neq D_{s-1}\restriction_u$ and we do not start an attack with some $x \in X$ at stage s, then we redefine $\gamma_s(n) = u$ with use u; and we redefine $\psi^e_s(x) = u$ for $x \in X$. If $s > r_1$ then we also define $\psi^e_{d,s}(y') = u$ for all $y' \in Y$.

ATTACKING WITH A FOLLOWER $x \in X$.

1. Let t_0 be the stage at which the attack begins. We define $\gamma_{t_0}(n) = v$.

2. We wait for a stage $t_1 > t_0$ at which $g^e_{t_1}(n) = \Gamma_{t_1}(D_{t_1}, n)$. While waiting, we leave $\psi^e_s(x')$ for $x' \in X$ undefined. Note that $\psi^e_{d,s}(y)$ for $y \in Y$ will be undefined throughout the attack with x.

If $\varphi_d(x) \in D_{t_1}$ then we interrupt the attack cycle, discard all associated followers and anchor, and cease all action for the agent.

Otherwise, we enumerate x into $V^e_{t_1}$. We define $\psi^e_{t_1}(x') = v$ for all $x' \in X$.

3. We wait for a stage $t_2 > t_1$ at which $\varphi_d(x) \in D_{t_2}$.

If $X \subseteq V^e_{t_2}$ then we discard X and start a new attack with some $y \in Y \setminus V^e_{d,t_2}$. Otherwise we commence a new attack with some $x' \in X \setminus V^e_{t_2}$.

The functionals $\Gamma(D, n)$ and $\Psi^e(D, x')$ are maintained as above.

Also as in the ω case, we ensure totality of functionals by defining them with use 0 on all inputs which are not used as anchors or followers.

6.2.3 Verification

We need to show that the construction can be performed as described. Fix an agent (d, c) for the eth construction.

First we observe that if an attack cycle begins at some stage w then all functionals are divergent at that stage. Namely:

- if an attack with $y \in Y$ begins at stage $w = r_0$ then $\Gamma_{w-1}(D_w, n)\uparrow$, and $\Psi^e_{d,w-1}(D_w, y')\uparrow$ for all $y' \in Y$; and
- if an attack with $x \in X$ begins at stage $w = t_0$ then also $\Psi^e_{w-1}(D_w, x')\uparrow$ for all $x' \in X$.

But as above these are ensured by the D-change encountered at the last stage of the previous cycle. If the set-up phase ended at stage w, then we just saw a change on $D\restriction_n$, and all uses are n; at the end of an attack with $y \in Y$, we just saw a change on $D\restriction_u$, and all uses are u; at the end of an attack with $x \in X$, we just saw a change on $D\restriction_v$, and the uses $\gamma(n)$ and $\psi^e(x')$ are v, while $\psi^e_d(y)$ are undefined throughout the attack with x.

We also obtain an analogue of Lemma 6.3: if Y is already defined at stage w then $Y \not\subseteq V^e_{d,w}$. Suppose that the set Y is chosen at some stage s_1, with $o^e_{s_1}(n) = \omega \cdot m^* + p$, so $|Y| = m^* + 2$. We argue that an attack with some follower in Y is started at most $m^* + 1$ many times. For $t < \omega$ let $o^e_t(n) = \omega \cdot m_t + p_t$. We claim that if two attacks with followers in Y start at stages $s < t$ then

$m_t < m_s$. This in turn is done by examining attacks started with elements of X. We have $|X_s| = p_s + 2$. The argument in the ω-case shows that if $w < r$ are stages in (s,t) at which we start an attack with an element of X_s then $o_r^e(n) < o_w^e(n)$. The fact that $p_s + 2$ many such attacks occur implies that $m_t < m_s$ as required.

Next we observe that $\Gamma(D)$ is total. The argument is similar to the one in the ω-case. Suppose that n is an anchor for some agent, and is never discarded. A computation $\Gamma_s(D_s, n)$ is defined at every stage $s > n$. The use is bounded. There are three possibilities. An attack may never begin; in this case $\gamma_s(n) = n$ for all s. Alternatively, an attack with some $y \in Y$ is never ended; we then eventually have $\gamma_s(n) = u$. Finally it is possible that an attack with some $x \in X$ for some version of X is never ended; we then eventually have $\gamma_s(n) = v$ (and v is never redefined).

We fix some e such that $\Gamma(D) = g^e$ and $\langle g_s^e, o_s^e \rangle$ is eventually ω^2-computable. We will show that the eth construction succeeds. We drop all superscripts e from now on.

The argument proving Lemma 6.4 shows that $\Psi(D) = V$. If $V \not\leq_m D$ then we are done. Assume this fails; fix some total computable function φ_d such that $\varphi_d^{-1}[D] = V$.

We argue that $\Psi_d(D) = V_d$. Observing that we only enumerate $y \in Y$ into V_d at stages at which $\Psi_d(D, y)$ diverges, again it suffices to show that $\Psi_d(D)$ is total. We focus on some y which is a follower in some set of followers Y for some agent (d, c). If no attack by the agent is every started (it is always in the set-up phase) then $\psi_d(y)\downarrow$ at every stage after y is appointed, with a bounded use n_k. Otherwise, the key is that since $\varphi_d^{-1}[D] = V$, every attack by this agent with a follower $x \in X$ must end. So there is an attack with some $y' \in Y$ by the agent which never ends. However the assumption that $\varphi_d^{-1}[D] = V$ implies that the attack is not stuck waiting for a stage r_1; φ_d is total. So we are eventually stuck waiting for a stage r_2; while waiting, we keep defining $\psi_d(y) = u$.

Finally, the argument of Lemma 6.5 shows that $V_d \not\leq_m D$. Fix some total φ_c. The simple permitting argument shows that the agent (d, c) will enter the attack phase; we just observed that an attack with some $y \in Y$ must succeed.

6.3 TOTALLY $<\omega^\omega$-C.A. DEGREES ARE NOT m-TOPPED

The general case follows the structure of the ω^2 case. Each construction guesses the m such that $\Gamma(D)$ is ω^m-c.a., and an appropriate approximation. It builds sets in m layers of nonuniformity.

6.3.1 Construction

We are given a c.e. set D whose Turing degree is totally $<\omega^\omega$-c.a. We use uniform lists $\langle g^{e,m} \rangle$ of all ω^m-c.a. functions, with tidy $(\omega^m + 1)$-computable approximations $\langle g_s^{e,m}, o_s^{e,m} \rangle$, for all $m < \omega$. We enumerate a Turing functional Γ with intended oracle D.

For every pair (e,m) we perform an (e,m)-construction. These constructions are independent of each other. For every m-tuple $\bar{d}=(d_0,\dots,d_{m-1})$, the (e,m)-construction will employ an agent $\bar{d}$. The construction enumerates c.e. sets $V_{\bar{c}}^{e,m}$ for all tuples $\bar{c}$ of numbers of length strictly smaller than m. For each such sequence $\bar{c}$, the construction also enumerates a functional $\Psi_{\bar{c}}^{e,m}$, as usual with the aim of having $\Psi_{\bar{c}}^{e,m}(D)=V_{\bar{c}}^{e,m}$, so as usual, to define a computation for one of these functionals, we only need to specify its use.

The action of agent $\bar{d}$ for the construction (e,m)

The agent aims to establish m sets of followers X_{m-1}, $X_{m-2},\dots$, X_0. The followers in X_k are targeted for $V_{\bar{d}\restriction_k}^{e,m}$. After receiving simple permission, the set X_{m-1} is fixed but the sets X_{m-2}, $X_{m-3},\dots$ are not fixed. When all followers in $X_{k-1},X_{k-2},\dots,X_0$ are used, we discard these sets and attack with a new follower from X_k.

Before we receive our simple permission though we need to appoint a sequence of candidates for X_{m-1}. These will be denoted by $Y_1,Y_2,\dots$.

The agent starts with setting up the first set Y_1.

Setting up Y_i.

1. Let s_0 be the stage at which this set-up cycle begins. We choose a large anchor n_i. Define $\gamma_{s_0}(n_i)=n_i$.

2. We wait for a stage s_1 at which $o_{s_1}^{e,m}(n_i)<\omega^m$. Suppose that $o_{s_1}^{e,m}(n_i)=\omega^{m-1}\cdot p+\beta$ (for some $\beta<\omega^{m-1}$). At stage s_1 we choose a set Y_i of $(p+2)$-many large followers. For each $y\in Y_i$ we define $\psi_{\bar{d}\restriction_{m-1},s_1}^{e,m}(y)=n_i$.

3. We wait for a stage $s_2>s_1$ at which $\varphi_{d_{m-1},s_2}(y)\downarrow$ for all $y\in Y_i$.

4. We then wait for a stage $s_3>s_2$ at which $D_{s_3}\restriction_{n_i}\neq D_{s_3-1}\restriction_{n_i}$. While waiting we (recursively) set up the set Y_{i+1}.

When such a stage s_3 is found, we interrupt all set-up cycles. We discard all anchors $n_{i'}$ and sets of followers $Y_{i'}$ for $i'\neq i$. We let $X_{m-1}=Y_i$ and $n=n_i$. We start an attack with some $x\in X_{m-1}$.

Throughout the set-up phase, if some anchor n_i is already chosen and $D_s\restriction_{n_i}\neq D_{s-1}\restriction_{n_i}$ then unless we start an attack at stage s we define $\gamma_s(n_i)=n_i$ and if also Y_i is defined, $\psi_{\bar{d}\restriction_{m-1},s}^{e,m}(y)=n_i$ for $y\in Y_i$.

Throughout the attack phase we let

$$o_s^{e,m}(n)=\omega^{m-1}p_{m-1,s}+\omega^{m-2}p_{m-2,s}+\cdots+\omega\cdot p_{1,s}+p_{0,s}.$$

When we start an attack with some element of X_k (for $k<m$) the sets X_{m-1}, $\dots,X_k$ are defined but $X_{k-1},\dots,X_0$ are not. If X_k is defined then so is

$u_k = 1 + \max\{\varphi^{e,m}_{d_k}(x) : x \in X_k\}$. During an attack with some $x \in X_k$, the computations $\Psi^{e,m}_{\bar{d}\restriction_{k'}}(y)$ for all $k' > k$ and $y \in X_{k'}$ are undefined.

Attacking with a follower $x \in X_k$ for $k > 0$.

1. Let r_0 be the stage at which the attack begins. We define $\gamma_{r_0}(n) = u_k$. We appoint a set X_{k-1} of $(p_{k-1,r_0} + 2)$-many large followers. For each $z \in X_{k-1}$ we define $\psi^{e,m}_{\bar{d}\restriction_{k-1},r_0}(z) = u_k$. For now we leave $\psi^{e,m}_{\bar{d}\restriction_k,s}(x')$ for all $x' \in X_k$ undefined.

2. We wait for a stage $r_1 > r_0$ at which $\varphi_{d_{k-1},r_1}(z)\downarrow$ for all $z \in X_{k-1}$. If $\varphi_{d_k}(x) \in D_{r_1}$ then we interrupt the attack cycle, discard all associated followers and anchor, and cease all action for the agent.

Otherwise we enumerate x into $V^{e,m}_{\bar{d}\restriction_k,r_1}$. For all $x' \in X_k$ we define $\psi^{e,m}_{\bar{d}\restriction_k,r_1}(x') = u_k$.

3. We wait for a stage $r_2 > r_1$ at which $\varphi_{d_k}(x) \in D_{r_2}$. At that stage we end the current attack and commence an attack with some $y \in X_{k-1}$.

As usual we respond to spontaneous $D\restriction_{u_k}$-changes by rectifying existing computations with the same use u_k.

Attacking with a follower $x \in X_0$.

1. Let t_0 be the stage at which the attack begins. We define $\gamma_{t_0}(n) = u_0$.

2. We wait for a stage $t_1 > t_0$ at which $g^{e,m}_{t_1}(n) = \Gamma_{t_1}(D_{t_1}, n)$. While waiting, we leave $\psi^{e,m}_{\langle\rangle,s}(y)$ for $y \in X_0$ undefined. If $\varphi_{d_0}(x) \in D_{t_1}$ then we interrupt the attack cycle, discard all associated followers and anchor, and cease all action for the agent.

Otherwise we enumerate x into $V^{e,m}_{\langle\rangle,t_1}$. For all $x' \in X_0$ we define $\psi^{e,m}_{\langle\rangle,t_1}(x') = u_0$.

3. We wait for a stage $t_2 > t_1$ at which $\varphi_{d_0}(x) \in D_{t_2}$. At that stage we end the current attack. Let $k \geqslant 0$ be the least such that $X_k \not\subseteq V^{e,m}_{\bar{d}\restriction_k,t_2}$. Discard $X_{k'}$ (and so $u_{k'}$) for all $k' < k$. Start a new attack with some $y \in X_k \setminus V^{e,m}_{\bar{d}\restriction_k,t_2}$.

As above, we maintain functionals, and define them on numbers that are not used by any construction.

6.3.2 Verification

These are similar to the previous verifications. First we need to ensure that the construction can be carried out as described. As above we observe that at the end of any cycle (set up or attack), all related computations are undefined.

We also prove that if X_{m-1} is defined at a stage s (for some agent $\bar{d}$ for a construction (e,m)) then $X_{m-1} \nsubseteq V^{e,m}_{\bar{d}\upharpoonright m-1}$. To see this, by induction on $k<m$ we observe that if $s<t$ are stages at which at attack with some $x \in X_k$ is started, then $o^{e,m}_s(n) - o^{e,m}_t(n) \geqslant \omega^k$.

The proof that $\Gamma(D)$ is total is as above. Fixing (e,m) which is a correct guess ($\Gamma(D) = g^{e,m}$ and $\langle g^{e,m}_s, o^{e,m}_s \rangle$ is eventually ω^m-computable), and dropping the superscripts (e,m), we argue that the (e,m) construction is successful. As above we argue that $\Psi(D) = V$. If V is not as required, we fix some d_0 such that $\varphi^{-1}_{d_0}[D] = V$. Then for any agent $\bar{c}$ such that $c_0 = d_0$, no attack with some $x \in X_0$ can succeed. This shows that $\Psi_{d_0}(D) = V_{d_0}$. If V_{d_0} is not as required then we fix some d_1 such that $\varphi^{-1}_{d_1}[D] = V_{d_0}$. Then for any agent $\bar{c}$ with $(c_0, c_1) = (d_0, d_1)$, no attack with some $x \in X_1$ can succeed. This shows that $\Psi_{d_0,d_1}(D) = V_{d_0,d_1}$. And so on ... This process must end at some $k<m$, giving some $V_{d_0,d_1,\ldots,d_k}$ which shows that $\deg_{\mathrm{T}}(D)$ is not m-topped.

Chapter Seven

Embeddings of the 1-3-1 lattice

ONE OF THE central and longstanding areas of classical computability theory concerns the structure of the degrees of unsolvability, and particularly the computably enumerable degrees. In the same way that studying symmetries in nature and solutions to equations leads to group theory, studies of the computational content of mathematics lead naturally to the structure of sets of integers under reducibilities. Understanding these structures should lead to insights into relative computability.

Notable in these studies is the question of embeddability into the c.e. degrees. We know that the c.e. degrees form an upper semilattice. Sacks [82] showed that this structure is a dense partial ordering. Lachlan [59] and Yates [105] proved that it is not a lattice, but some lattices could be embedded preserving meet and join. For example, Lachlan and Yates showed that the diamond could be embedded. Their constructions of minimal pairs mean that there are nontrivial c.e. problems whose only common information is precisely the computable sets. The Lachlan-Lerman-Thomason theorem (see [91, IX.2]) established that any countable distributive lattice could be embedded as a lattice into the computably enumerable degrees. It is natural to wonder precisely which lattices can be embedded. We note that this question is related to the longstanding question of a decision procedure for the two quantifier theory of the c.e. degrees. Unfortunately, we do not know a full characterisation of the finite lattices embeddable into the c.e. degrees. The most up-to-date state of our knowledge can be found in Lerman-Lempp-Solomon [64].

We do know that there are nondistributive lattices that can be embedded. As we mentioned in Chapter 1, both the nonmodular 5-element lattice and the 1-3-1 modular nondistributive lattice can be embedded in the c.e. degrees (Lachlan [61], fig. 1.1). The embedding of the 1-3-1 lattice was an amazing result, and introduced the "continuous tracing" technique into computability theory. The first inkling of quite how remarkable this technique is was the proof of Lachlan and Soare [62], who showed that it is not possible to embed 1-3-1 while making the top element branching, i.e., the bottom of a diamond in the c.e. degrees (see fig. 1.2). This was the first *non-embedding* result in the c.e. degrees.

There is a hidden message in the Lachlan-Soare technique. The non-embedding of S_8 was proved by a Lachlan game, in which one more or less assumes that the given embedding of the 1-3-1 lattice follows Lachlan's construction. Then the minimal pair machinery for the top diamond is shown to interact fatally with this methodology. This gives us the intuition that Lachlan's technique is not only sufficient but necessary for embedding the 1-3-1 lattice. This is in some sense the essence of the main result of this monograph, Theorem 1.6, which we prove in this chapter: the 1-3-1 lattice is embeddable in the c.e. degrees below a c.e. degree **d** if and only if **d** is not totally $<\omega^\omega$-c.a. This result shows that the class of totally $<\omega^\omega$-c.a. degrees is definable in the c.e. degrees.

Recall that Downey and Shore [35] showed that the 1-3-1 lattice can be embedded in the c.e. degrees below any non-low$_2$ degree. Our embedding of the 1-3-1 below any degree which is not totally $<\omega^\omega$-c.a. is an elaboration on their construction. In the introduction (in Section 1.3) we discussed the dynamics of this construction, and explained why they align with not totally $<\omega^\omega$-c.a. permitting. What we did not do is justify why indeed these are the dynamics one gets when embedding the 1-3-1 lattice. This is done below (in Subsection 7.1.1) once we state the requirements involved.

In the other direction, Downey [21] and Weinstein [103] showed that there are c.e. degrees which do not bound a (weak) critical triple (see fig. 1.3); Walk [101] showed that such degrees can be made array noncomputable. Toward proving the other direction of Theorem 1.6, we cannot simply adapt their constructions to an anti-permitting argument, as we know that there are totally $<\omega^\omega$-c.a. degrees which do bound critical triples, namely, all such degrees which are not totally ω-c.a. Thus we will need to find an elaboration on their constructions which can be adapted to such a proof.

7.1 EMBEDDING THE 1-3-1 LATTICE

We prove the first direction: if **d** is not totally $<\omega^\omega$-c.a. then the 1-3-1 lattice is embeddable below **d**.

7.1.1 Lachlan's construction

To prove this we use the construction of Downey and Shore's [35] which shows that the 1-3-1 lattice can be embedded below any non-low$_2$ degree. This is an elaboration on Lachlan's original embedding of the 1-3-1 lattice into the c.e. degrees. We briefly recall a version of the construction given by Stob (unpublished notes), using Lerman's pinball machine technique [65].[1]

In this construction we enumerate three c.e. sets A_0, A_1 and A_2. To ensure that their degrees form the middle section of an embedding of the 1-3-1 lattice

[1]This is one of the few infinite-injury constructions which does not benefit much from the use of a tree of strategies.

(with bottom **0**) we need to ensure that they are noncomputable, any two form a minimal pair (which also implies that they must be Turing incomparable), and each is computable from the join of the other two. The requirements to meet are:

P_e^i: $A_i \neq \Phi_e$,

where $\langle \Phi_e \rangle$ is an enumeration of all partial computable functions; and for $i \neq j$ from $\{0, 1, 2\}$,

$N_e^{i,j}$: If $\Theta_e(A_i)$ and $\Psi_e(A_j)$ are total and equal, then they are computable;

here $\langle \Theta_e, \Psi_e \rangle$ is an enumeration of all pairs of Turing functionals.

The global requirement that $A_i \leqslant_T A_j \oplus A_k$ when $\{i, j, k\} = \{0, 1, 2\}$ is met by the mechanism of appointing traces. A requirement P_e^i will appoint a follower x, targeted for A_i, and wait for it to be realised, which means $\Phi_e(x)\downarrow = 0$; as usual, when the follower is realised the requirement will want to enumerate it into A_i. Before x is realised, it is assigned a *trace* $y > x$, another number, which is targeted for either A_j or A_k. This is essentially the current $A_j \oplus A_k$-use of computing $A_i(x)$. The main rule is that we cannot enumerate x into A_i before we enumerate y into the set it is targeted for, A_j or A_k. Sometimes we will be able to enumerate y into the required set, but not be yet able to enumerate x into A_i; in this case, we will appoint a new trace y', again targeted for A_j or A_k, but not necessarily to the same set for which y was targeted. Indeed it is switching between A_j and A_k which is the key idea which makes the construction work.

Say that currently (at some stage s) x has a trace y, targeted for A_j. Another global requirement is $A_j \leqslant_T A_i \oplus A_k$. And so we need to repeat: the number y receives a trace $z > y$ of its own, targeted for either A_i or A_k. Overall, the follower x is accompanied by an *entourage* of traces $y, z, \ldots$, each element of the sequence being a trace for the number appearing before. At any stage, only numbers in a final segment of the entourage may be enumerated into the sets for which they are targeted. No two successive elements of the entourage are targeted for the same set. At stage s, the last element w of the entourage is a number of size at least s, and so does not yet require a trace. At the end of the stage, if $w < s+1$ then we will assign it a new, large trace. The construction will specify the set for which the new trace will be targeted. For simplicity of expression, we abuse the term a little by letting the word *entourage* refer to the entire sequence $x, y, z, \ldots$, including the follower.

All numbers we use in the construction as potential elements of the three sets A_0, A_1 and A_2 are represented as balls which will move in a pinball machine (see fig. 7.1). The main components of the machine are gates and holes. Some balls drop through holes to the main track of the machine. The balls move downwards. Along their journey they encounter gates. A gate may allow a ball to pass, or stop its movement. In the latter case, the ball is placed in a *corral* associated with the gate. Balls in the corral may later be released and allowed to resume their journey. When a ball arrives at the bottom of the machine we imagine

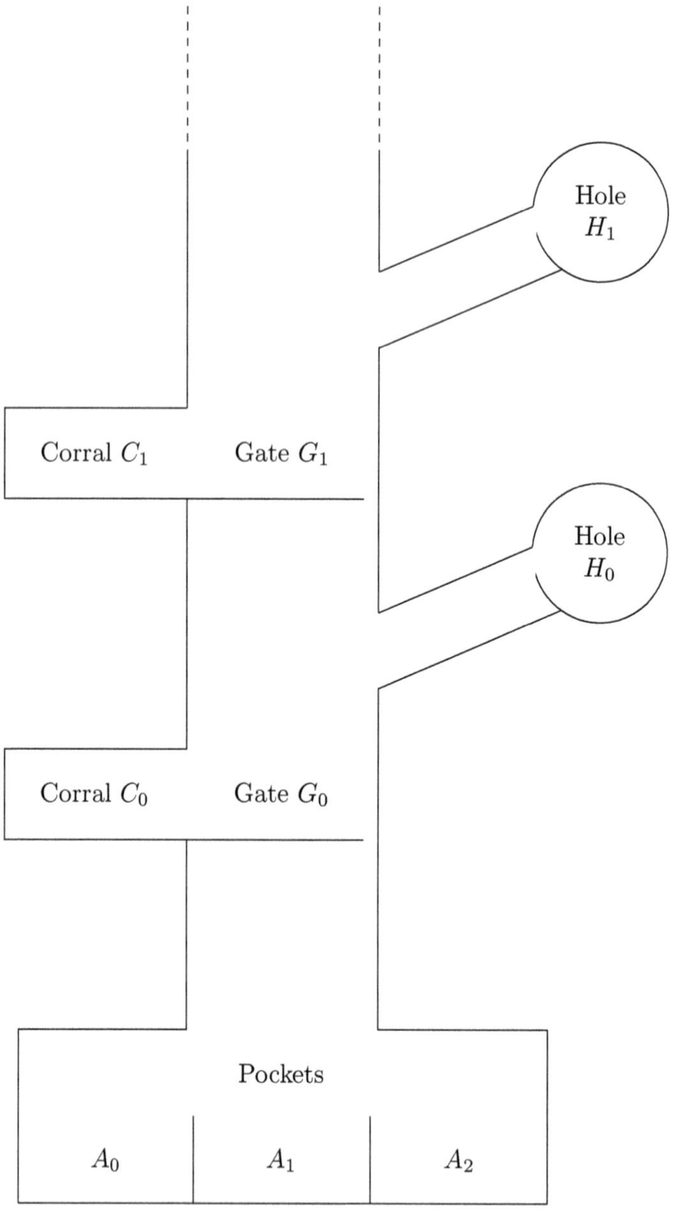

Figure 7.1. A pinball machine.

that it lands in one of the pockets associated with one of the sets A_i, namely the set the ball is targeted for. When a ball marked with the number x lands in the pocket associated with A_i, the number x is enumerated into the set A_i.

Holes $H_0, H_1, H_2, \ldots$ are associated with positive requirements P_e^i (much like strategies on a tree are assigned to requirements). As described above, the requirement appoints a follower $x = t_0$. While waiting for the follower to be realised, an entourage of traces $t_1, t_2, t_3, \ldots$ is appended to x. Once the follower is realised, the entourage $t_0, t_1, \ldots$ drops through the associated hole H_n and starts moving down through the machine. The entourage may be stopped by one of the gates G_m for $m \leqslant n$, in which case it enters the corral C_m. The last ball $y = t_\ell$ in the entourage rolls out of the corral and waits at the gate G_m. While waiting, the entourage is extended with more traces, all of which wait at the gate with y. At some point the gate opens and y and its sequence of traces (the final segment of the current entourage waiting at the gate) continue their journey down the machine. This sequence of balls may be stopped at a lower gate $G_{m'}$ (so $m' < m$). All of the balls enter the corral $C_{m'}$ and the last element $z = t_{\ell'}$ rolls out to the gate. Again while waiting, new traces are added to the entourage beyond z. When the gate opens, z and its traces continue their fall. Eventually some of these balls, in a final segment of the entourage, pass all of the gates and land in their pockets (with numbers enumerated into the sets they are targeted for). These balls are removed from the entourage. Say the final segment starting with t_k has just landed in the pockets, and $k > 0$. The ball t_{k-1} is now the last element of the entourage. It has been waiting in some corral C_n. It now rolls out to the gate G_n and waits for the gate to open. While it is waiting, new traces $t_k, t_{k+1}, \ldots$ are added to the entourage; they wait at the gate G_n together with t_{k-1}. The process continues ... In general the structure is as described in the following lemma.

Lemma 7.1. *Let x be a follower for some requirement P_e^i associated with the hole H_m. Suppose that x has already been dropped through its hole but has not yet been enumerated into A_i, so all balls in x's entourage are currently lying at various corrals and gates below the hole. The entourage is partitioned into segments $I_m < I_{m-1} < \cdots < I_0 < I^*$, where each I_k lies in the corral C_k and I^* waiting at some gate G_n. Some of the segments I_k may be empty; indeed all segments I_k for $k < n$ are empty. I^* however is nonempty.*

We need to address two issues:

(1) we need to describe when gates open and when balls are stopped at some gate; and
(2) we need to explain why the follower will eventually be enumerated into its set.

We first explain (1). The gates G_0, G_1, $G_2, \ldots$ are associated with negative requirements. Let G_n be a gate and suppose that it is associated with the requirement $N_e^{i,j}$. The requirement is met by following Lachlan's minimal pair strategy of freezing a computation on one side or the other until it recovers on the other side. Suppose that s is a stage and that $t < s$ was the previous stage at which the gate G_n was open. Then G_n opens at stage s if the length of agreement between $\Theta_e(A_i)$ and $\Psi_e(A_j)$ exceeds t. That is, if for all $x \leqslant t$, $\Theta_e(A_i, x){\downarrow} = \Psi_e(A_j, x){\downarrow}$ $[s]$. When open, the gate G_n cannot allow both balls targeted for A_i and balls targeted for A_j to drop below it. For this reason we need to ensure that if a final segment I^* of an entourage is waiting at the gate G_n at the beginning of stage s then either no ball in I^* is targeted for A_i or no ball in I^* is targeted for A_j. This is achieved by appointing new traces correctly: say that the first ball z in I^* rolled out to the gate G_n from the corral C_n at stage $r < s$. Suppose that z is targeted for A_i. Then the next trace w that we appoint for z will be targeted not for A_j but for A_k. And the next trace that we appoint for w will be targeted for A_i; and so on, so no ball waiting at the gate at stage s is targeted for A_j. The segment I^* is sometimes called an (i,k)-*stream*. If z is targeted for A_j then we build I^* to be a (j,k)-stream. Of course if z is targeted for A_k then we can build I^* to be either a (k,i)-stream or a (k,j)-stream.

The whole process can be thought of as *re-targeting* of traces. Say that the segment I_n waiting in C_n is an (i,j)-stream. Each ball in that segment waits until its trace, its successor in I_n, is enumerated into its set; we then appoint a new trace, targeted for A_k.

This brings us to question (2) above. We need to show that progress is made at every step. Let x be a follower. On the face of it, it would appear that because we keep extending the entourage, it is possible that balls in x's entourage move down at infinitely many stages (but x itself is never enumerated). This is not so. Consider as a first example the case of one gate: suppose that $x = t_0$ and its entourage $I = (t_0, t_1, \dots, t_\ell)$ at stage r arrive at the corral C_0 at that stage. The last ball t_ℓ in I rolls out to the gate G_0. We keep appointing traces and extend the entourage beyond t_ℓ, but when the gate opens, t_ℓ and all of these new balls fall to the pockets and are removed from the entourage. Next, the ball $t_{\ell-1}$ rolls out to the gate and the process resumes. We see that after $t_\ell + 1$ many iterations, the follower $x = t_0$ will be enumerated into the set it is targeted for, and the process will end.

Now consider two gates G_0 and G_1. At some stage r_0, x and its entourage $I = (t_0, \dots, t_\ell)$ arrive at the corral C_1. The ball t_ℓ rolls out to the gate. While waiting the entourage is extended to $I^* = (t_\ell, t_{\ell+1}, \dots, t_{\ell+p})$. At some stage the gate G_1 opens, this segment is allowed to proceed, but is placed in the corral C_0. As discussed above, after $p - \ell + 1$ many times at which G_0 opens, t_ℓ will be enumerated into its set and the ball $t_{\ell-1}$ will roll out to the gate G_1. After $\ell + 1$ many iterations of this longer process, the follower lands in C_0, and we are back in the first case. This kind of nested analysis can be extended to any finite number of gates.

This argument can be coded succinctly using ordinals below ω^ω. Say x is a follower, and let $I_m < I_{m-1} < \dots < I_0 < I^*$ be a partition of its entourage as in Lemma 7.1. Consider the ordinal $\omega^m |I_m| + \omega^{m-1} |I_{m-1}| + \dots + \omega^0 |I_0| + \omega^n$, where I^* lies at the gate G_n. The analysis above shows that each time a gate opens and part of x's entourage moves, this ordinal decreases. The well-foundedness of ω^ω guarantees that parts of x's entourage move only finitely many times. In the next subsection we will see that this "ordinal analysis" corresponds to the kind of permitting which is required to get the argument to work below a given c.e. degree.[2]

The main ideas of this construction have been described, but we mention a couple of aspects which we missed. In the analysis above we ignored the possibility that the last segment of an entourage is waiting at a gate which will never open again, because the hypothesis of the associated negative requirement fails. In this case the follower will not be enumerated into its set. For this reason, a positive requirement needs to appoint more followers and hope that one of them succeeds. We need to ensure that not all the followers will get stuck in this way. A good way to do this is to let each gate apprehend the entourage-segment of at most one follower. This is made possible by a process of cancellation. Followers are assigned priorities based on the time they were appointed. When a positive requirement receives attention (for example when appointing a new follower or when one of its followers receives attention), all followers for weaker requirements are cancelled. Thus the priority ordering between followers respects the ordering between requirements. When a follower receives attention (when its last entourage segment moves), all weaker followers, even for the same requirement, are cancelled. As usual, since new followers are appointed large, a follower x is stronger than a follower y if and only if $x < y$. Suppose that the last segment of x's entourage is waiting at a currently closed gate G_n, and that the segment of y's entourage is currently moving down. The gate can let y's segment pass even though it is not currently open and even though y's segment may contain balls targeted for both sets A_i and A_j that the gate cares about. The reason is the following. The fact that x's segment is still waiting at the gate when y's segment is moving (and so when y receives attention) shows that x is stronger than y; otherwise x would be cancelled at this stage. The last stage r at which x received attention is no earlier than the last stage u at which G_n was open. The computations currently protected by the gate have been observed at stage u. At stage r, followers weaker than x are cancelled, and so y was appointed later than stage u. It is therefore much too large to disturb any of the computations that the gate is currently trying to protect, and it (or part of its entourage) can pass without let or hindrance. Overall, this shows that if a positive requirement is using the hole H_n, then at most $n+1$ many of its followers could be permanently stuck at some gate. One of its followers will therefore either never get realised, or successfully enumerated into its set.

[2]We also remind the reader of Theorem 1.8, part of which relies on the fact that for most admissible ordinals $\alpha > \omega$, the 1-3-1 lattice cannot be embedded (at least with an incomplete top). The reason the argument fails is the instruction that the *last* ball of the entourage roll out to the gate. Since entourages might keep growing, it is perfectly possible that some will have order-type a limit ordinal. The only way to overcome this is if an α-c.e. degree can compute a bijection between α and ω. In that case the construction is essentially rearranged to resemble the standard ω-construction, with finite entourages at every stage.

We remark that the necessity for appointing more than one follower could be avoided if we put the construction on a tree of strategies. The tree now acts as the track of the machine, with positive nodes acting as holes and others as gates. A positive node on the true path guesses correctly which gates will open infinitely often and so its follower cannot get permanently stuck. However, when we add permitting in the next section we will need to let positive requirements appoint many followers; even if they do not get stuck at gates, they can wait in vain for a permission. For the permitting argument it seems that adding a tree of strategies does not help simplify the construction.

7.1.2 Embedding the 1-3-1 lattice with non-total $<\omega^\omega$-c.a. permitting

Recall the argument above for why every follower x receives attention only finitely many times. The ordinals used to show that the progress was well-founded correspond to the amount of permissions required to get the follower to its pocket. First note that for that argument, it is crucial that when part of x's entourage lands in the pockets, that the numbers are actually enumerated into their sets. We cannot appoint a new trace for the last element of the entourage still waiting in a corral without first enumerating its current trace. Further, before a gate opens again, we need to ensure that the numbers that it allowed to pass at the last time it was open are actually enumerated into the sets. Otherwise it may let balls potentially injuring the other side pass, and then both sides may get injured before the next time the gate opens. So the number of permissions we need to get until x is enumerated is close to the number of times the follower actually receives attention. The fact that x receiving attention corresponds to a decrease in the ordinal shows us that a bound on the ordinal also bounds the number of permissions required. For the hole H_{m-1} the bound is ω^m.

This can be explained in detail looking at the simple cases. In the one-gate case, once the follower is realised, we know the size of its entourage that enters the corral C_0, and so the number of times the gate G_0 needs to open until the follower arrives in its pocket. If the gate opens at some stage and releases one of the balls in the entourage, then we need a permission before the next such stage. So the number of permissions required is the same as the size of the entourage. This corresponds to non-total ω-c.a. permitting. (It is not array noncomputable permitting because we need to wait until the follower is realised before we know the eventual size of the entourage that enters C_0; we cannot tell it in advance.) When two gates are involved, when the follower is realised we know how many times we need G_1 to open. Each time it does open (and not before) we find out how many G_0-openings, and so how many permissions, we need until the next G_1-opening. Even though we don't need a permission between G_1-opening and the first time after that when G_0 opens, the size of the entourage in C_1 does tell us how many times we need to update the bound on the number of permissions required. This is precisely non-total ω^2-c.a. permitting. Weaker holes need to pass more and more gates, so overall for permitting we need a function which is not ω^n-c.a for any $n < \omega$.

This analysis shows that to pass m gates, *a single ball* requires ω^m permissions. However, the situation becomes more complicated when more than one ball is involved. As usual, a positive requirement will issue many followers, because some of them may get stuck at gates that don't open, and some of them will get stuck waiting for permissions. When one ball receives attention, weaker balls for the same requirement are cancelled. In many other α-c.a. permitting constructions, if a follower x cancels a follower y then x takes over the "permitting number" of y. That is, from that point on, every y-permission should be also counted as an x-permission. We cannot do this in this construction. The reason is that in order to increase x's permission number we first need to actually receive x-permission (with the old number). Otherwise the whole process of requiring permissions does not help us show that the permitting degree bounds all the sets being constructed. However in the 1-3-1 construction below a non-low$_2$ degree we cannot require permission during every movement of a follower; this is only possible with high permitting. (This has to do with the question of what happens when a gate opens but the corresponding follower is waiting for permission to move.) In a non-low$_2$ or weaker construction we can only require permissions when attempting to enumerate numbers into sets. So whenever x receives attention but does not try to enumerate numbers into sets, weaker followers y will be cancelled, but their permitting numbers cannot be taken over by x.

Our solution is to abandon the technique of taking over permitting numbers. Essentially this means that if y is a follower with permitting number k, and y is cancelled, then the next follower y' appointed gets the permitting number k as well (technically this is not quite so, but for nonessential reasons). However the first ordinal we compute for y' may be much larger than the ordinals we observed for y while y was still alive. When arguing that the positive requirement is met we need to threaten to give an ω^n-computable approximation (for some n) for a function which doesn't have one. During this approximation we are not allowed to increase the ordinals. However we notice that y was cancelled because a stronger follower x received attention. This means that x's ordinal count went down. Multiplying x's ordinal by the bound ω^m (on the left) and adding to y's ordinals we see that a single drop in x's ordinal allows us to increase the y-ordinal to the y'-ordinal. Overall, to pass m gates, we need ω^{2m}-permission. The details are given in the proof of Lemma 7.5.

The permitted embedding cannot be done while preserving the least element. Our embedding will have a bottom degree $\mathbf{b} > \mathbf{0}$. This is similar to the non-low$_2$ construction of Downey and Shore's [35]. The reason is an aspect of the construction that we glossed over in the previous section. Let G_n be a gate, working for requirement $N_e^{i,j}$, and suppose that the requirement's hypothesis holds: $\Theta_e(A_i) = \Psi_e(A_j)$. We need to show how to compute this common function. We look at a stage s at which the gate opens; we need to argue that if no balls targeted for A_j (say) drop from the gate at this stage, then the computation

$\Psi_e(A_j)[s]$ up to the length of agreement will survive until the next stage t at which the gate opens. This is not actually always true, the reason being that small balls targeted for A_j are currently waiting at a gate G_m below G_n and may be enumerated between stages s and t. We only certify the computation at stage s if we know that no small balls targeted for A_j that are at stage s waiting at gates below G_n will ever be enumerated into A_j. Note that some such balls may be stuck forever at a gate below G_n. So G_n cannot wait for a stage at which there are no small balls targeted for A_j at any gate below. It only needs to ensure that such balls will not enter A_j. How can G_n tell? Well, there are only finitely many gates below G_n, and each can have at most one segment as a permanent resident. The information about which of the gates below has permanent residents can be given to G_n nonuniformly, and we can wait for stages at which, below G_n, only gates with permanent members are occupied.[3]

In the permitted construction, many more balls can get stuck below G_n: those which passed all the gates, are lying in their pockets, but are still waiting for permission to be enumerated (the pockets act as a "permission bin"). Over all the construction, there will be infinitely many such balls. We need some uniform way to tell G_n which of those are dangerous. For this reason we introduce the new c.e. set B. To ensure that $\deg_{\mathrm{T}}(A_0 \oplus B)$, $\deg_{\mathrm{T}}(A_1 \oplus B)$, and $\deg_{\mathrm{T}}(A_2 \oplus B)$ form the middle of an embedding of the 1-3-1 lattice with bottom $\deg_{\mathrm{T}}(B)$ we need to meet the modified requirements:

P^i_e: $A_i \neq \Phi_e(B)$; and

$N^{i,j}_e$: If $\Theta_e(A_i, B)$ and $\Psi_e(A_j, B)$ are total and equal, then they are computable from B.

The global requirements are now $A_i \leqslant_{\mathrm{T}} A_j \oplus A_k \oplus B$.

When an entourage segment lands in the pockets, we attach a new trace to the end of the entourage; this new trace is targeted for B. When permitted, the balls in that segment, together with the new trace, are enumerated into their sets. A gate G_n now can look at the pockets and consulting B can tell which entourage segments will be enumerated in the future into their sets, and so find whether a computation it is examining may be injured by balls waiting in the pockets.

Note that a number targeted for B does not need a trace of its own. We may be tempted to close off entourages with a trace targeted for B before they land in the pockets. We cannot appoint such a trace while the ball is waiting to be realised: since we are now diagonalising against B, a positive requirement will protect the B-computation which realises the follower; it can certainly not plan to enumerate a number into B before it sees the use of such

[3] Again, a tree of strategies is equivalent to nonuniformly giving this advice to G_n; but as we will now see, this advice will be insufficient in the permitted construction.

a computation. Suppose that the follower dropped through the hole, is moving down the machine, and its entourage has a final segment I^* waiting at a gate. When the gate will open it will want to protect a computation on one side. However now both sides use B, so again, the gate cannot allow the appointing of a small number targeted for B before it sees the use of these computations. Thus only an entourage segment which passed all the gates and is waiting in the pockets can appoint a trace targeted for B.

The reader may want to compare this construction with the permitted construction of a critical triple below a non-totally ω-c.a. degree in [25]. In that construction the gates do not look at computations involving the "centre" B, and so a B-trace can be appointed at the node working for the positive requirement, once the B-computation realising the follower is discovered.

Toward the construction

Let $\mathbf{d}$ be a c.e. degree which is not totally $<\omega^\omega$-c.a. Let $g \in \mathbf{d}$ be a function which is not ω^n-c.a. for any $n < \omega$. As in the argument in Chapter 5, since $\mathbf{d}$ is c.e., we may replace g by its modulus, and obtain an approximation $\langle g_s \rangle$ which is non-decreasing and such that changes in $g(n)$ force changes in $g(m)$ for all $m \geqslant n$.

List both kinds of requirements in order-type ω; associate the hole H_n with the nth positive requirement P_e^i and the gate G_n with the nth negative requirement $N_e^{i,j}$.

As discussed, each positive requirement appoints followers. Each follower x for a positive requirement will be assigned a permitting number $a(x)$. We say that a follower x for the requirement P_e^i is *realised* at stage s if $\Phi_{e,s}(B_s, x)\downarrow = 0$. An uncancelled follower may, at a given stage, either still reside above its hole; occupy some gate or corral; lie in a pocket; or already be enumerated into the set A_i. We say that a follower x is *permitted at stage* s if $g_{s+1}(a(x)) \neq g_s(a(x))$. We say that the requirement is *satisfied* at stage s if there is a follower x for P_e^i which is still realised and has already been enumerated into A_i.

Also as discussed, followers are linearly ordered by priority. When a follower x receives attention, all weaker followers are cancelled. When a follower is cancelled, all of its entourage is cancelled with it. We allow cancellation of followers which are already enumerated into the sets for which they are targeted.[4]

At each stage, a gate may be *occupied* by a final segment of some entourage. We will ensure the following.

Lemma 7.2. *Let G_n be a gate, associated with the requirement $N_e^{i,j}$. At a stage s the gate may be occupied by at most one final segment of an entourage.*

[4]The point is that if x is enumerated into A_i but a stronger follower acts, then this action may cause an enumeration into B which destroys the computation which made x realised. We then need to cancel x, and the requirement to which x belonged will need to start again.

That entourage segment does not contain both a ball targeted for A_i *and a ball targeted for* A_j.

The associated corral may contain segments of more than one entourage. However, if the gate is occupied by the final segment of the entourage of some follower x*, then* x *is weaker than any other follower which has a segment of its entourage in the corral.*

We also ensure the following:

Lemma 7.3. *Let* x *be a follower for a requirement associated with the hole* H_m*. Suppose that at stage* s*,* x *is on the machine. Then* x*'s entourage at stage* s *is increasing and is partitioned into intervals* $I_m < I_{m-1} < \cdots < I_0 < I^*$ *such that:*

- *for each* $k \leqslant m$*,* I_k *is in the corral* C_k*; and*
- I^* *is nonempty, and either occupies a gate* G_k *for some* $k \leqslant m$*, or is lying in the pockets. If* I^* *is at gate* G_k *then* $I_n = \varnothing$ *for all* $n < k$.

Every ball in the entourage, except possibly for the last one, is targeted for one of the sets A_0*,* A_1 *or* A_2*, with no two successive balls in the entourage targeted for the same set. The last ball of the entourage is targeted for* B *if and only if* I^* *lies in the pockets.*

Construction

At stage s a gate G_n, associated with the requirement $N_e^{i,j}$, *opens* if for all $y \leqslant t$, $\Theta_e(B, A_i, y) = \Psi_e(B, A_j, y)\,[s]$, where t is the previous stage at which the gate opened, $t = 0$ if there was no such stage.

At stage s, a follower x *requires attention* if one of the following holds:

(1) x is still waiting above its hole, and is now realised;
(2) x is on the machine, and the final segment I^* of its entourage (as in Lemma 7.3) is waiting at a gate G_n, which is now open; or
(3) x is on the machine, the final segment I^* of its entourage is waiting in the pockets, and x is permitted at stage s.

A positive requirement P_e^i requires attention if either one of its followers requires attention, or if it is not currently satisfied, and no follower for this requirement is currently waiting above the hole.

Let P_e^i be the strongest requirement which requires attention at stage s. We cancel the followers for all weaker requirements. If no follower for P_e^i requires attention at this stage, then we appoint a new, large follower x for P_e^i, and place it over the hole. Define $a(x)$ to be large.

Otherwise, let x be the strongest follower for P_e^i which requires attention at stage s. We cancel all weaker followers for P_e^i.

Let I^* be the final segment of x's entourage given by Lemma 7.3; if x currently lies above its hole let I^* be all of x's current entourage. In cases (1) and (2), the segment I^* drops to the highest gate below its current location which is now unoccupied (this is measured after the cancellation of weaker followers). The segment I^* is put in the corresponding corral, and the last ball in I^* rolls out to wait at the gate.

However, if there are no unoccupied gates below I^*'s current location, then the balls in the segment I^* are put into the pockets. A new, large trace, targeted for B, is appended to this segment.

In case (3), all of the balls in I^* are enumerated into the sets for which they are targeted; they are removed from x's entourage. If I^* consisted of the entirety of x's entourage then x has just been enumerated and the requirement is now satisfied; we can cancel all other followers for P^i_e. Otherwise, the last ball in the remaining entourage is waiting in some corral. That last ball now rolls out of the corral and waits at the gate.

At the end of the stage, for any follower z which is still uncancelled, if the last ball w in z's entourage is smaller than $s+1$, and is not targeted for B, then we appoint a new, large trace and append it to the end of z's entourage. The location on the machine of the new trace is the same as the location of the previously last ball w. Say w is targeted for a set A_i. The new trace is targeted for one of the two sets A_j or A_k (where $\{i,j,k\}=\{0,1,2\}$) so that Lemma 7.2 still holds.

Verification

Before we embark on the verification, we need to ensure that the construction can actually be carried out as described. We need to show that Lemmas 7.2 and 7.3 hold at every stage. These two lemmas are proved together by induction on the stage. Most parts are immediate. We verify two parts of Lemma 7.2:

(1) if x and y are distinct followers, and at stage s, part of x's entourage lies in the corral C_n and part of y's entourage waits at the gate G_n, then y is weaker than x; and
(2) if a ball z rolls out to a gate G_n at stage s, then at that time, the gate is unoccupied.

For (1), consider the stage $r<s$ at which the segment of y's entourage which occupies the gate G_n at the beginning of stage s arrived at the gate. Between stages r and s the gate is occupied so no new entourage segments are added to the gate or corral. Hence x's entourage segment already lay in the corral at the beginning of stage r. Since x was not cancelled at stage r, it must be stronger than y.

For (2), let x be the follower of whose entourage z is a member. The follower x receives attention at stage s. If at that stage the final segment of x's entourage arrives at the corral C_n, then by the instructions, G_n is empty when that segment moves. Otherwise, balls in a final segment of x's entourage are enumerated into their sets at stage s. The new final segment (of which z is the last element) has been waiting in the corral C_n at the beginning of the stage. Suppose that G_n was occupied at the beginning of the stage. Then we know it was occupied by the final segment of the entourage of some other follower y. By (1), x is stronger than y. And so all the balls in y's entourage are cancelled at stage s (as x receives attention), and the gate becomes unoccupied.

Let x be a follower which at stage s has already been issued from the hole H_m but is not yet cancelled or enumerated into the set it is targeted for. Let $I_{m,s}(x) < I_{m-1,s}(x) < \cdots < I_{0,s}(x) < I_s^*(x)$ be the decomposition of x's entourage at that stage given by Lemma 7.3. We define an ordinal $\beta_s(x)$. Let

$$\bar{\beta}_s(x) = \omega^m \cdot 2|I_{m,s}(x)| + \cdots + \omega^1 \cdot 2|I_{1,s}(x)| + \omega^0 \cdot 2|I_{0,s}(x)|.$$

If $I_s^*(x)$ resides at gate G_k at stage s then we let $\beta_s(x) = \bar{\beta}_s(x) + \omega^k$. If $I_s^*(x)$ resides in the pockets then we let $\beta_s(x) = \bar{\beta}_s(x)$.

Considering various cases, we observe:

Lemma 7.4. *Suppose that x is on the machine at stage s and is not cancelled at stage s, nor is it enumerated into its set at stage s. Then $\beta_{s+1}(x) \leqslant \beta_s(x)$; if x receives attention at stage s then $\beta_{s+1}(x) < \beta_s(x)$.*

It follows that every follower receives attention only finitely many times.

Lemma 7.5. *Every positive requirement P_e^i receives attention finitely many times, and is met.*

Proof. Suppose that the requirement P_e^i is associated with the hole H_{m-1}.

To begin, we note that if x is a follower for P_e^i which is realised at some stage r and is still not cancelled at a stage $s > r$ then $\Phi_e(B,x)\downarrow = 0[s]$ by the same computation which was present at stage r. This is standard: suppose that a number $b < \varphi_{e,s}(B_s, x)$ enters B at stage s. The number b is the last element of an entourage of some follower y. If y is stronger than x then x is cancelled at stage s. Otherwise, the trace b is chosen after stage r, and so is greater than $\varphi_{e,r}(B_r, x)$, which by induction equals $\varphi_{e,s}(B_s, x)$.

By induction, all positive requirements stronger than P_e^i eventually cease all action; in particular, they stop cancelling followers for P_e^i. Let r^* be the last stage at which a requirement stronger than P_e^i receives attention.

If some follower for P_e^i enters A_i after stage r^* then the lemma holds. This is also the case if some follower x for P_e^i is never cancelled but never realised. We will show that one of these two cases must hold. Suppose otherwise, for a contradiction. We will give an ω^{2m}-computable approximation for g.

Suppose that x is a follower for P_e^i which is never cancelled. By assumption, it is realised at some stage. By Lemma 7.4 the follower receives attention finitely many times. We assumed that x is not enumerated into A_i. This means that the final configuration for x (given by Lemma 7.3) contains an ever-increasing final segment $I^*(x)$ which is either a permanent resident of some gate G_n, or a permanent resident of the pockets. In the first case, we say that x's entourage is *stuck* at the gate G_n; in the second case, that it is stuck in the pockets.

There are only finitely many followers for P^i_e whose entourage gets stuck at some gate. Indeed there are at most m many. This is because each gets stuck at some gate G_n for some $n < m$, and each gate contains at most one segment as a permanent resident.

We let $r^{**} > r^*$ be the last stage at which a follower, whose entourage is eventually stuck at some gate, receives attention; $r^{**} = r^*$ if there is no such stage. Every follower which receives attention after stage r^{**} was also appointed after stage r^{**}. Every such follower is either eventually cancelled, or eventually its entourage is stuck in the pockets, awaiting permission which is never given.

Infinitely many followers are appointed for P^i_e, and of those, infinitely many are never cancelled. The argument is again standard: for any stage t consider the strongest follower x which requires attention after stage t. Then x is never cancelled, and after the last stage at which x receives attention, a new follower is appointed, and eventually receives attention as it is eventually realised.

Let $p > r^{**}$. To approximate $g(p)$ we let, for $s > p$, $X_s(p)$ be the set of followers $y > r^{**}$ for P^i_e which are uncancelled at stage s such that $a(y) \leqslant p$. This set is naturally ordered (in an increasing fashion). If $s < t$ then $X_t(p)$ is an initial segment of $X_s(p)$; some followers in $X_s(p)$ may get cancelled; new permitting numbers are always assigned to be large.

Let $S(p)$ be the set of stages $s > r^{**}, p$ such that:

- at stage s there is some follower $x > r^{**}$ for P^i_e such that $a(x) > p$; and
- if $x = x_s(p)$ is the least such follower, then the final segment $I^*_s(x)$ of x's entourage is waiting in the pockets at stage s.

The set $S(p)$ is infinite, indeed it is cofinite. The sets $X_s(p)$ stabilise to some $X(p)$; let s be the last stage at which any follower in $X(p)$ receives attention. The next follower x, appointed at stage $s+1$, is never cancelled and $a(x) > p$, so $x = x_t(p)$ for all $t > s$; x's entourage is eventually stuck in the pockets.

Let $s \in S(p)$; let $y_{1,s}, y_{2,s}, \dots, y_{\ell(s),s}$ be the increasing enumeration of $X_s(p)$. We let

$$\sum_{y \in X_s(p)} \beta_s(y) = \beta_s(y_{1,s}) + \beta_s(y_{2,s}) + \dots + \beta_s(y_{\ell(s),s})$$

and

$$\gamma_s(p) = \omega^m \cdot \left(\sum_{y \in X_s(p)} \beta_s(y) \right) + \beta_s(x_s(p)).$$

Since $\beta_s(x) < \omega^m$ for all x, we see that $\gamma_s(p) < \omega^{2m}$.

Let $s \in S(p)$, and let t be the next stage in $S(p)$ after stage s. We show that $\gamma_t(p) \leqslant \gamma_s(p)$, and that if $g_t(p) \neq g_s(p)$ then $\gamma_t(p) < \gamma_s(p)$.

Suppose that $x_t(p) \neq x_s(p)$. Then the follower $x_s(p)$ must be cancelled by stage t. This means that one of the followers in $X_s(p)$ received attention between stages s and t; let z be the least such follower. Then z is the last (greatest) element of $X_t(p)$. By Lemma 7.4, $\beta_t(z) < \beta_s(z)$. This shows that $\sum_{y \in X_t(p)} \beta_t(y) < \sum_{y \in X_s(p)} \beta_s(y)$. Even though $\beta_t(x_t(p))$ may be much larger than $\beta_s(x_s(p))$, it is smaller than ω^m, and this shows that $\gamma_t(p) < \gamma_s(p)$.

So we assume that $x_t(p) = x_s(p)$; let $x = x_s(p)$. For all $y \in X_s(p) = X_t(p)$, $\beta_t(y) \leqslant \beta_s(y)$, and $\beta_t(x) \leqslant \beta_s(x)$, so $\gamma_t(p) \leqslant \gamma_s(p)$. Suppose that $g_t(p) \neq g_s(p)$. Since x is not cancelled between stages s and t and $a(x) > p$, it follows that x is permitted at some stage between s and t. At the first such stage, x's final entourage segment is still waiting in the pockets, and so x receives attention between stages s and t. Lemma 7.4 guarantees that $\beta_t(x) < \beta_s(x)$, and this implies that $\gamma_t(p) < \gamma_s(p)$ as required. □

Lemma 7.6. *All sets A_0, A_1, A_2 and B are computable from* $\mathbf{d}$.

Proof. To determine if a number z is an element of one of these sets or not, we first go to stage z. We then see if z has already been chosen as a follower or a trace; and if so, to which set it is targeted. If not, then z does not enter any set, since new followers and traces are chosen to be large.

Suppose that z is an element of an entourage of a follower x (possibly $x = z$) at some stage $t \leqslant z$. The number $a(x)$ is already determined by stage z. With oracle g we can find a stage after which the follower x is never permitted. The function g can thus calculate a stage after which z cannot enter any set. □

The verification concludes with the following two lemmas, which are standard, but are added for completeness.

Lemma 7.7. *If $\{i, j, k\} = \{0, 1, 2\}$ then $A_i \leqslant_{\mathrm{T}} A_j \oplus A_k \oplus B$.*

Proof. We ensured that if y is targeted for A_i then at all stages $s > y$ at which y is on the machine, y has a trace z targeted to one of the sets A_j, A_k or B, and y does not enter A_i unless the trace z enters the set it is targeted for. Further, y is either cancelled or eventually receives a trace which is never cancelled; this is due to Lemma 7.4. □

Lemma 7.8. *Every negative requirement $N_e^{i,j}$ is met.*

Proof. Suppose that $\Theta_e(A_i, B) = \Psi_e(A_j, B)$ are total. Let G_n be the gate associated with the requirement $N_e^{i,j}$.

By Lemma 7.5, let r^* be the last stage at which either:

- any positive requirement which is associated with a hole H_m for some $m < n$ receives attention; or
- any follower whose entourage is eventually stuck at some gate G_m for some $m < n$ receives attention.

Let M be the set of $m \leqslant n$ such that the gate G_m does not have a permanent resident. We assumed that the hypothesis of $N_e^{i,j}$ holds; this implies that G_n opens infinitely often, and so $n \in M$.

We let S^* be the set of stages $s > r^*$ at the beginning of which:

- for all $m \in M$, the gate G_m is unoccupied;
- if $I_s^*(x)$ is the final segment of an entourage of a follower x which lies in the pockets, then x will not receive attention at stage s or after stage s.

The set S^* is computable from B; this is because entourage segments in the pockets end with traces targeted for B. We note that if $s \in S^*$ and x is a follower, part of whose entourage resides anywhere below the gate G_n, then x does not receive attention after stage s; the last segment of x's entourage is either permanently at a gate or in the pockets.

The set S^* is infinite. Let t be a large stage. As usual, let x be the strongest follower which ever receives attention after stage t; say x last receives attention at stage $s-1 > t$. All balls on the machine at the beginning of stage s will never move again; if a gate G_m is occupied at the beginning of stage s then the residents of G_m at stage s are permanent. Hence $s \in S^*$.

Let $p < \omega$. We let $s(p)$ be the least stage $s \in S^*$ such that $s > p$, G_n was open at some stage in the interval (p, s), and $\Theta_e(A_i, B, p)\downarrow = \Psi_e(A_j, B, p)\downarrow [s]$. Such a stage exists because we assume that the hypothesis of $N_e^{i,j}$ holds. Let $a = \Theta_e(A_i, B, p)[s(p)]$. We claim that $a = \Theta_e(A_i, B, p)$. To show this we prove by induction that for all $s > s(p)$, either $\Theta_e(A_i, B, p)\downarrow [s] = a$ or $\Psi_e(A_j, B, p)\downarrow [s] = a$.

Let $s > s(p)$ and suppose that the claim is already established for all stages in the interval $[s(p), s)$. Let x be the strongest follower which receives attention at any stage in the interval $[s(p), s)$ (if no follower receives attention then the computations which were observed at stage $s(p)$ were not destroyed by stage s).

Since $s(p) \in S^*$, no part of x's entourage lies below G_n at stage $s(p)$. Suppose that no part of x's entourage crosses the gate G_n at any stage in the interval $[s(p), s)$. In this case let $t < s$ be the last stage before stage s at which x received attention. By induction either $\Theta_e(A_i, B, p)\downarrow [t] = a$ or $\Psi_e(A_j, B, p)\downarrow [t] = a$; without loss of generality, assume the former. No numbers are enumerated into sets during stage t. If a number from some follower y's entourage is enumerated into any set between stages t and s, then y is weaker than x, and so was appointed after stage t, and so is greater than the use $\theta_{e,t}(p)$. Thus the computation $\Theta_e(A_i, B, p)[t]$ is preserved until stage s.

Suppose then that parts of x's entourage do cross the gate G_n at some stages in the interval $[s(p), s)$. Let t be the last stage in that interval at which any part of x's entourage crosses the gate. We note that whenever x receives attention, all other followers that were appointed after stage $s(p)$ are cancelled. In particular, G_n becomes unoccupied. We conclude that no segments of x's entourage ever pass by the gate without stopping first. Hence, at stage t, the gate opens, and part of x's entourage that was waiting at the gate is allowed to proceed downwards.

This implies two things: the first, that $\Theta_e(A_i, B, p)\downarrow[t] = \Psi_e(A_i, B, p)\downarrow[t]$; by induction, the common value is a. The second is that the segment of x's entourage which is released from the gate at stage t does not contain both balls targeted for A_i and balls targeted for A_j. Without loss of generality, suppose it does not contain any balls targeted for A_j. We claim that the computation $\Psi_e(A_j, B, p)[t]$ is not destroyed by stage s.

For suppose that some number u below the use $\psi_{e,y}(p)$ of that computation is enumerated into A_j or B at some stage in the interval $[t, s)$. Let y be the follower to whose entourage u belongs. By the choice of x, either $y = x$ or y is weaker than x. If y is weaker than x then y is appointed after stage t, and so y, and all of the balls in its entourage, are greater than the use $\psi_{e,t}(p)$. But $y = x$ is impossible too: u must be appointed before stage t, and so is already an element of x's entourage at stage t. But it does not cross the gate at stage t: no balls targeted for either A_j or B proceed from the gate at stage t. All other balls in x's entourage at stage t remain above the gate until stage s. □

7.1.3 The 1-4-1 lattice

The embedding technique used above actually shows:

Theorem 7.9. *If c.e. degree* $\mathbf{d}$ *is not totally* $<\omega^\omega$*-c.a. c.e. degree then for all* $n \geqslant 3$*, the 1-*n*-1 lattice can be embedded into the c.e. degrees below* $\mathbf{d}$.

Take for example the case $n = 4$. We enumerate sets A_0, A_1, A_2 and A_3, and a bottom set B. The requirements are as above, except for the pairwise joins: if i, j, k are distinct indices from $\{0, 1, 2, 3\}$ then $A_i \leqslant_{\mathrm{T}} A_j \oplus A_k \oplus B$. The rule for traces now is that if $\{i, j, k, l\} = \{0, 1, 2, 3\}$ then every number targeted for A_i needs to have two traces, for two of the sets A_j, A_k and A_l.

It would seem that an entourage in this construction will be a binary branching tree, but we can actually make do with linear entourages as in the construction above; the two balls following a ball in a (linear) sequence of balls are considered its traces. That is, if the follower is t_0 and the entourage is $t_0, t_1, t_2, \dots, t_\ell$ then for all $i \leqslant \ell - 2$, t_{i+1} and t_{i+2} are the traces for t_i. For the tracing to work we need to require that for any such i, no two of the three balls t_i, t_{i+1} and t_{i+2} are targeted for the same set. Given two previous balls t_{i-2} and t_{i-1}, this still leaves two options for choosing a target for the next ball t_i, and this allows us to re-target followers at gates. A sequence of balls waiting at

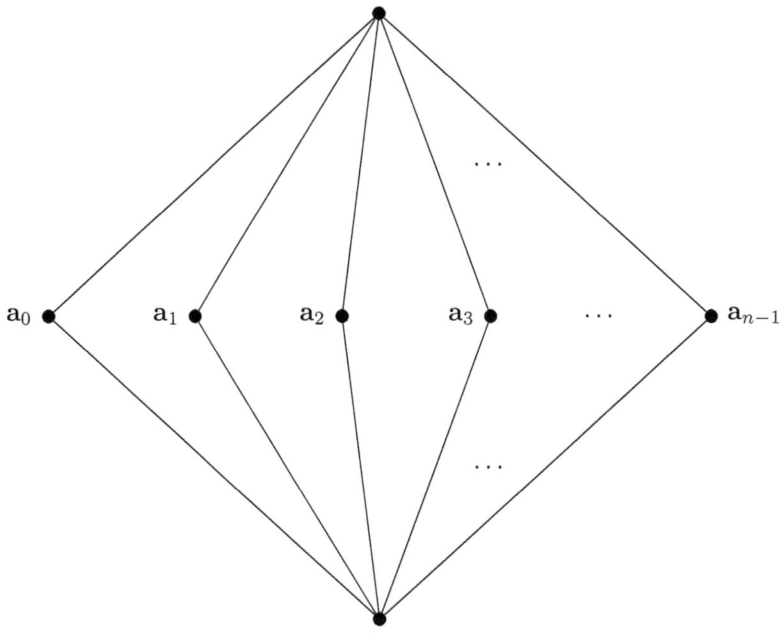

Figure 7.2. The 1-n-1 lattice.

a gate working for $N_e^{i,j}$ will be an (i,k,l)-stream or a (j,k,l)-stream. The rest of the construction is identical.

In fact, we can string together these constructions to obtain an embedding of the 1-ω-1 lattice; the nth follower appointed (across all requirements) and its entourage will only concern itself with the first n middle sets; reductions $A_i \leqslant_{\mathrm{T}} A_j \oplus A_k \oplus B$ will be nonuniform.

7.2 NON-EMBEDDING CRITICAL TRIPLES

As discussed in the introduction, a *critical triple* in an upper semilattice consists of three incomparable elements $\mathbf{a}_0, \mathbf{a}_1$ and $\mathbf{b}$ such that $\mathbf{a}_i \leqslant \mathbf{b} \vee \mathbf{a}_{1-i}$ for $i=0,1$, and such that any $\mathbf{e}$ lying below both $\mathbf{a}_0$ and $\mathbf{a}_1$ lies below $\mathbf{b}$ as well. That is, $\mathbf{a}_0 \wedge \mathbf{a}_1 \leqslant \mathbf{b}$, except that we don't actually require the meet $\mathbf{a}_0 \wedge \mathbf{a}_1$ to exist. The element $\mathbf{b}$ is called the *centre* of the triple.

In [25] the authors show that a c.e. degree bounds a critical triple (in the c.e. Turing degrees) if and only if it is not totally ω-c.a. The proof shows that the same holds for weak critical triples. The proof that no totally ω-c.a. c.e. degree bounds a weak critical triple is an "anti-permitting" elaboration on an argument from [13] that constructs a c.e. degree which bounds no weak critical triple. That argument in turn is a simplification of an argument from [103], which constructs a c.e. degree that bounds no weak critical triple. Toward the proof of the second half of Theorem 1.6, we now give an anti-permitting elaboration on Downey's original argument in [21]. It is somewhat more complicated than Weinstein's weak critical triple argument, and gives a weaker result. But it will be the argument that we need to generalise in order to prove our theorem. To avoid an extra step of simple permitting we work with array computable degrees rather than totally ω-c.a. That is, in this section we prove:

- no array computable c.e. degree bounds a critical triple in the c.e. degrees.

7.2.1 Layering

The fundamental notion from [21] is that of protecting computations by layers. In our setting, let D be a c.e. set whose Turing degree is array computable; and let $A_0, A_1, B \leqslant_{\mathrm{T}} D$ be sets whose degrees potentially form a critical triple. To show that they in fact do not form a critical triple we will build a c.e. set $Q \leqslant_{\mathrm{T}} A_0, A_1$ such that $Q \nleqslant_{\mathrm{T}} B$; or we may fail to do so, but in that case we will show that A_0 is computable from B. We fix functionals Λ, Φ_0 and Φ_1 such that $\Lambda(D) = (B, A_0, A_1)$, and such that $\Phi_i(B, A_{1-i}) = A_i$ for $i = 0, 1$.

The general idea of the construction is as follows. We define an auxiliary function $\Delta(D)$, and as in the anti-permitting arguments in the previous chapters, nonuniformly we know an id-computable approximation for $\Delta(D)$. We enumerate the set Q, together with reductions Γ_i of Q to A_i. For each $d < \omega$, to ensure that $\Psi_d(B) \neq Q$ we appoint a follower x, and after it is realised ($\Psi_d(B, x)\downarrow = 0$) we hope for double permission—changes in both A_0 and A_1 below the uses of reducing $Q(x)$ to these sets—so that we can enumerate x into Q. The natural two questions are: (a) why would we get double permission? (b) if we do, how do we protect the realisation of the follower—i.e., how do we ensure that indeed $\Psi_d(B, x) = 0$?

The idea is to have a backup strategy. We build a functional Ξ_d; if the dth requirement fails, that is, if $\Psi_d(B) = Q$, then we will ensure that $\Xi_d(B) = A_0$. Suppose that x is a follower. When we see that x is realised, we set up a computation of $A_0\restriction_x$ from B, with use at least $\psi_d(B, x)$. If later we attack with x and then x becomes unrealised, then we will be able to cancel x, because any incorrect computation of $A_0\restriction_x$ from B can be discarded as well. This solves the problem (b) above. However, this process introduces two analogous problems (assuming that indeed $\Psi_d(B) = Q$): (b') how do we protect the correctness of a computation $\Xi_d(B) = A_0\restriction_x$ (when x is not cancelled); and (a') how do we ensure that infinitely many followers are not cancelled so that $\Xi_d(B)$ is total?

This is where anti-permitting comes in. We associate a follower x with an anchor n, an input for $\Delta(D)$. As long as we keep $\Delta(D)$ total, having guessed the correct approximation, we know there will be no more than n many changes to $D\restriction_{\delta(n)}$. If we can arrange $\delta(n)$ to be large enough, beyond $\lambda(u)$, then we can ensure that there are at most n many changes to $A_i\restriction_u$ or $B\restriction_u$ (recall that Λ is the functional computing A_0, A_1 and B from D).

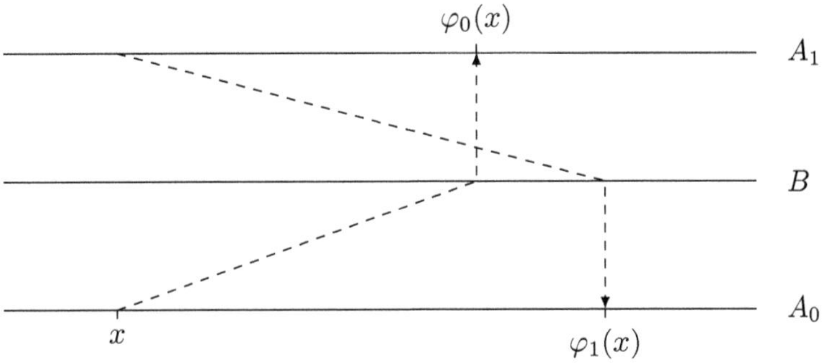

Figure 7.3. One layer.

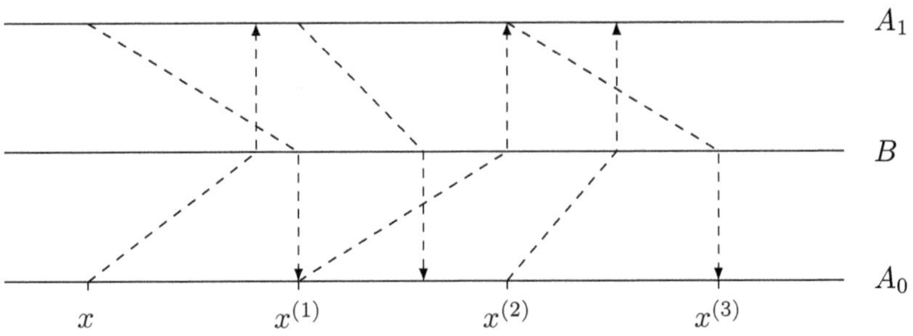

Figure 7.4. Three layers.

A single layer above x is the length $u > x$ required to ensure that a change in one of the sets A_0 or A_1 below x necessitates a change in at least one other set among A_0, A_1 and B below u. Formally we define

$$x^{(1)} = \max\{\varphi_0(B, A_1, x), \varphi_1(B, A_0, x)\}$$

(see fig. 7.3). We then let

$$x^{(n+1)} = (x^{(n)})^{(1)}$$

(see fig. 7.4).

When we set up x, we define the use of reducing $Q(x)$ to the sets A_i to be $x^{(n)}$; and set $\delta(n) = \lambda(x^{(n)})$. When x is realised, we set the use $\xi_d(x)$ of reducing $A_0 \upharpoonright_x$ to B to be $\max\{x^{(n)}, \psi_d(x)\}$. We consider what the next change could be. Assuming that x remains realised, we are concerned about A_i-changes. The key, again, is that the number $x^{(n)} = (x^{(n-1)})^{(1)}$ is chosen so that a change in some A_i below $x^{(n-1)}$ forces a change in either B or A_{1-i} below $x^{(n)}$. So now there can be two kinds of A_i-changes. If one A_i changes below $x^{(n-1)}$, then (again assuming that x remains realised, so B does not change) there must be a change in A_{1-i} below $x^{(n)}$. But $x^{(n)} = \gamma_i(x) = \gamma_{1-i}(x)$, the uses of reducing $Q(x)$ to A_i and

A_{1-i}; so in this case we get the double change we wished for, and we can attack with x: enumerate it into Q, and hopefully win the dth requirement $\Psi_d(B) \neq Q$. Otherwise, the A_i-change that concerns us happens below $x^{(n)}$ but above $x^{(n-1)}$. We say that the nth layer is *peeled.* Since $\delta(n) = \lambda(x^{(n)})$, the A_i-change allows us to redefine $\Delta(n)$ and extract one $D\restriction_{\delta(n)}$-change from our opponent. And the opponent's capital is bounded: at most n changes are possible. The nth layer is gone, but we now repeat the argument with the $(n-1)^{\text{st}}$ layer instead: a change below $x^{(n-2)}$ leads to an attack; a change below $x^{(n-1)}$ but not below $x^{(n-2)}$ means that the next layer is peeled, and another change in $\Delta(D, n)$ is paid by the opponent. Since we have set up sufficiently many layers, if an attack never occurs, the opponent cannot peel all of the layers, which in particular means that no changes to $A_0\restriction_x$ are possible—ensuring the correctness of the reduction $\Xi_d(B)$ on x.

Finally, the anchor n is also used to solve problem (a'): if we can ensure that each time that we cancel x, D changes below $\delta(n)$, then we can cancel x and appoint a new follower x', but keep the same anchor n. For each anchor n, at most n followers can be cancelled, and so one will be permanent. There are some delicate details involved, though, and we discuss them below.

7.2.2 Four procedures

Let us give more details and fix notation. For every $e < \omega$ we will perform an eth construction. All constructions together define a functional Δ, and ensure that $\Delta(D)$ is total. Let $\langle f^e, o^e \rangle$ be an effective enumeration of all id-c.a. functions (with tidy (id+1)-computable approximations). The eth construction guesses that $\Delta(D) = f^e$. The eth construction enumerates a c.e. set Q^e. For each d, an agent d for the eth construction tries to ensure that $\Psi_d(B) \neq Q^e$. The construction builds two functionals Γ^e_0 and Γ^e_1, with the aim of ensuring that $\Gamma^e_i(A_i) = Q^e$. The dth agent also enumerates a functional Ξ^e_d.

We adopt the conventions of Notation 6.2; for example, we write $\xi^e_{d,s}(x)\downarrow$ to indicate that $\Xi^e_d(B, x)\downarrow$ $[s]$, and when we define the computation we just assign a value to the use; we know that we always define $\Xi^e_d(B, x) = A_0\restriction_x$ $[s]$, $\Gamma^e_i(A_i, x) = Q^e(x)$ $[s]$, and $\Delta_s(D_s, n) = s$.

We go one step further and omit mentioning the stage number during the construction; so we just write $\xi^e_d(x)\downarrow$ and understand that this is to be evaluated at the present, i.e., at the stage currently under consideration. To further simplify the notation we omit the superscript e.

As discussed above, we are given functionals Λ and Φ_i such that for $i = 0, 1$, $\Phi_i(B, A_{1-i}) = A_i$, and $\Lambda(D) = (B, A_0, A_1)$. At a given stage of the construction we may refer to uses such as $\lambda(u)$ for some number u. When we do this we understand that we are speeding up the enumerations of the sets and functionals which are given to us so that we see a convergence of the relevant computation (in the example, $\Lambda(D, u)$). Applying this to the uses φ_i, this allows us to refer to numbers such as $x^{(n)}$ defined above.

At each stage, agent d will appoint a new anchor n (using the next unused number). Each anchor will start a process which will be independent of all other processes for all agents and all constructions. The process cycles between four procedures (or *phases*):

Set-up: Appointing a follower x; defining a parameter $u = x^{(n)}$, and defining $\delta(n) = \lambda(u)$; waiting for $\Delta(D, n) = f^e(n)$. Once this is observed, defining $\gamma_i(x) = u$.

Realisation: Waiting for $\Psi_d(B, x)\downarrow$. When convergence is obtained, defining $\xi_d(x) = \max\{u, \psi_d(x)\}$.

Maintenance: Waiting for double permission: both $\gamma_i(x)\uparrow$. (While waiting, demanding payment for layers being peeled.)

Attack: When double permission is received, enumerating x into Q. Then, monitoring the correctness of the realising computation $\Psi_d(B, x)$.

To understand the construction we need to explain under what circumstances we move from one procedure to another, and how we react to changes when we see them. We discuss some of the principles involved.

Cancelling a follower

We cancel a follower x if both $\delta(n)\uparrow$ and $\xi_d(x)\uparrow$, except during the set-up procedure. We need $\delta(n)\uparrow$ so that we will be free to redefine $\delta(n) = \lambda((x')^{(n)})$ for a new follower x' which will be appointed once x is cancelled. We need $\xi_d(x)\uparrow$ as while $\Xi_d(x)\downarrow$ we need to maintain the correctness of this computation. We are not allowed to cancel the follower during the set-up phase, because during set-up we are still waiting for our opponent to make a payment; each cancellation will be charged against a change in $f^e(n)$, and during set-up we have not seen this change yet.

Why would we need to cancel x? While we are in set-up, both $\gamma_i(x)$ are undefined, and so any change to any of the sets A_i or B below u will cause us to simply recalculate a new value for $u = x^{(n)}$ and restart the set-up procedure; note that this change forces $\delta(n)\uparrow$. However once we exit set-up, a change in B below u might cause many layers to disappear but it is still possible that one of $\gamma_i(x)$ remains defined; so we cannot return to a fresh set-up for x. And certainly, once we have attacked, if realisation is destroyed ($\Psi_d(B, x)\uparrow$) then we need to get rid of x, as we cannot extract it from Q.

The value of u

As discussed above, during the set-up phase, any changes to sets A_i or B may increase the value of $x^{(n)}$; we need to keep track of these changes and update the value of u. Once we leave set-up we cannot update the value of u anymore; peeling the layers one by one would result in increases to $x^{(n)}$, but at least one of $\gamma_i(x)$ is still defined, so we cannot increase this use to be the new $x^{(n)}$. Once we leave set-up, the value of u is fixed (until the follower x is cancelled).

Actually, one could ask why we ever need to give up on any layer. When the last layer is peeled—say $A_0\restriction_u$ changes but not $A_1\restriction_u$—why shouldn't we just redefine $\gamma_0(x)$ to be the new $x^{(n)}$ and leave $\gamma_1(x) = u$? And later if $A_1\restriction_u$ changes we could update $\gamma_1(x)$ as well. However the change causes $x^{(n)} > \xi_d(x)$. A change now in A_1 below $x^{(n-1)}$ would cause a change in B (rather than A_0) below the new $x^{(n)}$ but not below $\xi_d(x)$; we cannot cancel x, so we are peeling another layer even though we tried to resurrect the last layer. In other words, there is no way to actually revive the last layer: one change means it is gone.

The value of $\delta(n)$

To keep $\Delta(D)$ total, as usual, we need to ensure that $\delta(n)\downarrow$ at every stage (even if the guess $\Delta(D) = f^e$ is wrong), and we need to ensure that the value of this use is bounded. When exiting set-up we have $\delta(n) = \lambda(u)$; when we see that x is realised we will likely have $\psi_d(x) > u$ so will not have $\delta(n) \geqslant \lambda(\xi_d(x))$. This means that during maintenance it is quite possible that a B-change causes the realising computation $\Psi_d(B, x)$ to disappear, but D does not change below $\delta(n)$. In this case we need to go back to the realisation procedure and cannot cancel x.

However, once we attack, it is important that $\delta(n) \geqslant \lambda(\xi_d(x))$; the reason is that if $B\restriction_{\xi_d(x)}$ changes we must be able to cancel x, as it is already enumerated into Q. However the double change in $A_i\restriction_u$ that enabled that very attack caused $\delta(n)\uparrow$, and this allows us to redefine $\delta(n)$ to be at least $\lambda(\xi_d(x))$ as required.

Further, during maintenance, if we see one layer peeled then we must update $\delta(n)$ to be $\lambda(\xi_d(x))$. The reason is that while waiting for the opponent to pay for this peeling we may see that $\Psi_d(B, x)\uparrow$. We would then like to cancel x: if we do not do so, while waiting we may see more layers unravel, so we would like to attack, but obviously cannot do so if x is no longer realised.

7.2.3 Construction

We detail how to react to changes during each procedure for an anchor n for an agent d (for construction e). Recall that during the construction, at each stage, every agent for every construction appoints a new anchor n and starts cycling through the procedures for n. The following description of these procedures therefore describes the entire construction.

SET-UP.

1. Appoint a new follower x. Define $\delta(n) = \lambda(x^{(n)})$. Wait for $\Delta(D, n) = f^e(n)$.

- While waiting, if D changes below $\delta(n)$, we redefine $\delta(n)$ using the current value of $x^{(n)}$.

2. Once we see that $\Delta(D,n)=f^e(n)$, we define $u=x^{(n)}$ and $\gamma_i(x)=u$, and move to realisation.

REALISATION.

1. Wait for $\Psi_d(B,x)\downarrow=0$.

- If, while waiting, we see that D changes below $\delta(n)$, then we cancel x and return to set-up.

2. Once we see that $\Psi_d(B,x)\downarrow=0$, we define $\xi_d(x)=\max\{u,\psi_d(x)\}$, and move to maintenance.

MAINTENANCE.

We wait for a change in D below $\delta(n)$ or in B below $\xi_d(x)$. When we see such a change we react according to the first case which applies:

(a) *Cancellation:* If both $\delta(n)\uparrow$ and $\xi_d(x)\uparrow$ then we cancel x and return to set-up.
(b) *Realisation:* If $\xi_d(x)\uparrow$ (but $\delta(n)\downarrow$), we return to the realisation phase.
(c) *Attack:* If both $\gamma_i(x)\uparrow$ (but $\xi_d(x)\downarrow$) then we move to the attack phase.
(d) *Layer peeled:* If only one $\gamma_i(x)\uparrow$ then we redefine $\delta(n)=\lambda(\xi_d(x))$ and wait for $\Delta(D,n)=f^e(n)$.

- While waiting, if one of the cases (a), (c), or (e) applies, we react accordingly. (b) cannot happen anymore.

When we see the required agreement we redefine $\gamma_i(x)=u$, $\delta(n)=\lambda(u)$, and stay at the maintenance phase.

(e) *Trivial change:* If only $\delta(n)\uparrow$ then we redefine $\delta(n)=\lambda(\xi_d(x))$ and stay at the maintenance phase.

ATTACK.

1. We enumerate x into Q. We define $\delta(n)=\lambda(\xi_d(x))$.

2. We wait for $\xi_d(x)\uparrow$. When this is observed, we cancel x and return to set-up.

- While waiting, if we see that $\delta(n)\uparrow$, we redefine $\delta(n)=\lambda(\xi_d(x))$, and keep waiting.

7.2.4 Verification

Lemma 7.10. *Let e be a construction, d an agent for e, and n an anchor for d. There is a follower which is appointed for n and is never cancelled.*

Proof. Let s_0 be a stage after which $f^e(n)$ does not change. Suppose that at some stage $s_1 > s_0$ a follower x is appointed for n. Then the set-up phase is never exited, and so x is never cancelled. □

Lemma 7.11. *$\Delta(D)$ is total.*

Proof. Let $n<\omega$ be an anchor for some agent d (for construction e). We note that $\delta(n)$ is never left undefined at the end of a stage, so we just need to show that the value of $\delta(n)$ is bounded (over all stages).

By Lemma 7.10, let x be the last follower appointed for n. There are several possibilities for where we can end up with x.

(1) It is possible to get stuck forever waiting for realisation. In this case, we know that $\delta(n)$ can never get undefined after starting the realisation run, as that would cancel x.
(2) An attack with x is performed. We would never end this attack. The value $\xi_d(x)$ is constant during the attack. During the attack we let $\delta(n)=\lambda(\xi_d(x))$. Since $\Lambda(D)$ is total, the value $\lambda(v)$ stabilizes for all v.
(3) It is possible to be left in the set-up cycle, never getting a correct f^e guess. The value of $x^{(n)}$ may change a number of times, but since $\Phi_i(B, A_{1-i})$ are both total, it eventually stabilises. We always define $\delta(n)=\lambda(x^{(n)})$, and so again since $\Lambda(D)$ is total, this value is eventually constant.
(4) After we enter the maintenance phase, $D\restriction_{\delta(n)}$ never changes. In this case obviously $\delta(n)$ is constant after we enter maintenance.
(5) We enter maintenance with x, and at some stage s_1 after that we see a $D\restriction_{\delta(n)}$-change. We then define $\delta(n)=\lambda(\xi_d(x))$. After stage s_1 there cannot be a change in $B\restriction_{\xi_d(x)}$—such a change would cause us to cancel x. We will therefore remain at maintenance and always define $\delta(n)=\lambda(\xi_d(x))$; again, this reaches a limit. □

We fix some e such that $\Delta(D)=f^e$, and continue with omitting the superscript e.

Lemma 7.12. *Q is computable from both A_0 and A_1.*

Proof. The construction ensures that a follower x never enters Q unless both $\Gamma_0(A_0, x)\uparrow$ and $\Gamma_1(A_1, x)\uparrow$. We always define $\Gamma_i(A_i, x)$ to agree with $Q(x)$; so we just need to show that $\gamma_i(x)\downarrow$, or x is cancelled, or is enumerated into Q. Suppose that x is a follower (for some anchor n for some agent d) which is never cancelled and is never enumerated into Q. We show that $\gamma_i(x)$ is defined at infinitely many stages, and that the value is bounded. (As usual we assume that if x is cancelled, or never chosen as a follower, or is enumerated into Q, then we eventually define both computations $\Gamma_i(A_i, x)$ with use 0.)

Since the guess $\Delta(D)=f^e$ is correct, we successfully exit the set-up phase for x. After set-up, the parameter u is fixed, and $\gamma_i(x)$, when defined, is always defined to equal u, and is thus bounded. The only time after set-up at which $\gamma_i(x)$ is undefined is when a layer is peeled, and we wait for agreement between $\Delta(n)$ and $f^e(n)$; such agreement will eventually be found, and then $\gamma_i(x)$ will be redefined. Since whenever $\gamma_i(x)\uparrow$ we also get $\delta(n)\uparrow$, any other context at which $\gamma_i(x)\uparrow$ causes x to be cancelled (or attacked with). □

If $Q \not\leqslant_{\mathrm{T}} B$ then we are done. Otherwise, we fix some d such that $\Psi_d(B)=Q$; we will show that $\Xi_d(B)$ computes A_0 successfully. We made sure that if a follower x for agent d is ever cancelled, then $\xi_d(x)\uparrow$ when we do so. The agent d appoints a new anchor at every stage; by Lemma 7.10, for each one there is a follower which is never cancelled. So it suffices to show that if x is a follower for agent d which is never cancelled, then eventually a permanent computation $\Xi_d(B,x)$ is defined, and this computation correctly computes $A_0\restriction_x$. Fix a never-cancelled follower x for an anchor n for agent d.

Since e's guess that $\Delta(D)=f^e$ is correct, we exit the set-up phase with x. Since $\Psi_d(B)=Q$, every time we enter the realisation phase with x we will also exit it. Further, the use $\psi_d(x)$ reaches a limit, which implies that the use $\xi_d(x)$ reaches a limit; whence we eventually define a permanent computation $\Xi_d(B,x)$. We need to verify its correctness. We note that since x is never cancelled, we do not enter the attack phase with x. And so after the permanent computation $\Xi_d(B,x)$ is defined, we will forever be in maintenance with x, potentially observing layers being peeled. Again, since e is correct, after each peeling we will observe agreement between $\Delta(D,n)$ and $f^e(n)$.

Let s^* be the stage at which the permanent computation $\Xi_d(B,x)$ is defined. We need to show that $A_0\restriction_x = A_{0,s^*}\restriction_x$. This is the heart of the argument: showing that setting up sufficiently many layers protects the correctness of $\Xi_d(B,x)$. First we observe again that between set-up and last realisation we do not see $D\restriction_{\delta(n)}$-changes. That is, if t^* is the stage at which set-up of x is exited, then $D_{s^*}\restriction_{\delta(n)} = D_{t^*}\restriction_{\delta(n)}$; otherwise, we would increase $\delta(n)$ to be $\xi_d(x)$, and then at some stage before stage s^*, x would be cancelled. This implies that $A_{i,s^*}\restriction_u = A_{i,t^*}\restriction_u$ and $B_{s^*}\restriction_u = B_{t^*}\restriction_u$; since $u=x^{(n)}$ as calculated at stage t^*, we have $u=x^{(n)}$ at stage s^* as well.

For $k\leqslant n$ we let $v_k=x^{(k)}$ as calculated at stage s^* (or t^*); and we let $s_1 < s_2 < s_3 < \cdots < s_m$ be the stages at which a layer for x is peeled (stages at which we observe case (d) of the maintenance cycle for x). So for some $i<2$, $A_{i,s_k+1}\restriction_u \neq A_{i,s_k}\restriction_u$.

Since $o^e_0(n)\leqslant n$ and during the set-up stage we force one change in $\Delta(D,n)$, we have $o^e_{s_1}(n)\leqslant n-1$. Every time a layer is peeled we force one more change in $\Delta(D,n)$; this implies that for all k, $o^e_{s_k}(n)\leqslant n-k$. It follows that $m\leqslant n$.

Lemma 7.13. *For all $k\leqslant m$, for both $i=0,1$,*

$$A_{i,s_k}\restriction_{v_{n-k+1}} = A_{i,s^*}\restriction_{v_{n-k+1}}\,. \tag{7.1}$$

Proof. The stage s_1 is the least stage after stage s^* at which we see any change in either A_i below u. In other words, $A_{i,s_1}\restriction_u = A_{i,s^*}\restriction_u$; since $u = v_n$, the equalities (7.1) hold for $k=1$.

Now by induction let $k<m$ and suppose that Eq. (7.1) holds for k (for both $i<2$). We note that for all $s>s^*$ and $r\leqslant n$, if $A_{i,s}\restriction_{v_r} = A_{i,s^*}\restriction_{v_r}$ for both i then $x^{(r)} = v_r$ when calculated at stage s. Fix i such that $A_{i,s_k+1}\restriction_u \neq A_{i,s_k}\restriction_u$. Since at the beginning of stage s_k, $x^{(n-k+1)} = v_{n-k+1}$, the fact that $A_{1-i}\restriction_u$ does not change at stage s_k implies that the change in A_i at that stage is necessarily above v_{n-k}. Now, by induction on $s\in(s_k, s_{k+1})$ we show that for both $j<2$, $A_{j,s}\restriction_{v_{n-k}} = A_{j,s_k}\restriction_{v_{n-k}}$.

Let $t_k > s_k$ be the stage at which we exit the peeling subroutine (d) of the maintenance cycle that we enter at stage s_k. Suppose that $s\in(s_k,t_k)$. Between stages s_k and t_k we see no changes in $A_{1-i}\restriction_u$ as such a change would open an attack. Recall that we are assuming that $B\restriction_{\xi_d(x)}$, and hence $B\restriction_u$, is correct from stage s^* onwards. This, and the fact that $A_{1-i,s}\restriction_u = A_{1-i,s_k}\restriction_u$, implies that $\varphi_i(B, A_{1-i}, v_{n-k})[s]\leqslant v_{n-k+1}$, and that $A_{i,s}\restriction_{v_{n-k}} = A_{i,s_k}\restriction_{v_{n-k}}$.

After stage t_k and before stage s_{k+1} we see no changes in $A_j\restriction_u$ for either $j<2$; this follows from the definition of s_{k+1}. It follows that for both $j<2$, $A_{j,s_{k+1}}\restriction_{v_{n-k}} = A_{j,s_k}\restriction_{v_{n-k}} = A_{j,s^*}\restriction_{v_{n-k}}$ as required. □

7.3 DEFEATING TWO GATES

We go up one level in our hierarchy; in this section we show:

- a uniformly totally ω^2-c.a. c.e. degree does not bound a copy of the 1-3-1 lattice in the c.e. degrees.

Of course the main difference between this and the previous section must come from the fact that some uniformly totally ω^2-c.a. degrees do bound critical triples (those which are not totally ω-c.a.). We observe that if $\mathbf{a}_0$, $\mathbf{a}_1$ and $\mathbf{a}_2$ are the middle elements of the 1-3-1 lattice then each of the $\mathbf{a}_i$ is the centre of a critical triple (consisting of these three elements). Given a c.e. set D of uniformly totally ω^2-c.a. degree and $B_0, B_1, A\leqslant_{\mathrm{T}} D$ we show that either B_0 is not the centre of a critical triple B_1, B_0, A; or B_1 is not the centre of a critical triple B_0, B_1, A. As expected, this adds one more level of nonuniformity.

The main idea is the following. We enumerate a c.e. set $Q = Q^e$ which will be computable from A and B_0, and try to ensure that $Q\nleqslant_{\mathrm{T}} B_1$. If we fail, say $\Psi_d(B_1) = Q$, then we enumerate a backup set $Q_d = Q_d^e$, this time computable from A and B_1, and hope that $Q_d\nleqslant_{\mathrm{T}} B_0$. If we fail then we will ensure that $B_1\leqslant_{\mathrm{T}} B_0$.

The number of times that $D\restriction_{\delta(n)}$ could change will be at most ωn. We will appoint two followers x and y; the latter targeted for Q, the former for Q_d. We will ensure that if the remaining number of changes is $\omega m + k$ then $y > u_x \geqslant x^{(m)}$ and $u_y \geqslant y^{(k)}$, where u_x and u_y are our analogues of u of the previous construction. The peeling as above will happen from outside in: first, y layers will be peeled by successive A- and B_0 changes, while B_1 remains unchanged. When all the y-layers have been peeled, one or two x-layers will be peeled. But peeling the x-layers happens in successive A- and B_1-changes, not B_0-changes. Such a B_1-change will allow us to cancel our follower y (while keeping x), and set up a new version of y, with however many new layers we might need (the new ordinal is now $\omega(m-1)+k'$, with k' as large as our opponent may like).

An overall intuition is that the alternation between A, B_0-peeling and A, B_1-peeling reflects the re-targeting of traces in two gates of the pinball machine used for constructing an embedding of the 1-3-1 lattice. Speaking vaguely, we say that a degree which is not totally ω-c.a. has enough power to pass one gate, but may run out of gas when trying to pass two gates.

7.3.1 Discussion

We start with some details. Let D be a c.e. set whose Turing degree is uniformly totally ω^2-c.a. Let $A, B_0, B_1 \leqslant_T D$; fix a functional Λ such that $\Lambda(D) = (A, B_0, B_1)$. We further suppose that any two of these sets compute the third; we fix functionals Φ, Φ_0 and Φ_1 such that $\Phi(B_0, B_1) = A$, $\Phi_0(A, B_1) = B_0$ and $\Phi_1(A, B_0) = B_1$. For $x < \omega$ we define

$$x^{(1)} = \max\left\{\varphi(B_0, B_1, x), \varphi_0(A, B_1, x), \varphi_1(A, B_0, x)\right\}$$

and $x^{(n+1)} = (x^{(n)})^{(1)}$.

Again the idea is that a change in one of the sets A, B_0 or B_1 below x necessitates a change in one other of these sets below $x^{(1)}$.

Let $h(n) = \omega n$; let $\langle f^e, o^e \rangle$ be an effective listing of all h-c.a. functions (with tidy $(h+1)$-computable approximations). We will define a functional Δ; the eth construction will guess that $\Delta(D) = f^e$.

The eth construction will enumerate a c.e. set $Q = Q^e$ and functionals $\Gamma = \Gamma^e$ and $\Theta = \Theta^e$ with the aim of having $\Gamma(A) = Q$ and $\Theta(B_0) = Q$. Further, for each $d < \omega$ the construction will enumerate a c.e. set $Q_d = Q^e_d$ and functionals $\Gamma_d = \Gamma^e_d$ and $\Theta_d = \Theta^e_d$ with the aim of having $\Gamma_d(A) = Q_d$ and $\Theta_d(B_1) = Q_d$. The action for the construction will be done by agents indexed by pairs of natural numbers. An agent (d, c) for the eth construction will enumerate a functional $\Xi_{d,c} = \Xi^e_{d,c}$ with the aim of computing B_1 from $\Xi_{d,c}(B_0)$.

As mentioned, each anchor for each agent will try to appoint a pair of followers x and y. The movement between the four procedures is now complicated by the fact that each x can have several y's. In other words we will sometimes cancel y but not x (we always cancel y if we cancel x). So for example we may need to return to the set-up procedure to set up a new y; but a change may cause us to interrupt the set-up and either cancel x or attack with it.

How should we set up our uses? On top of the principles applied in the simpler construction above, we have the following. Recall that the idea is to set up $x < u_x < y < u_y$ and to arrange that if at the current stage we have $o^e(n) = \omega m + k$ then $u_y \geqslant y^{(k)}$ and $u_x \geqslant x^{(2m)}$. We need to think about the possible changes and at which times they occur.

The follower y behaves similarly to the follower in the previous construction. It is targeted for Q; we will define $\gamma(y) = \theta(y) = u_y$ once we leave the set-up procedure (and define $\delta(n) \geqslant \lambda(u_y)$). After y is realised ($\Psi_d(B_1, y)\downarrow = 0$), when both A and B_0 change below u_y we will be able to attack with y: enumerate it into Q. Changes in B_1 below $\psi_d(y)$ will either cause us to return to the realisation phase or to cancel y; when a single layer is peeled (either $\gamma(y)\uparrow$ or $\theta(y)\uparrow$) then we redefine $\Delta(D, n)$ and wait for the opponent to catch-up.

The follower x is targeted for Q_d; we will be able to attack with x if we see that $\Psi_c(B_0, x)\downarrow = 0$ and then both $\gamma_d(x)$ and $\theta_d(x)$ are undefined. As discussed, the idea is that if two layers below u_x are peeled and x is still realised (no change in B_0) then we are guaranteed a change in B_1 (and in A); so we would be able to cancel y and set up many layers for the new y.

One role of x in the simpler construction is taken up in this construction by x and not by y: we will define $\xi_{d,c}(x) \geqslant \psi_c(x)$, and will use the peeling of x-layers to protect the computation $\Xi_{d,c}(B_0, x) = B_1 \restriction_x$. The role of the y-layers is secondary; they protect the x-layers. As before, we can only cancel x if it becomes unrealised ($\psi_c(x)\uparrow$)—otherwise we need to keep protecting the correctness of the $\Xi_{d,c}$-computation. However, we will also only be allowed to cancel y if it is unrealised ($\psi_d(y)\uparrow$); while it is realised, it needs to keep protecting the outermost x-layers.

A threat

The success of this process relies on the layers between y and u_y to be peeled one at a time, so that when the two layers below u_x are peeled, we will have already seen $o^e(n)$ drop below the next limit ordinal (we see $\omega m' + k'$ for some $m' < m$). Consider though the situation in which layers between y and u_y are still unpeeled, but the last layer below u_x is peeled due to an A-change. Of course there is a change in either B_0 or B_1 on the first y-layer; the former would allow us to attack with y. The latter would allow us to cancel y. However, our opponent will pay by decreasing the ordinal, but not below the limit ordinal ωm; rather, to $\omega m + k'$, for $k' < k$. We are now left with insufficiently many x-layers.

In this situation what we would really like to do is attack with x. For this reason we will define the use $\theta_d(x)$ to be at least u_y, not u_x.

In fact, we will want to define $\theta_d(x) \geqslant \psi_d(y)$ as well. This is done to prepare the ground for the new follower. When y is cancelled we appoint a new one, say y', and then we would like to define $\theta_d(x) \geqslant u_{y'}$. For us to be able to do so, we need $\theta_d(x)\uparrow$ when y is cancelled. The cancellation of y of course follows from $\psi_d(y)\uparrow$.

This requirement in turn means that while we are waiting for y to be realised, we must leave $\theta_d(x)$ undefined. This is ok because we only need to use the set Q_d if our first attempt with Q has failed; we only need $\Theta_d(B_1)=Q_d$ if $\Psi_d(B_1)=Q$.

Similarly, if during an attack with y we see that $\gamma_d(x)\uparrow$, then we leave it undefined for the duration of the attack. The attack is prompted by changes in A and in B_0, but B_1 remains fixed; in particular, $\theta_d(x)\downarrow$. The A-change below $u_x=\gamma_d(x)$ causes an x-layer to be peeled; the opponent has not paid for this by successive peeling of y-layers. If the attack later fails (B_1 changes below $\psi_d(y)\leqslant\theta_d(x)$) then the fact that $\gamma_d(x)\uparrow$ will allow us to attack with x instead.

7.3.2 Construction

At every stage, every agent (d,c) for a construction e appoints a new anchor n and starts a new set-up procedure for n. We then cycle through the four procedures for n as soon described. For brevity:

- We say that x is *realised* if $\xi_{d,c}(x)\downarrow$. We say that y is realised if $\theta_d(y)\downarrow$.
- We say that a follower is *confirmed* if we have already exited the set-up cycle during which it was appointed.
- We *may cancel* a follower if it is confirmed, unrealised and $\delta(n)\uparrow$.
- We *may attack* with x if it is realised, and both $\gamma_d(x)\uparrow$ and $\theta_d(x)\uparrow$. We may attack with y if it is realised, and both $\gamma(y)\uparrow$ and $\theta(y)\uparrow$.

We stipulate that throughout the construction, including the set-up cycle, if we may cancel x or attack with it then we do so; *in either case we cancel* y. Otherwise, if we may cancel y or attack with it we do so, except during the set-up of y. If we cancel a follower but are not attacking, then we return to the set-up cycle. These instructions override all other instructions during the construction.

We now describe the procedures. For each procedure we also list (in small font) facts about divergence of functionals at the beginning of each procedure, to be verified later.

SET-UP: $\delta(n)\uparrow$ AND $\theta_d(x)\uparrow$.

1. If x is not currently defined, appoint a new follower x. In either case, appoint a new follower $y>x^{(2n)}$. Define $\delta(n)=\lambda(y^{(k)})$, where currently $o^e(n)=\omega m+k$. Wait for $\Delta(D,n)=f^e(n)$. Note that if x is already defined, then it is realised, and we choose $y>\xi_{d,c}(x)$, so $\delta(n)\geqslant\lambda(\xi_{d,c}(x))$.

While waiting, we react to changes as follows.

- If x was appointed during this set-up cycle, and one of A, B_0 or B_1 changes below $x^{(2n)}$, we cancel y, appoint a new y, and redefine $\delta(n)$ accordingly.
- Otherwise, if $\delta(n)\uparrow$ then we redefine $\delta(n)=\lambda(y^{(k)})$ (using the current value of $y^{(k)}$).

2. Once we see that $\Delta(D,n)=f^e(n)$, we define $u_y=y^{(k)}$ and $\gamma(y)=\theta(y)=u_y$. If x was appointed during this cycle, then we define $u_x=x^{(2n)}$. If $\gamma_d(x)\uparrow$ then we define $\gamma_d(x)=u_x$. As discussed, we leave $\theta_d(x)$ undefined.

We move to realisation.

REALISATION: $\theta_d(x)\uparrow$ OR $\xi_{d,c}(x)\uparrow$.

1. If y is unrealised, wait for $\Psi_d(B_1,y)\downarrow=0$. Once this is observed, define $\theta_d(x)=\max\{u_y,\psi_d(y)\}$.

2. If x is unrealised, wait for $\Psi_c(B_0,x)\downarrow=0$. Once this is observed, define $\xi_{d,c}(x)=\max\{u_x,\psi_c(x)\}$; move to maintenance. We could have defined $\xi_{d,c}(x)\geqslant u_y$ but this cannot be maintained, since we may later cancel y but be unable to move $\xi_{d,c}(x)$.

MAINTENANCE: ALL FUNCTIONALS DEFINED.

We wait for a change in D below $\delta(n)$ or for x or y to become unrealised. When this occurs:

(a) If x or y are unrealised, move to realisation.
(b) If a layer is peeled—either $\gamma(y)\uparrow$ or $\theta(y)\uparrow$, but not both—redefine $\delta(n)=\lambda\left(\max\{\theta_d(x),\xi_{d,c}(x)\}\right)$. Wait for $\Delta(D,n)=f^e(n)$. While waiting, if $\delta(n)\uparrow$ (but no attack or cancellation are possible) then we just redefine it by the same formula. When $\Delta(D,n)=f^e(n)$ is observed we redefine all the markers $\gamma(y),\theta(y),\gamma_d(x)$ which are undefined, with value u_y or u_x as appropriate.
(c) If only $\delta(n)\uparrow$ then we redefine $\delta(n)=\lambda\left(\max\{\theta_d(x),\xi_{d,c}(x)\}\right)$ and stay at the maintenance phase.

ATTACK WITH y: $\theta(y)\uparrow$, $\gamma(y)\uparrow$, $\delta(n)\uparrow$.

We enumerate y into Q. We define $\delta(n)=\lambda\left(\max\{\theta_d(x),\xi_{d,c}(x)\}\right)$. We wait for changes. If $\delta(n)\uparrow$ we redefine it according to the formula above. As discussed, if $\gamma_d(x)\uparrow$ we leave it undefined.

ATTACK WITH x: $\theta_d(x)\uparrow$, $\gamma_d(x)\uparrow$, $\delta(n)\uparrow$.

We enumerate x into Q_d. We define $\delta(n)=\lambda\left(\xi_{d,c}(x)\right)$. If $\delta(n)\uparrow$ we redefine it according to the same formula.

7.3.3 Verification

First, we observe that functionals discussed indeed diverge as promised at the beginning of each cycle. For example, we indeed have $\delta(n)\uparrow$ at the beginning

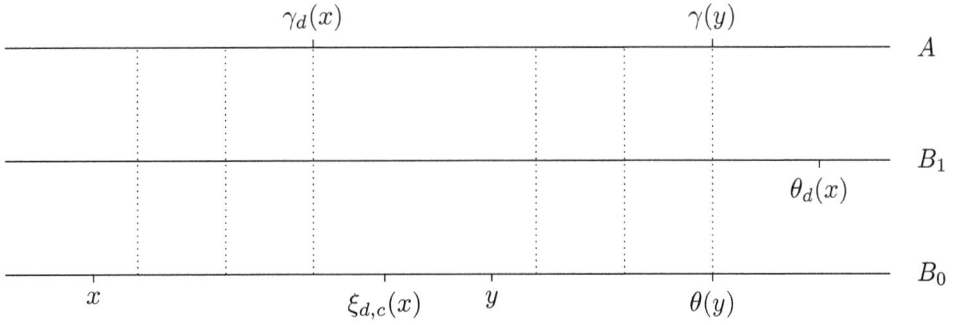

Figure 7.5. Two gates: a typical configuration.

of an attack because we always define $\delta(n) \geqslant \lambda(u_y)$ (which in turn is at least $\lambda(u_x)$), and $\gamma(y) = u_y$ and $\gamma_d(x) = u_x$ whenever they are defined. Similarly, when we return to a set-up and x is not cancelled, it is because y is cancelled; y became unrealised, which means that $\theta_d(x)\uparrow$.

We also observe that the instructions described cover all possible occurrences. Consider for example the maintenance cycle. We stipulated that if x or y can be either cancelled or attacked with then we do so (with x having precedence over y in that respect). Suppose that $\delta(n)\uparrow$ during maintenance. If x or y are unrealised, then they are cancelled. Otherwise, at most one of $\gamma(y)\uparrow$ or $\theta(y)\uparrow$, in which case a y-layer is peeled; and possibly $\gamma_d(x)\uparrow$ but as y is realised, $\theta_d(x)\downarrow$, so an x-layer is peeled.

Also observe that during an attack with x, if x becomes unrealised then it is cancelled, as $\delta(n) \geqslant \lambda(\xi_{d,c}(x))$. Similarly, during an attack with y, if either x or y becomes unrealised then it is cancelled. And similarly, if $\delta(n)\uparrow$ during maintenance then we never return to the realisation cycle without passing through set-up again.

We note that if we attack with x then we may indeed cancel y, as $\theta_d(x)\uparrow$ implies that y is unrealised, and $\gamma_d(x)\uparrow$ implies that $\delta(n)\uparrow$.

Finally note that $\theta_d(x) \geqslant \psi_d(y)$ so if y is realised then $\Psi_d(B_1, y)\downarrow = 0$; if x is realised then $\Psi_c(B_0, x)\downarrow = 0$.

We extend Lemma 7.10.

Lemma 7.14. *Let e be a construction, d an agent for e, and n an anchor for d. There is a follower x for n which is never cancelled. There is a last follower y for n which is ever appointed; it is only cancelled if we attack with x.*

Proof. As in the proof of Lemma 7.10, let s_0 be a stage after which $f^e(n)$ does not change. Suppose that at some stage $s_1 > s_0$ we are in the set-up cycle. The follower x at that time will never be cancelled. The follower y may be cancelled, but only if one of the sets A, B_0 or B_1 change below $x^{(2n)}$. Eventually, the value of $x^{(2n)}$ stabilizes. □

Lemma 7.15. *$\Delta(D)$ is total.*

Proof. Let $n<\omega$ be an anchor for some agent d (for a construction e). Again we note that $\delta(n)$ is never left undefined at the end of a stage, so we just need to show that the value of $\delta(n)$ is bounded (over all stages).

By Lemma 7.14, let x and y be the last followers appointed for n. There are several possibilities for where we can end up.

(1) It is possible to get stuck forever waiting for realisation for either x or y. In this case, we know that $\delta(n)$ can never get undefined after starting the realisation run, as that would cancel x or y.
(2) An attack with x or with y is performed. The attack with y can be exited only if we start an attack with x (otherwise, y is cancelled). The attack with x cannot be exited. The value $\theta_d(x)$ is constant during an attack with y; the value $\xi_{d,c}(x)$ is constant during an attack with y or with x. And $\Lambda(D)$ is total.
(3) It is possible to be left in the set-up cycle, never getting a correct f^e guess. The value of $o^e(n)$ and so of $y^{(k)}$ eventually stabilizes; we again then use the totality of $\Lambda(D)$.
(4) After we enter the maintenance phase, $D\upharpoonright_{\delta(n)}$ never changes. In this case obviously $\delta(n)$ is constant after we enter maintenance.
(5) We enter maintenance with x, and at some stage s_1 after that we see a $D\upharpoonright_{\delta(n)}$-change. After that stage, x and y are always realised. □

As above we fix e such that $\Delta(D)=f^e$.

Lemma 7.16. *Q is computable from both A and B_0.*

Proof. The proof is pretty much identical to the proof of Lemma 7.12: if y is a permanent follower for some anchor n for some agent for e, then u_y is eventually defined; if we never attack with y then we only leave $\gamma(y)$ or $\theta(y)$ undefined when waiting for agreement between $\Delta(D,n)$ and $f^e(n)$ (after a layer is peeled). □

If $Q\not\leq_{\mathrm{T}} B_1$ then we are done. Otherwise fix some d such that $\Psi_d(B_1)=Q$.

Lemma 7.17. *Q_d is computable from both A and B_1.*

Proof. The proof is slightly more elaborate; let x be a follower for an anchor n for an agent (d,c), and suppose that x is neither cancelled nor attacked with. We consider stages during which $\gamma_d(x)\uparrow$ or $\theta_d(x)\uparrow$.

We possibly have $\gamma_d(x)\uparrow$ while waiting for agreement between $\Delta(D,n)$ and $f^e(n)$. As for y, during a realisation cycle, if $\gamma_d(x)\uparrow$ then $\delta(n)\uparrow$ and then we cancel x or y; eventually this stops happening. We may also have $\gamma_d(x)\uparrow$ during an attack with some y. But such an attack must end, as $\Psi_d(B_1)=Q$. So $\gamma_d(x)$ is defined at all but finitely many stages, and its value is constant u_x.

Usually, when $\theta_d(x)\uparrow$ we can cancel y. Otherwise, we can have $\theta_d(x)$ while we are waiting for some y to be realised (here it is important that if both y and x are unrealised, we first realise y, then x); but $\Psi_d(B_1)=Q$ implies that every y is eventually realised or cancelled. There will be a last y appointed, and never cancelled (as we assumed that we do not attack with x); and the value $\psi_d(x)$ will eventually stabilise. This implies that the values of $\theta_d(x)$ are bounded. □

If $Q_d \not\leqslant_T B_0$ then we are done. Otherwise fix some c such that $\Psi_c(B_0)=Q_d$. We will show that with $\Xi_{d,c}$, B_0 correctly computes B_1. As in the simpler construction, we need to show that if x is a follower for some anchor for the agent (d,c), and x is never cancelled, then eventually we define a computation $\Xi_{d,c}(B_0,x)$ which always converges, and that $B_1\restriction_x$ is constant from the stage at which this computation is defined. Fix such x. The argument of the simpler construction shows that $\xi_{d,c}(x)$ is bounded and defined at infinitely many stages. We only need to notice that if y is the last follower appointed for x's anchor, then every realisation cycle that we enter after appointing y must be exited, as both $\Psi_d(B_1)=Q$ and $\Psi_c(B_0)=Q_d$.

So it all comes down to correctness, which as above is the heart of the argument. Let s^* be the stage at which the permanent computation $\Xi_{d,c}(B_0,x)$ is defined. For $k\leqslant 2n$ let $v_k=x^{(k)}$ as calculated at stage s^*. As x is not cancelled, $\delta(n)\downarrow$ at all stages from the end of the set-up of x and stage s^*; it follows that $u_x=v_{2n}$.

The key observation is that the peeling of the x-layers has to alternate between A and B_1. For $k\leqslant n$ let s_k be the least stage $s\geqslant s^*$ such that $B_{1,s+1}\restriction_{v_{2k}}\neq B_{1,s}\restriction_{v_{2k}}$; otherwise let $s_k=\infty$. By induction on $s\in[s^*,s_k]$ we see that $v_{2k-1}=x^{(2k-1)}$ at stage s and that $A_s\restriction_{v_{2k-1}}=A_{s^*}\restriction_{v_{2k-1}}$; but $A_{s_k}\restriction_{v_{2k+1}}\neq A_{s^*}\restriction_{v_{2k+1}}$. Let t_k be the least stage $t\geqslant s^*$ such that $A_{t+1}\restriction_{v_{2k-1}}\neq A_{s^*}\restriction_{v_{2k-1}}$; the fact that we never attack with x implies that $s_n<t_n<s_{n-1}<t_{n-1}<\cdots$.

Lemma 7.18. *For all $k<n$ such that $s_k<\infty$,*

$$o_{s_k}(n)<\omega k \tag{7.2}$$

(where $o=o^e$).

The inequality will imply that s_0 must equal ∞, and so $B_{1,s^*}\restriction_x=B_1\restriction_x$ as required.

Proof. Since we start with $o_0(n)=\omega n$ and we redefine $\Delta(D,n)$ when setting x up, we have $o_{s_n}(n)<\omega n$; so Eq. (7.2) holds for $k=n$.

We prove Eq. (7.2) by induction on k. Fix $k\leqslant n$ such that $s_{k-1}<\infty$, and suppose that $o_{s_k}<\omega k$. Since $v_{2k}\leqslant u_x\leqslant\theta_d(x)$, y is unrealised at stage s_k and

$\delta(n)\uparrow$ at that stage; so we cancel y at stage s_k. At stage t_k we must have $\theta_d(x)\downarrow$, since otherwise we attack with x at that stage. So there is some last stage $r_k \in (s_k, t_k)$ at which we realise a follower $y = y_k$. The familiar argument shows that at stage r_k we have $u_y = y^{(m)}$ where $o_{r_k}(n) = \omega(k-1) + m'$ for some $m' < m$ (we may assume that $o_{r_k}(n) \geqslant \omega(k-1)$, otherwise we are done for this inductive step). The follower y_k is not cancelled before stage t_k. An important point is that we do not attack with y_k before or at stage t_k. To see this, observe that every attack with y_k must eventually fail, and y_k is then cancelled; so this failure does not happen before stage t_k. But then, as $\gamma_d(x)\uparrow$ at stage t_k, it remains undefined until the attack with y fails—and then we would attack with x.

At stage t_k we do not start an attack with x so at that stage $\theta_d(x)\downarrow$ (and recall that $\theta_d(x) \geqslant u_y$). We do not start an attack with y at that stage, whereas $\gamma(y)\uparrow$ at t_k; so $\theta(y)\downarrow$ at t_k. So $y^{(1)} > u_y$ at stage t_k. The only way this could happen is that between stages r_k and t_k, all the layers between y and u_y were peeled. Each time this happens we extract another $\Delta(D, n)$ change; we have m such changes, which drives the ordinal $o_{t_k}(n)$ below $\omega(k-1)$ as required. $\square$

7.4 THE GENERAL CONSTRUCTION

No new ideas are required for the general construction. The general idea is that if we guess that $\Delta(D)$ is ω^m-c.a. then we set up m many followers. We go straight to the details. We are presented with a c.e. set D of totally $<\omega^\omega$-c.a. degree, three c.e. sets A, B_0 and B_1, and reductions $\Lambda(D) = (A, B_0, B_1)$, $\Phi(B_0, B_1) = A$, and $\Phi_i(A, B_{1-i}) = B_i$ for $i = 0, 1$. For $x < \omega$ we define $x^{(1)}$ and $x^{(n)}$ as in the previous section.

For $m < \omega$ define $h_m(n) = \omega^m \cdot n$. Every function computable from D is h_m-c.a. for some m. Fix (uniformly in m) an effective list $\langle f^{e,m}, o^{e,m}\rangle$ of all h_m-c.a. functions, with the usual tidy approximations. For simplicity of notation we will only use odd m's. We enumerate a functional Δ; a construction (e, m) for $e < \omega$ and odd m will guess that $\Delta(D) = f^{e,m}$. Agents for the $(e, m)^{\text{th}}$-construction are indexed by $m+1$-tuples $\bar{d} = (d_0, d_1, \ldots, d_m)$ of natural numbers. For each sequence $\bar{c}$ of length at most m the construction enumerates a c.e. set $Q_{\bar{c}} = Q^{e,m}_{\bar{c}}$ and functionals $\Gamma_{\bar{c}} = \Gamma^{e,m}_{\bar{c}}$ and $\Theta_{\bar{c}} = \Theta^{e,m}_{\bar{c}}$; we plan for $\Gamma_{\bar{c}}(A) = Q_{\bar{c}}$ and for $\Theta_{\bar{c}}(B_{[\bar{c}]}) = Q_{\bar{c}}$, where we let $[\bar{c}] = |\bar{c}| \mod 2$. For simplicity we will also write $[k]$ for $k \mod 2$. Each agent $\bar{d}$ defines a functional $\Xi_{\bar{d}} = \Xi^{e,m}_{\bar{d}}$, hoping that $\Xi_{\bar{d}}(B_0) = B_1$ (if m were even we would need to exchange B_0 and B_1, all the rest would be identical). We write θ, γ, Q for $\theta_{\langle\rangle}, \gamma_{\langle\rangle}, Q_{\langle\rangle}$.

An agent $\bar{d}$ will appoint anchors n, inputs for $\Delta(D)$. Each anchor will try to appoint a sequence of followers $x_m < x_{m-1} < \cdots < x_1 < x_0$, with x_k targeted for $Q_{\bar{d}\upharpoonright k}$. When a follower x_k is cancelled or attacked with, we cancel all the larger followers $x_{k'}$ for $k' < k$. The main idea will be to ensure that if $o^{e,m}(n) = \omega^m p_m + \omega^{m-1} p_{m-1} + \cdots + \omega p_1 + p_0$ then x_{k-1} will bound at least p_k many layers above x_k.

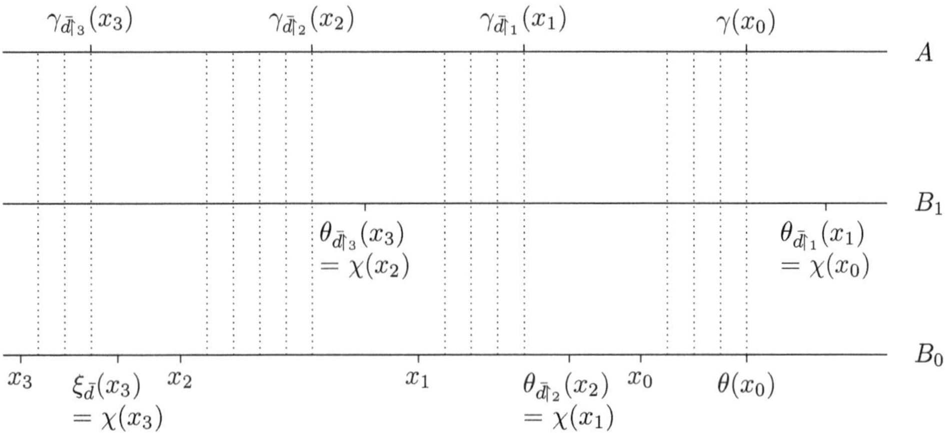

Figure 7.6. Four gates: a typical configuration.

To streamline the description of the construction we define, for $k=0,\ldots,m-1$, $\chi(x_k)=\theta_{\bar{d}\upharpoonright_{k+1}}(x_{k+1})$; and define $\chi(x_m)=\xi_{\bar{d}}(x_m)$. See Figure 7.6. We will say that the follower x_k is *realised* if $\chi(x_k)\downarrow$.

As before, we say that we *may attack* with a follower x_k if it is realised, and both $\gamma_{\bar{d}\upharpoonright_k}(x_k)\downarrow$ and $\theta_{\bar{d}\upharpoonright_k}\downarrow$. We say that a follower x_k is *confirmed* if the set-up cycle at which it was appointed has already finished. We *may cancel* a confirmed follower x_k if it is unrealised and $\delta(n)\uparrow$. Throughout the construction, if we may cancel a follower or attack with it then we do so, always choosing the smallest follower (the one with largest index) with which to attack or cancel. If we cancel a follower and do not start an attack, then we return to the set-up cycle.

We now describe the procedures undertaken by an anchor n.

SET-UP.

1. Say that $x_m, x_{m-1},\ldots,x_{k+1}$ are defined and confirmed. We appoint new followers $x_k<x_{k-1}<x_{k-2}<\cdots<x_0$ so that $x_k>u_{k+1}$, and for all $j<k$, $x_j>x_{j+1}^{(2p_{j+1})}$, where currently $o^{e,m}(n)=\omega^m p_m+\cdots+\omega p_1+p_0$. We then define $\delta(n)=\lambda(x_0^{(p_0)})$, and wait for $\Delta(D,n)=f^{e,m}(n)$. While waiting, we update the values of x_j for $j<k$ and of $\delta(n)$ to keep the desired inequalities. We do so in a conservative way: only cancel x_j if there is a change in A, B_0 or B_1 below $x_{j+1}^{(2p_{j+1})}$.

2. Once we see that $\Delta(D,n)=f^{e,m}(n)$ we define for all $j=1,\ldots,k$, $u_j=x_j^{(2p_j)}$, and define $u_0=x_0^{(p_0)}$. For each j such that $\gamma_{\bar{d}\upharpoonright_j}(x_j)\uparrow$ we define this marker to equal u_j. We also define $\theta(x_0)=u_j$.

We move to realisation.

REALISATION.

For each $k \leqslant m$, if x_k is unrealised, wait for $\Psi_{d_k}(B_{1-[k]}, x_k)\downarrow = 0$. Once this is observed we define $\chi(x_k) = \max\{u_k, \psi_{d_k}(x_k)\}$. That is, we define $\theta_{\bar{d}\upharpoonright_{k+1}}(x_{k+1})$ or $\xi_{\bar{d}}(x_m)$ depending if $k = m$ or $k < m$.

Note that the search is done in parallel, and we define $\chi(x_k)$ immediately when the realising computation is discovered. Once all followers are realised we move to maintenance.

MAINTENANCE.

We wait for a change in D below $\delta(n)$ or for some follower to become unrealised. When this occurs:

(a) If a follower is unrealised, move to realisation. As above, this assumes that $\delta(n)\downarrow$, otherwise we would cancel the follower.

(b) If either $\gamma(x_0)\uparrow$ or $\theta(x_0)\uparrow$, but not both, redefine

$$\delta(n) = \lambda\left(\max\{\chi(x_k) : k \leqslant m\}\right).$$

Wait for $\Delta(D, n) = f^{e,m}(n)$. While waiting, if $\delta(n)\uparrow$ (but no attack or cancellation are possible) then we just redefine it by the same formula. When $\Delta(D, n) = f^{e,m}(n)$ is observed we redefine all the markers $\gamma(x_k)$ and $\theta(x_0)$ which are undefined (with value u_k).

(c) If only $\delta(n)\uparrow$ then we redefine $\delta(n) = \lambda\left(\max\{\chi(x_k) : k \leqslant m\}\right)$ and stay at the maintenance phase.

ATTACK WITH x_k.

We enumerate x_k into $Q_{\bar{d}\upharpoonright_k}$. We define $\delta(n) = \lambda\left(\max\{\chi(x_j) : j \geqslant k\}\right)$. We wait for changes. If $\delta(n)\uparrow$ we redefine it according to the formula above. If $\gamma_d(x_j)\uparrow$ for some $j < k$ we leave it undefined.

7.4.1 Verification

The verification is identical to the two-gate case and so we omit it.

7.4.2 A conjecture

There are two known obstacles for embedding finite lattices into the c.e. degrees. One is structural, involving the impossibility for a re-targeting procedure past a number of gates; this results in the failure to embed lattices such as L_{20} (Lempp and Lerman [63]). Computational strength (highness) is irrelevant here. The other is the interference of a meet requirement with continuous tracing, preventing lattices such as S_8 being embeddable (Lachlan and Soare [62]; see fig. 1.2). Perhaps these are the only obstacles. We thus conjecture:

Conjecture 7.19. If a finite lattice is embeddable into the c.e. degrees then it is embeddable below any non-totally $<\omega^\omega$-c.a. c.e. degree.

A counterexample to the conjecture would need significant new insight into lattice embeddings into the c.e. degrees.

Chapter Eight

Prompt permissions

IN THIS CHAPTER we consider prompt versions of the permitting notions we investigated in this monograph. These prompt notions of permission allow us to perform constructions that are closer to the original construction we considered, rather than their variations when adopted for permitting. For example, in the usual embedding of the 1-3-1 lattice one gets the bottom element to be **0**. When we then added permitting, we showed that the 1-3-1 lattice can be embedded below any not totally $<\omega^\omega$-c.a. degree, but we did not get an embedding with bottom **0**; this seems necessary. However, below any degree which is promptly not totally $<\omega^\omega$-c.a., we can get an embedding of the 1-3-1 lattice with bottom **0**.

8.1 PROMPT CLASSES

Recall that a c.e. set A *permits promptly* if it has an enumeration $\langle A_s \rangle$ such that for some computable function $p \geqslant \mathrm{id}$, for any e, if W_e is infinite then there is some n which enters W_e at some stage s such that $A_s \upharpoonright_n \neq A_{p(s)} \upharpoonright_n$. This notion is invariant under Turing equivalence; a degree permits promptly if and only if it contains a promptly simple set; see [2]. Prompt permitting is the prompt version of simple permitting; a set which permits promptly is in some sense promptly non-computable.

For considering the prompt version of non-total α-c.a. permitting, fix an effective listing $\langle f^{e,\alpha} \rangle$ of all α-c.a. functions, each equipped (uniformly) with tidy $(\alpha+1)$-computable approximations $\langle f^{e,\alpha}_s, o^{e,\alpha}_s \rangle$ as in Proposition 2.7. We will shortly use more properties of this list. However to motivate these properties we first give our definitions.

Definition 8.1. Call a function g *self-modulating* if there is a computable approximation $\langle g_s \rangle$ of g such that:

- for all s and n, $g_s(n) \leqslant s$;
- for all s and n, if $g_s(n) \neq g_{s-1}(n)$ then $g_s(n) = s$ and in fact for all $m \geqslant n$, $g_s(m) = s$.

It follows that for all s, $g_s \leqslant g_{s+1}$ (pointwise) and that if $g_s(n) \neq g_{s-1}(n)$ then $g_s(m) \neq g_{s-1}(m)$ for all $m \geqslant n$. The idea is that g is the modulus of the approximation $\langle g_s \rangle$. Above we used the fact that if $\mathbf{d}$ is c.e. but not totally α-c.a. then there is a self-modulating function $g \in \mathbf{d}$ which is not α-c.a. Note that every self-modulating function has a c.e. degree. Below we assume that each self-modulating function g "comes with" the approximation $\langle g_s \rangle$ of which it is the modulus.

Definition 8.2. A *speed-up function* is a non-decreasing, computable function p such that $p(n) \geqslant n$ for all n.

Definition 8.3. Let g be a self-modulating function and let p be a speed-up function. Let $n < \omega$. Let $\langle f_s, o_s \rangle$ be a tidy $(\alpha+1)$-computable approximation. We say that g promptly p-escapes $\langle f_s, o_s \rangle$ on input n if for all s, if $o_s(n) < \alpha$ and $f_s(n) = g_s(n)$ then $g_{p(s)}(n) \neq g_s(n)$. We say that g promptly p-escapes $\langle f_s, o_s \rangle$ if it promptly p-escapes it on some input.

A self-modulating function g is *promptly not α-c.a.* if there is some speed-up function p such that g promptly p-escapes each $\langle f_s^{e,\alpha}, o_s^{e,\alpha} \rangle$.

A c.e. degree $\mathbf{d}$ is *promptly not totally α-c.a.* if there is a self-modulating function $g \leqslant_{\mathrm{T}} \mathbf{d}$ which is promptly not α-c.a.

Note that if an approximation $\langle f_s^{e,\alpha}, o_s^{e,\alpha} \rangle$ is not eventually α-computable then vacuously, for almost all n, g promptly p-escapes this approximation on n; the power of promptness is when it is applied to "total" approximations (approximations which are eventually α-computable).

8.1.1 Slow-down lemma

Recall how prompt permitting is used in constructions. Suppose for example that we want to show that a promptly permitting degree $\mathbf{d}$ is not half of a minimal pair. Let $D \in \mathbf{d}$ and let B be c.e. and non-computable. We build a c.e. set Q computable from both D and B and plan to make Q noncomputable. To diagonalise against the eth computable set, a requirement appoints a follower x and waits for it to be realised ($\varphi_e(x)\downarrow = 0$). When it is realised we wait for simple permitting from B; $B_{s+1}\restriction_x \neq B_s\restriction_x$. When we see this we ask for prompt permission from D, namely $D_{p(s)}\restriction_x \neq D_s\restriction_x$. If both are granted then we can enumerate x into Q and meet the requirement. Why will permission be granted? Of course we potentially appoint infinitely many followers. Since B is noncomputable, infinitely many of them will be permitted by B. Let $U_e = W_{g(e)}$ be the c.e. set of followers for this requirement which will be permitted by B. Applying prompt permission to this set U_e guarantees prompt permission from D for one of the followers in U_e.

This sketch of an argument involved a little cheating. While indeed we know, by the recursion theorem, an index $g(e)$ for U_e, the effective enumeration of $W_{g(e)}$ may be different from our enumeration of U_e. We put x into U_e at the stage at which B permits x. It is conceivable that x is enumerated into $W_{g(e)}$ at an earlier stage; so the prompt permission for x was given in the past, and is useless for us now. We need to find $g(e)$ such that not only $W_{g(e)} = U_e$ but every number enters $W_{g(e)}$ not before we put it into U_e.

This "slow-down lemma" can be obtained by a more sophisticated use of the recursion theorem (see [91, Thm.XII.1.5]). This elaborate use of the recursion theorem is actually not quite necessary. Interpret the eth partial computable function φ_e as a function of two variables. We can transform this function into an effective enumeration of a c.e. set (call it W_e) such that if φ_e is an effective enumeration $\langle V_{e,s}\rangle$ of a c.e. set V_e (that is, φ_e is total and for all s, $\varphi_e(-,s)$ is the characteristic function of $V_{e,s}$) then $W_e = V_e$ and further, for all s, $W_{e,s} \subseteq V_{e,s}$. Namely, we put x into W_e at stage s if at that stage we have seen sufficiently much convergence from φ_e to see that $x \in V_e$. The slow-down lemma can now be obtained by using the recursion theorem to obtain an index $g(e)$ such that $\varphi_{g(e)}$ is our enumeration of U_e; we then apply prompt permitting to $W_{g(e)}$.

In our usage of prompt permitting of the form given by Definition 8.3 we need a similar form of a slow-down lemma. Namely, to force changes we will define, for some requirement, an α-computable approximation $\langle h_s\rangle$ attempting to trail the function g given by the definition, and ask for immediate changes in g. To do this we will need to find one of the functions $f^{e,\alpha}$ on the list such that for all n, for all s there is some $t \geqslant s$ such that $f^{e,\alpha}_t(n) = h_s(n)$. To obtain this we follow the construction proving Proposition 2.7. Using the notation of the proof of that proposition, we think of φ_e as giving the sequence $\langle h_s, m_s\rangle$ which we transform into the partial approximation $\langle f^{e,\alpha}_s, o^{e,\alpha}_s\rangle$, making sure that as long as $\langle h_s, m_s\rangle$ appears to be a tidy $(\alpha+1)$-computable approximation, we copy every value that shows up. It is this sequence of approximations that we use in Definition 8.3. This sequence will be acceptable in a strong way.

Call a pair $\langle h_s, m_s\rangle$ of partial computable functions a *partial tidy* $(\alpha+1)$-*computable approximation* if for all x and s, $h_s(x)\downarrow \Leftrightarrow m_s(x)\downarrow$ and if so, for all $y \leqslant x$ and $r \leqslant s$, $h_r(y)\downarrow$ and the array $\langle h_r(y), m_r(y)\rangle_{r\leqslant s, y\leqslant x}$ satisfies the conditions for being an initial segment of such an approximation: that is, $m_0(y) = \alpha$, $m_r(y) \leqslant \alpha$, $h_0(y) = 0$, $m_r(y) \leqslant m_{r-1}(y)$, and if $h_r(y) \neq h_{r-1}(y)$ then $m_r(y) < m_{r-1}(y)$. The sequence $\langle f^{e,\alpha}_s, o^{e,\alpha}_s\rangle$ is acceptable in the following sense:

- if $\langle h^e_s, m^e_s\rangle_{e,s<\omega}$ is a sequence of (uniformly) partial tidy $(\alpha+1)$-computable approximations then there is a computable function k (obtained uniformly from an index for the sequence) such that for all e, x and $s > 0$, if $h^e_s(x)\downarrow$ then there is some $t \geqslant s$ such that $o^{k(e),\alpha}_t(x) = m^e_s(x)$ and $f^{k(e),\alpha}_t(x) = h^e_s(x)$.

In particular, for each e, if $\langle h^e_s, m^e_s\rangle$ is a (total) α-computable approximation, then $\left\langle f^{k(e),\alpha}_s, o^{k(e),\alpha}_s\right\rangle$ is eventually α-computable and further, for all n and s there is some $t \geqslant s$ such that $h^e_s(n) = f^{k(e),\alpha}_t(n)$.

Finally, in some arguments it would be useful to assume that like the enumeration of the sets W_e, at each stage s we have only said finitely much about all functions. Formally,

- for all s, e and n, $f^{e,\alpha}_s(n) \leqslant s$, and $o^{e,\alpha}_s(n) < \alpha$ implies $e, n < s$.

8.1.2 Counting down α

The functions $f_s^{e,\alpha}$ are not really important for promptness; it is the ways $o_s^{e,\alpha}$ of counting down α that we need to escape.

Definition 8.4. A *counting down* α is a sequence of uniformly computable functions $\langle o_s \rangle$ from ω to $\alpha+1$ such that for all n, $o_0(n)=\alpha$; $o_s(n)=\alpha$ if $s \leqslant n$; $o_s(n) \leqslant o_{s-1}(n)$ for all n and s; and if $o_s(n) < \alpha$ then $o_s(n-1) < \alpha$ as well.

In other words, $\langle o_s \rangle$ is a counting down α if it appears as the ordinal part in a tidy $(\alpha+1)$-computable approximation $\langle f_s, o_s \rangle$.

Definition 8.5. Let g be a self-modulating function and let p be a speed-up function; let $\langle o_s \rangle$ be a counting down α. We say that g promptly p-escapes $\langle o_s \rangle$ on an input n if for all $s>0$, if $o_s(n) \neq o_{s-1}(n)$ then $g_{p(s)}(n) \neq g_s(n)$. We say that g promptly p-escapes $\langle o_s \rangle$ if it does so on some input.

Lemma 8.6. *Let g be a self-modulating function. Then g is promptly not α-c.a. if and only if there is a speed-up function q such that g promptly q-escapes each $\langle o_s^{e,\alpha} \rangle$.*

One direction is short.

Lemma 8.7. *Let $\langle f_s, o_s \rangle$ be a tidy $(\alpha+1)$-computable approximation such that $f_s(n) \leqslant s$ for all s and n. Suppose that a self-modulating function g promptly p-escapes $\langle o_s \rangle$ on input n. Then it also promptly p-escapes $\langle f_s, o_s \rangle$ on input n.*

Proof. Suppose that $o_s(n) < \alpha$ and that $f_s(n) = g_s(n)$. Let $t \leqslant s$ be the least such that $o_t(n) = o_s(n)$. So $f_s(n) = f_t(n)$. Since $f_t(n) \leqslant t$ we see that $g_s(n) \leqslant t$; since g is self-modulating, this implies that $g_t(n) = g_s(n)$. By assumption, $o_t(n) \neq o_{t-1}(n)$, and so $g_{p(t)}(n) \neq g_t(n) = g_s(n)$. But $p(t) \leqslant p(s)$ and g is non-decreasing so $g_{p(s)}(n) \geqslant g_{p(t)}(n) > g_s(n)$ as required. ∎

Proof of Lemma 8.6. One direction is provided by Lemma 8.7 and one of our conditions on the listing of approximations $\langle f_s^{e,\alpha}, o_s^{e,\alpha} \rangle$. In the other direction suppose that p witnesses that g is promptly not α-c.a. For brevity we write f_s^e and o_s^e for $f_s^{e,\alpha}$ and $o_s^{e,\alpha}$. For each e we define an approximation $\langle h_s^e \rangle$ which chases g as much as o^e allows it. Namely, we define

$$h_s^e(n) = \begin{cases} 0, & \text{if } s=0; \\ h_{s-1}^e(n), & \text{if } s>0 \text{ and } o_s^e(n) = o_{s-1}^e(n); \text{ and} \\ g_s(n), & \text{otherwise.} \end{cases}$$

The approximation $\langle h_s^e, o_s^e\rangle$ is $(\alpha+1)$-computable and tidy. By the α-slow-down lemma find some computable function k such that for all e, n and s there is some $t=t(e,n,s)\geqslant s$ such that $o_t^{k(e)}(n)=o_s^e(n)$ and $f_t^{k(e)}(n)=h_s^e(n)$. For $s<\omega$ define $t^*(s)=\max\{t(e,n,s) : e,n\leqslant s\}$, and let $q(s)=p(t^*(s))$.

Fix e. There is some n such that g promptly p-escapes $\left\langle f_s^{k(e)}, o_s^{k(e)}\right\rangle$ on input n. We claim that g promptly q-escapes $\langle o_s^e\rangle$ on input n. For let $s>0$ be a stage such that $o_s^e(n)\neq o_{s-1}^e(n)$. Then $h_s^e(n)=g_s(n)$; so $f_t^{k(e)}(n)=g_s(n)$ for $t=t(e,n,s)$. We need to show that $g_{q(s)}(n)\neq g_s(n)$. Note that $o_s^e(n)<\alpha$ implies that $e,n<s$, so $t\leqslant t^*(s)$. If $g_t(n)\neq g_s(n)$ then we are done, as $q(s)\geqslant t$. Otherwise $f_t^{k(e)}(n)=g_t(n)$ (and $o_t^{k(e)}(n)=o_s^e(n)<\alpha$) so by our assumption, $g_{p(t)}(n)\neq g_t(n)$; but $q(s)\geqslant p(t)$. □

Therefore for the purposes of promptness we from now on ignore the function part f_s. We state the slow-down lemma in this context. As expected, define a *partial counting down* α to be a partial computable sequence $\langle o_s\rangle$ such that for all s and x: (a) if $o_s(x)\downarrow$ then $o_s(x)\leqslant\alpha$ and $o_t(y)\downarrow$ for all $t\leqslant s$ and $y\leqslant x$; (b) if $o_0(x)\downarrow$ then $o_0(x)=\alpha$; (c) if $s>0$ and $o_s(x)\downarrow$ then $o_s(x)\leqslant o_{s-1}(x)$; (d) if $o_s(x)\downarrow$, $y<x$ and $o_s(y)=\alpha$ then $o_s(x)=\alpha$.

Lemma 8.8. *Suppose that $\langle m_s^e\rangle$ is a uniform sequence of partial countings down α. There is a computable function k such that for all e, s and x, if $m_s^e(x)\downarrow$ then there is some $t\geqslant s$ such that $o_t^{k(e),\alpha}(x)=m_s^e(x)$. The function k can be obtained effectively.*

We can conclude that promptness does not really depend on the choice of list $\langle o_s^{e,\alpha}\rangle$ (as long as it is acceptable).

Corollary 8.9. *Suppose that $\langle m_s^e\rangle$ is a uniformly computable sequence of (total) countings down α such that for all s the set $\{(e,x) : m_s^e(x)<\alpha\}$ is bounded, computably in s. Suppose that a function g is promptly not α-c.a. Then there is a speed-up function q such that g promptly q-escapes $\langle m_s^e\rangle$ for each e.*

Proof. As in the proof of Lemma 8.6 let $t^*(s)$ be a bound on stages $t=t(e,x,s)\geqslant s$ such that $o_t^{k(e)}(x)=m_s^e(x)$ for all e,x such that $m_s^e(x)<\alpha$ (where k is given by the slow-down Lemma 8.8; as above $o^e=o^{e,\alpha}$). Suppose that p witnesses that g is promptly not α-c.a.; let $q(s)=p(t^*(s))$. To see that this works, suppose that g promptly p-escapes $\left\langle o_s^{k(e)}\right\rangle$ on an input x. Let $s>0$ and suppose that $m_s^e(x)\neq m_{s-1}^e(x)$. Then $m_s^e(x)<\alpha$, so $t^*(s)\geqslant t(e,x,s)$. Let u be the least such that $o_u^{k(e)}(x)=m_s^e(x)$; so $u\leqslant t(e,x,s)$. But also $u>t(e,x,s-1)\geqslant s-1$ so $u\geqslant s$. By assumption, $g_{p(u)}(x)\neq g_u(x)$, and $q(s)\geqslant p(u)$. □

We can escape infinitely many inputs.

Lemma 8.10. *Suppose that g is promptly not α-c.a. Then there is some speed-up function q such that for all e there are infinitely many x such that g promptly q-escapes $\langle o_s^{e,\alpha}\rangle$ on input x.*

Proof. Note that the first attempt that comes to mind to prove this does not work. Nonuniformly we could guess an initial segment of g and change an approximation to make sure that permission is not given on the first n locations. But there are infinitely many possible initial segments of a fixed finite length, and we cannot define our speed-up taking into account all of them (see the proof of [91, Thm.XII.1.7(iii)]). What we do is shift by n.

Namely, for all e and n define $m_s^{e,n}(x) = o_s^e(x+n)$ (for brevity let $o_s^e = o_s^{e,\alpha}$). Note that $m_s^{e,n}(x) < \alpha$ implies $e, x, n < s$. Let q be given by Corollary 8.9. Suppose that g promptly q-escapes $\langle m_s^{e,n}\rangle$ on input x; we conclude that g promptly q-escapes $\langle o_s^e\rangle$ on input $x+n$, the reason being that if $g_{q(s)}(x) \neq g_s(x)$ then $g_{q(s)}(y) \neq g_s(y)$ for all $y > x$. □

The proof of this lemma shows that we can effectively, given a uniform list $\langle m_s^e\rangle$ of tidy $(\alpha+1)$-computable approximation and a speed up-function p such that g promptly p-escapes each $\langle m_s^e\rangle$, find a speed-up function q such that g promptly q-escapes each $\langle m^{e,s}\rangle$ on infinitely many inputs.

8.1.3 Powers of ω

Let $\alpha \leqslant \varepsilon_0$. For brevity let $\mathsf{PN}(\alpha)$ denote the class of degrees which are promptly not totally α-c.a.

Lemma 8.11. *If $\beta < \alpha$ then every function which is promptly not α-c.a. is also promptly not β-c.a.*

Hence $\mathsf{PN}(\alpha) \subseteq \mathsf{PN}(\beta)$.

Proof. Define $m_t^e(x) = o_t^{e,\beta}(x)$ if this value is smaller than β; otherwise let $m_t^e(x) = \alpha$. Now apply Corollary 8.9. □

Proposition 8.12. *Suppose that g is promptly not α-c.a. Then for all $m < \omega$, g is promptly not $\alpha \cdot m$-c.a.*

So $\mathsf{PN}(\gamma) = \mathsf{PN}(\alpha)$ for all $\gamma \in [\alpha, \alpha \cdot \omega)$. As for the non-prompt case, this means that each prompt class is $\mathsf{PN}(\alpha)$ for an ordinal α which is a power of ω. Below we will see that this is sharp.

Proof. We need to uniformise Lemma 3.7. We define a list $\langle m_s^{e,k}\rangle_{e<\omega, k<m}$ of countings down α. We claim that by the recursion theorem we have a speed-up

function q such that g promptly q-escapes each $\langle m_s^{e,k}\rangle$ (and further we require that this happens on infinitely many inputs).

Actually this relies on a property of the construction. By stage s we will have already defined $m_r^{e,k}$ for all $r < s$ (for all e and $k < m$). The finiteness condition of Corollary 8.9 will be obtained by ensuring that $m_r^{e,k}(x) = \alpha$ unless $e, x < r$. During stage s we define the functions $m_s^{e,k}$, but in the process of doing so we only consult q on values strictly smaller than s. Then the fact that $m_s^{e,k}$ is defined for all s implies that $q(s)$ is defined (as in the proof of Corollary 8.9), and the construction can proceed to the next stage.

The counting $m_s^{e,k}$ guesses that $k = k^*$ (in the notation of Lemma 3.7). However it is not sufficient for g to escape $\langle o_s^{e,\alpha m}\rangle$ on some input only from the stage at which $o_s^{e,\alpha m}(x) < \alpha(k+1)$; we need it to escape earlier as well. So it looks for inputs which have already been escaped up to that point (using q) and only copies them. Inductively, Lemma 8.10 says there will be infinitely many such inputs.

Now to the details. To define $m_s^{e,k}(x)$ we search for some $y \geqslant x$ such that:

- $o_s^{e,\alpha m}(y) \in [\alpha k, \alpha(k+1))$ but $o_{s-1}^{e,\alpha m}(y) \geqslant \alpha(k+1)$ (note that this implies $y < s$); and
- for all $t < s$ at which $o_t^{e,\alpha m}(y) \neq o_{t-1}^{e,\alpha m}(y)$ we have $g_{q(t)}(y) \neq g_t(y)$.

If such y is found then we declare $y = y^{e,k}(x)$ and $s = s^{e,k}(x)$. If such y is never found we let $s^{e,k}(x) = \omega$. Now we can define:

$$m_t^{e,k}(x) = \begin{cases} \alpha, & \text{if } t < s^{e,k}(x); \\ \beta, & \text{if } t \geqslant s^{e,k}(x) \text{ and } o_t^{e,\alpha m}(y^{e,k}(x)) = \alpha k + \beta; \text{ and} \\ 0, & \text{if } t \geqslant s^{e,k}(x) \text{ and } o_t^{e,\alpha m}(y^{e,k}(x)) < \alpha k. \end{cases}$$

Fix e. For $k \leqslant m$ we let $I_k = I_k^e$ be the set of inputs x such that for all s such that $o_s^{e,\alpha m}(x) \neq o_{s-1}^{e,\alpha m}(x)$ and $o_s^{e,\alpha m}(x) \geqslant \alpha k$, we have $g_{q(s)}(x) \neq g_s(x)$. Vacuously we have $I_m = \omega$; and our aim is to show that I_0 is nonempty. In fact we show by decreasing induction on $k = m, m-1, \ldots, 0$ that each I_k is infinite.

Let $k < m$ and suppose that we know that I_{k+1} is infinite. There are two cases. It is possible that for almost all $x \in I_{k+1}$, for all s, $o_s^{e,\alpha m}(x) \geqslant \alpha(k+1)$. Each such x is in I_k (in fact in I_0). Otherwise, for all $x < \omega$, $s^{e,k}(x)$ (and $y^{e,k}(x)$) are defined. There are infinitely many x on which g promptly q-escapes $\langle m_s^{e,k}\rangle$. Let x be such an input and let $y = y^{e,k}(x)$, $s^* = s^{e,k}(x)$. So $y \in I_{k+1}$ and we claim that in fact $y \in I_k$: if $s \geqslant s^*$, $o_s^{e,\alpha m}(y) \geqslant \alpha k$ and $o_s^{e,\alpha m}(y) \neq o_{s-1}^{e,\alpha m}(y)$ then $m_s^{e,k}(x) \neq m_{s-1}^{e,k}(x)$ and so $g_{q(s)}(x) \neq g_s(x)$; since $y \geqslant x$, $g_{q(s)}(y) \neq g_s(y)$. ☐

8.1.4 Relation to prompt simplicity

A counting $\langle o_s\rangle$ down the ordinal 1 is essentially a computable function. Namely let $h(n)$ be the unique stage s such that $o_s(n) = 0$ but $o_{s-1}(n) = 1$. The domain of h is an initial segment of ω. As mentioned above, the property $\mathsf{PN}(1)$ can be thought of as being "promptly noncomputable": it forces that $g(n) \neq h(n)$, and this is observed promptly.

Lemma 8.13. *A c.e. degree is promptly simple if and only if it is in* $\mathsf{PN}(1)$.

Proof. Suppose that A permits promptly; let $\langle A_s \rangle$ be an enumeration of A which witnesses this fact. Let g be the modulus of the enumeration of A: $g_s(n) = t$ if $t \leqslant s$ is least such that $A_s \upharpoonright_n = A_t \upharpoonright_n$.

For each e and n let $h^e(n) = s$ if $o_s^{e,1}(n) = 0$ but $o_{s-1}^{e,1}(n) = 1$. If $h^e(n) = s$ then enumerate n into a c.e. set U^e at stage s. By the promptly simple slowdown lemma there is a non-decreasing computable function q such that for all e, if U^e is infinite then there is some n which enters U^e at some stage s such that $A_s \upharpoonright_n \neq A_{q(s)} \upharpoonright_n$, so g promptly q-escapes $\langle o_s^{e,1} \rangle$ on n. We only care about the case $U^e = \omega$.

In the other direction suppose that $\deg_{\mathrm{T}}(A) \in \mathsf{PN}(1)$, witnessed by some g (which recall comes with an approximation $\langle g_s \rangle$). Let Γ be a functional such that $\Gamma(A) = g$. Let $\langle A_s \rangle$ be some enumeration of A such that for all s, $\operatorname{dom} \Gamma_s(A_s) \geqslant s$. Define a subsequence $0 = s(0) < s(1) < \ldots$ such that for all k, $\Gamma_{s(k)}(A_{s(k)}) \upharpoonright_k = g_{s(k)} \upharpoonright_k$.

For each $x < \omega$ search for an index $k = k^e(x) > x, e$ such that some number n enters W_e at stage k and the use $\gamma_{s(k)}(x)$ is smaller than n. We then define $h^e(x) = s(k)$. The domain of h^e is an initial segment of ω. We translate this to a counting down the ordinal 1: $m_t^e(x) = 1$ iff $t < h^e(x)$ (or $h^e(x)\uparrow$). Note that the counting $\langle m_t^e \rangle$ is total even if h^e is partial. Further, $m_t^e(x) = 0$ implies $e, x < t$. So by Corollary 8.9 find a computable function q such that for all e, if h^e is total then there is some x such that $g_{q(h^e(x))}(x) \neq g_{h^e(x)}(x)$.

Fix e. If W_e is infinite then h^e is total. Suppose that g escapes h^e on x (as described above). If $h^e(x) = s(k)$ then find the $k' > k$ such that $q(s(k)) \in (s(k'-1), s(k')]$. Define $p(k) = k'$. Let n be a number which enters W_e at stage k such that $\gamma_{s(k)}(x) < n$. The fact that $g(x)$ changes between stages $s(k)$ and $p(s(k))$ means that $A_{s(k')} \upharpoonright_{\gamma_{s(k)}(x)} \neq A_{s(k)} \upharpoonright_{\gamma_{s(k)}(x)}$. We conclude that the enumeration $\langle A_{s(k)} \rangle$ and the function p witness that A permits promptly. $\square$

8.1.5 A prompt hierarchy theorem

Let $\mathsf{N}(\alpha)$ denote the class of c.e. degrees which are not totally α-c.a. The class $\mathsf{N}(1)$ consists of the nonzero degrees.

Figure 8.1 details the containment relations between the classes. The following theorem implies that no further implications hold between these classes.

Theorem 8.14. *Suppose that* $\alpha \leqslant \beta \leqslant \varepsilon_0$ *are powers of* ω. *Then there is a c.e. degree* $\mathbf{d}$ *such that:*

- $\mathbf{d} \in \mathsf{PN}(\gamma)$ *if and only if* $\gamma < \alpha$; *and*
- $\mathbf{d} \in \mathsf{N}(\gamma)$ *if and only if* $\gamma < \beta$.

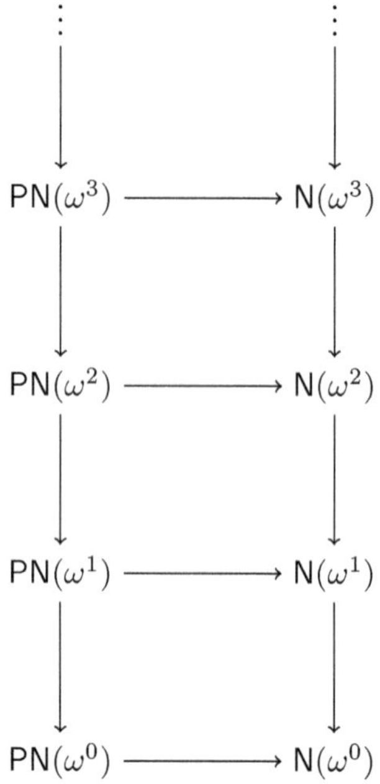

Figure 8.1. Prompt and regular classes. Arrows indicate containment.

For example, by choosing $\alpha = \beta$ we obtain:

Corollary 8.15. *Let $\delta \leqslant \varepsilon_0$. There is a degree which is promptly not totally ω^δ-c.a. but is totally $\omega^{\delta+1}$-c.a.*

On the other hand by choosing $\beta > \alpha$ we see:

Corollary 8.16. *Let $\delta \leqslant \varepsilon_0$. There is a degree which is not totally ω^δ-c.a., but not promptly so (i.e., not in $\mathsf{PN}(\omega^\delta)$), but is promptly not γ-c.a. for all $\gamma < \omega^\delta$. (In particular if $\delta > 0$ then the degree is promptly simple.)*

In this subsection we prove Theorem 8.14.

We define an approximation $\langle g_s \rangle$ witnessing that $g = \lim_s g_s$ is self-modulating, and intend to let $\mathbf{d} = \deg_{\mathrm{T}}(g)$.

For the positive side, for each $\gamma < \beta$ and $e < \omega$ we need to meet the requirement

$P^{e,\gamma}$: There is some $p < \omega$ such that:

- $g(p) \neq f^{e,\gamma}(p)$;
- and if $\gamma < \alpha$ then in fact g promptly id-escapes $\langle o_s^{e,\gamma} \rangle$ on the input p.

For the negative side, we need to meet the usual requirements

N^e: If $\Phi_e(g)$ is total then it is β-c.a.

where as usual $\langle \Phi_e \rangle$ is an effective listing of functionals. However, in addition we now have new requirements ensuring that $\mathbf{d}$ is not in $\mathsf{PN}(\alpha)$. Let $\langle \Gamma^j, \psi^j, h^j \rangle$ be an effective list of all triples of functionals, partial computable functions, and partial computable approximations. We will build a family m^j_s of (total) countings down α. We will need to meet the following requirements for each $j < \omega$:

M^j: If $\langle h^j(n,s) \rangle$ is a (total) approximation of a self-modulating function $\Gamma^j(g)$, and if ψ^j is total, then $\Gamma^j(g)$ (equipped with the approximation h^j) does not promptly ψ^j-escape m^j_s on any input.

We then appeal to Corollary 8.9 to see that $\mathbf{d} \notin \mathsf{PN}(\alpha)$.

The plan to meet this requirement is the following. One n at a time we:

(1) wait for a stage at which we see $\Gamma^j(g,n)\downarrow$, say with value q; until the end of the module for n we restrain g from changing below the use;
(2) wait for a stage s at which we see that $h^j(n,r)\downarrow = q$ for some $r \leqslant s$;
(3) define $m^j_s(n) \neq m^j_{s-1}(n)$;
(4) Wait until we see that $\psi^j(s)\downarrow$ and $h^j(n, \psi^j(s))\downarrow = q$; when this is observed we end the module for n, lift the restraint, and move to $n+1$.

The main conflict is between the actions that must be done promptly and those that must wait until they become accessible again. We argued above that to meet N^e we must use a tree of strategies. However to meet $P^{e,\gamma}$ for $\gamma < \alpha$ we need to change $g(p)$ immediately when we see that $o^{e,\gamma}_s(p)$ changes. The main observation here is that while action with existing followers must be immediate, the *appointment* of followers need not be: it can respect the priority tree. We will argue that this is sufficient to resolve the conflict between $P^{e,\gamma}$ and N^e.

Another conflict is between M^j and $P^{e,\gamma}$ for $\gamma \geqslant \alpha$, in particular when M^j is stronger. When $\gamma < \alpha$ we can allow action for $P^{e,\gamma}$ to injure the action for M^j. We restart the module above (for the same n). If we started with a large enough ordinal $m^e(n)$ then we have room to keep decreasing it. We just need to distribute priorities so that for all n, only finitely many $P^{e,\gamma}$ can disturb the module for n. If $\gamma \geqslant \alpha$ then we cannot allow $P^{e,\gamma}$ to injure M^j. However, if $\gamma \geqslant \alpha$ then we do not need to act promptly for $P^{e,\gamma}$. And between ending the module for n and starting the module for $n+1$, M^j can drop all restraint. On a tree, this is enough to ensure that $P^{e,\gamma}$ eventually succeeds.

Construction

On the tree of strategies we apportion to each requirement all nodes of some level of the tree. The outcomes for nodes working for N^e and M^j are $\infty < \mathtt{fin}$; nodes working for $P^{e,\gamma}$ have a single outcome.

We start with g_0 being the constant function 0. At a stage $s > 0$ we define g_s. This is done by determining a number p^*_s and letting $g_s(p) = s$ for $p \geqslant p^*_s$, and

$g_{s-1}(p) = g_s(p)$ for $p < p_s^*$. If the stage is ended without determining p_s^* then we let $g_s = g_{s-1}$.

Nodes σ working for some $P^{e,\gamma}$ will appoint followers. If a node σ is initialised then its follower is cancelled.

Nodes ρ working for some M^j will define a counting $\langle m_s^\rho \rangle$ down α. We start with m_0^ρ being the constant function α. At stage $s > 0$ we define m_s^ρ for all ρ. If ρ is initialised then we throw the counting $\langle m_s^\rho \rangle$ out and start a new one (we complete the old counting trivially, say with zeros everywhere, so that at the end we do get a uniformly computable sequence of total countings). If ρ is initialised at stage s then we (re)define m_t^ρ to be the constant function α for all $t \leqslant s$. If ρ is not initialised at stage s but is not accessible at stage s then we define $m_s^\rho = m_{s-1}^\rho$.

At each stage s, each node ρ working for M^j will be trying to meet the subrequirement M_n^ρ for some n; we denote this n by $n_s(\rho)$. We set $n_0(\rho) = 0$, and reset $n_s(\rho) = 0$ if ρ is initialised at stage s. Unless otherwise stated, we let $n_s(\rho) = n_{s-1}(\rho)$.

At stage s we first tend to promptness requirements. We ask if there is some node σ, working for some $P^{e,\gamma}$ for $\gamma < \alpha$, which has a follower p defined, and $o_s^{e,\gamma}(p) \neq o_{s-1}^{e,\gamma}(p)$. If so, we let σ be the strongest such node; we determine $p_s^* = p$, and initialise all nodes weaker than σ. No node is accessible, and we move to the next stage.

If there is no such node σ then we build the path of accessible nodes.

Suppose that a node τ, working for some N^e, is accessible at stage s. We let t be the greatest stage before s at which $\tau\hat{\ }\infty$ was accessible, $t = 0$ if there was no such stage. If $\operatorname{dom} \Phi_{e,s}(g_{s-1}) \geqslant t$ then we let $\tau\hat{\ }\infty$ be next accessible. Otherwise we let $\tau\hat{\ }\mathtt{fin}$ be next accessible.

Suppose that a node σ, working for some $P^{e,\gamma}$, is accessible at stage s. If σ has no follower then it appoints a new, large follower, initialises all weaker nodes, and ends the stage. If σ already has a follower p, $\gamma \geqslant \alpha$ and $f_s^{e,\gamma}(p) = g_{s-1}(p)$ then we determine that $p_s^* = p$, initialise all weaker nodes, and end the stage. Otherwise, we let the unique successor of σ on the tree be next accessible.

Suppose that a node ρ, working for some M^j, is accessible at stage s. Let $n = n_{s-1}(\rho)$. The subrequirement M_n^ρ is currently seen to be *satisfied* if there is some stage $r < s$ such that $m_r^\rho(n) \neq m_{r-1}^\rho(n)$, $\psi^j(r)\downarrow$ by stage s, and $h^j(n, \psi^j(r))\downarrow = h^j(n, r)$. If this subrequirement is currently seen to be satisfied then we let $n_s(\rho) = n+1$, and let $\rho\hat{\ }\infty$ be next accessible; we let $m_s^\rho(k) = 0$ for $k \leqslant n$ and $m_s^\rho(k) = \alpha$ for $k > n$.

Suppose that this is not the case. If $\Gamma_s^j(g_{s-1}, n)\uparrow$, let $m_s^\rho = m_{s-1}^\rho$ and let $\rho\hat{\ }\mathtt{fin}$ be next accessible. Suppose that $\Gamma_s^j(g_{s-1}, n)\downarrow = q$, and let $\gamma_s^j(n)$ be the use. If there is some node $\sigma \succcurlyeq \rho\hat{\ }\mathtt{fin}$, working for some $P^{d,\gamma}$, which has a follower $p < \gamma_s^j(n)$, then we initialise all nodes to the right of $\rho\hat{\ }\infty$, let $m_s^\rho = m_{s-1}^\rho$, and end the stage.

Otherwise, if there is no $r < s$ such that currently we see that $h^j(n, r) = q$ then again we let $m_s^\rho = m_{s-1}^\rho$ and let $\rho\hat{\ }\mathtt{fin}$ be next accessible. If there is such r, let t be the last stage at which $\rho\hat{}\infty$ was accessible, $t=0$ if there was no such stage. Let $\sigma_1, \sigma_2, \dots, \sigma_k$ be the list, with descending priority, of the nodes extending $\rho\hat{}\infty$, working for some $P^{e,\gamma}$ for some $\gamma<\alpha$, which currently have a follower p; let p_i be the follower for node σ_i and say that σ_i works for P^{e_i,γ_i}. We let

$$m^\rho_s(n) = \sum_{i\leqslant k} o^{e_i,\gamma_i}_s(p_i);$$

we let $m^\rho_s(n')=0$ for all $n'<n$ and $m^\rho_s(n')=\alpha$ for all $n'>n$. We let $\rho\hat{}\mathtt{fin}$ be next accessible.

Verification

Let ρ be a node, working for some M^j. Our first task is to prove:

Lemma 8.17. *$\langle m^\rho_s\rangle$ is a counting down α.*

Let $s<\omega$, and let r^* be the last stage prior to stage s at which ρ was accessible. We need to show that the conditions for m^ρ for being a counting have not been violated by stage s. We observe:

- if $r^*\leqslant t\leqslant s$ then $n_t(\rho)\leqslant n_s(\rho)$;
- for all $n'<n_s(\rho)$, $m^\rho_s(n')=0$;
- for all $n'>n_s(\rho)$, $m^\rho_s(n')=\alpha$.

So the only question is what happens on $n=n_s(\rho)$. Let $u^*\geqslant r^*$ be the least stage such that $n_{u^*}(\rho)=n$. For $t\in(u^*,s]$ let $\sigma^t_1, \sigma^t_2, \dots, \sigma^t_{k(t)}$ be the list, with descending priority, of the nodes extending $\rho\hat{}\infty$, working for some $P^{e,\gamma}$ for some $\gamma<\alpha$, which at stage t have a follower. Since $\rho\hat{}\infty$ is not accessible on the interval $(u^*,s]$, we in fact know that the node σ^t_i does not depend on t, so we write $\sigma_1, \sigma_2, \dots, \sigma_{k(t)}$; and the follower p_i for σ_i does not change. Say σ_i works for P^{e_i,γ_i}; for brevity let, for $t>u^*$,

$$\eta^\rho_t = \sum_{i\leqslant k} i\leqslant k o^{e_i,\gamma_i}_t(p_i).$$

So η^ρ_t is non-increasing, and if some node σ_i acts at a stage $w\in(u^*,s]$ then as $o^{e_i,\gamma_i}_w(p_i)<o^{e_i,\gamma_i}_{w-1}(p_i)$ we have $\beta^\rho_w<\beta^\rho_{w-1}$. Further, since α is closed under addition and each γ_i is smaller than α, we have $\beta^\rho_t<\alpha$ for all $t\in(u^*,s]$. Now let $t<s$. Either $m^\rho_t(n)=\alpha$, in which case certainly $m^\rho_s(n)\leqslant m^t_\rho(n)$; or there are stages $t'\leqslant t$ and $s'\geqslant t'$ such that $m^\rho_t(n)=\beta^\rho_{t'}$ and $m^\rho_s(n)=\beta^\rho_{s'}$; so we get $m^\rho_s(n)\leqslant m^\rho_t(n)$ as required. This proves Lemma 8.17.

Keeping with the same notation, say that ρ *acts* at a stage $t>u^*$ if it is accessible at stage t and ends the stage (initialising all extensions of $\rho\hat{}\mathtt{fin}$).

Lemma 8.18. *Suppose that ρ acts at two stages $s>t$, that $n_s(\rho)=n_t(\rho)$, and that ρ is not initialised at any stage in the interval $[t,s]$. Then $\beta^\rho_{s-1}<\beta^\rho_t$.*

Proof. Let $n = n_t(\rho)$. The action of ρ at stage t ensures that the computation $\Gamma^j_t(g_{t-1}, n)$ is injured between stage t and stage s. This action, and the fact that ρ itself is not initialised between stages t and s, means that some node σ extending $\rho\hat{}\infty$ acts at some stage $w \in (t, s)$ and changes g below the use of the computation. Since $\rho\hat{}\infty$ is not accessible at that interval, σ must work for some $P^{e,\gamma}$ where $\gamma < \alpha$. We observed that this means that $\beta^\rho_w < \beta^\rho_{w-1}$. □

Lemma 8.19. *The true path is infinite, and the construction is fair to every node on the true path.*

Proof. As usual, if p is a follower for some node σ then σ acts for p only finitely often. This shows that there are infinitely many stages at which we build the path of accessible nodes. Hence the root node lies on the true path, and of course is never initialised. Also this shows that a node that lies to the left of the true path can act at most finitely often.

Further, the usual arguments show that if a node working for either $P^{e,\gamma}$ or N^e is on the true path and is initialised only finitely many times, then some immediate successor of the node on the tree lies on the true path, and is only initialised finitely many times.

So we consider a node ρ on the true path, working for some M^j. The node ρ never initialises nodes extending $\rho\hat{}\infty$, so if $\rho\hat{}\infty$ is accessible infinitely often then we are done. Suppose that this is not the case. Then we can let t^* be the stage at which the last value n^* for $n_s(\rho)$ is set (either the last stage at which $\rho\hat{}\infty$ is accessible, or the last stage at which ρ is initialised). Now Lemma 8.18 implies that ρ acts only finitely many times after stage t^*. □

It is not difficult to see that every positive requirement is met. Further, following the proof of Theorem 3.6 we can see that each requirement N^e is met. As we mentioned above, it is not actually important that a computation $\Phi_e(g, x)$, already certified by a node τ on the true path, is injured only during $\tau\hat{}\infty$-stages; it is only important that the node injuring the computation extends $\tau\hat{}\infty$. We are left therefore with verifying that each M^j is met. Fix j, let ρ be a node on the true path working for M^j, and suppose that ψ^j is a total speed-up function, h^j is a (total) approximation witnessing that $\Gamma^j(g)$ (which is total) is self-modulating. We show that every subrequirement M^ρ_n is satisfied: for every n there is some stage r such that $m^\rho_r(n) \neq m^\rho_{r-1}(n)$ and $h^j(n, \psi^j(r)) = h^j(n, r)$. Of course if the subrequirement is ever seen to be satisfied then it is indeed satisfied. So by induction we show that $\langle n_s(\rho)\rangle$ is unbounded, equivalently that $\rho\hat{}\infty$ lies on the true path.

Suppose that this is not the case; let $n = \lim_s n_s(\rho)$; let t^* be the least stage (not before the last stage at which ρ was initialised) such that $n_{t^*}(\rho) = n$. The fact that $\Gamma^j(g, n)\downarrow$ and that $\lim_s h^j(n, s) = \Gamma^j(g, n)$ implies that $\lim_s m^\rho_s(n) = \lim_s \beta^\rho_s(n)$; let δ be that common value. Let s be the least stage at which $m^\rho_s(n) = \delta$; since $\delta < \alpha$, $m^\rho_s(n) \neq m^\rho_{s-1}(n)$. Also, by our instructions, $m^\rho_s(n) = \beta^\rho_s(n)$ so $\beta^\rho_s(n) = \delta$ and in fact $\beta^\rho_t(n) = \delta$ for all $t > s$.

Suppose that the computation $\Gamma^j_s(g_{s-1}, n) = q$ is correct. There is some $r < s$ such that $h^j(n, r) = q$; since h^j correctly approximates $\Gamma^j(g)$, and in a non-decreasing way, it must be that $h^j(n, w) = q$ for all $w \geqslant r$. But then since ψ^j is total, we will eventually see that M^ρ_n is satisfied, contrary to our hypothesis. Hence the computation $\Gamma^j_s(g_{s-1}, n)$ is injured at some stage $w > s$. The fact that ρ does not act at stage s implies, as in the arguments above, that some node σ extending $\rho\hat{\ }\infty$ does this injury, and that it must work for $P^{e,\gamma}$ for some $\gamma < \alpha$; this implies that $\beta^\rho_w < \beta^\rho_s$. This is the desired contradiction, showing that M^j is met, and concluding the proof of Theorem 8.14.

8.1.6 Uniform prompt classes

The uniform layers in our hierarchy also have prompt versions. Let $\alpha \leqslant \varepsilon_0$ be an infinite power of ω. Recall the definition of an α-order function h and of h-computable approximations (Definition 3.16). Recall also that we have a uniform listing $\langle f^{e,h}_s, o^{e,h}_s \rangle$ of tidy $(h+1)$-computable approximations of all h-c.a. functions. To avoid technical annoyances we define:

Definition 8.20. A self-modulating function g is *promptly not h-c.a.* if there is a speed-up function p such that g promptly p-escapes each counting $\langle o^{e,h}_s \rangle$ on infinitely many inputs.

An elaboration on the argument giving Lemma 3.17 yields the following.

Lemma 8.21. *The following are equivalent for a c.e. degree* $\mathbf{d}$*:*

(1) for some α-order function h, some $g \leqslant_{\mathrm{T}} \mathbf{d}$ is promptly not h-c.a.;
(2) for every α-order function h, some $g \leqslant_{\mathrm{T}} \mathbf{d}$ is promptly not h-c.a.

If these conditions hold then we say that $\mathbf{d}$ is *promptly not uniformly α-c.a.* When $\alpha = \omega$ we say that $\mathbf{d}$ is *promptly array noncomputable.*

Proof. Let h and $\bar{h}$ be α-order functions; let f be a function which is promptly not h-c.a. As in the proof of Lemma 3.17 partition ω into an increasing sequence of finite intervals $I^* < I_0 < I_1 < I_2 < \dots$ such that for all n, for all $x \in I_n$ we have $h(x) \geqslant \bar{h}(n)$.

Define a self-modulating function g by setting $g_s(n) = s$ if $f_s(x) = s$ for some $x \in I_m$ for some $m \leqslant n$.

For each e, define a counting $\langle m^e_s(x) \rangle$ down h by letting

$$m^e_s(x) = \begin{cases} 0, & \text{if } x \in I^*; \\ h(x), & \text{if } x \in I_n \text{ and } o^{e,\bar{h}}_s(n) = \bar{h}(n); \text{ and} \\ o^{e,\bar{h}}_s(n) & \text{if } x \in I_n \text{ and } o^{e,\bar{h}}_s(n) < \bar{h}(n). \end{cases}$$

The slow-down lemma holds for h and so an analogue of Corollary 8.9 ensures that there is a speed-up function p such that f promptly p-escapes each $\langle m_s^e \rangle$ on infinitely many inputs.

Fix e and suppose that f promptly p-escapes $\langle m_s^e \rangle$ on an input $x \notin I^*$; say $x \in I_n$. Then g promptly p-escapes $\left\langle o_s^{e,\bar{h}} \right\rangle$ on the input n. □

We can also define the prompt version of the class of not totally $<\alpha$-c.a. functions; the definition carries no surprises. The techniques used above allow us to prove hierarchy theorems for these classes; we do not elaborate here.

8.2 MINIMAL PAIRS OF SEPARATING CLASSES

To demonstrate the dynamic power encapsulated by prompt classes we discuss separating classes. For disjoint sets A_0 and A_1, we let $\mathsf{Sep}(A_0, A_1)$ denote the class of separators of A_0 and A_1—the *separating class* of A_0 and A_1. If A_0 and A_1 are c.e. sets, then $\mathsf{Sep}(A_0, A_1)$ is a Π^0_1 class. If $A_0 \cup A_1$ is co-infinite then $\mathsf{Sep}(A_0, A_1)$ is perfect, otherwise it is finite. In the literature at the time (for example [30, 52]), computing the extendible strings in a Π^0_1 class $\mathcal{P}$ (the strings σ such that $[\sigma] \cap \mathcal{P} \neq \varnothing$) was referred to as "computing the class $\mathcal{P}$." In the case of a separating class $\mathsf{Sep}(A_0, A_1)$, this is equivalent to computing both A_0 and A_1; here we only discuss computing separating classes.

Downey, Jockusch, and Stob [30] proved that a c.e. degree is array noncomputable if and only if it computes two separating classes $\mathcal{P}$ and $\mathcal{Q}$ which are incomparable in the sense that any element of $\mathcal{P}$ is Turing incomparable with any element of $\mathcal{Q}$.[1] Here we prove:

Theorem 8.22. *Every c.e. degree which is promptly array noncomputable computes two separating classes $\mathcal{P}$ and $\mathcal{Q}$ such that any element of $\mathcal{P}$ forms a minimal pair with any element of $\mathcal{Q}$.*

8.2.1 The Jockusch-Soare construction

To prove the theorem, we first recall how to construct two separating classes $\mathcal{P}$ and $\mathcal{Q}$ such that every element of $\mathcal{P}$ forms a minimal pair with every element of $\mathcal{Q}$. This was first done by Jockusch and Soare in [52]. We are not aware of a modern presentation of this construction, so we discuss it in some detail.

We wish to enumerate four c.e. sets A_0, A_1, B_0 and B_1 with the intention of letting $\mathcal{P} = \mathsf{Sep}(A_0, A_1)$ and $\mathcal{Q} = \mathsf{Sep}(B_0, B_1)$. The minimality requirements we need to meet are:

[1] In one direction they showed that any separating class computed by an array computable degree has an element of degree $\mathbf{0}'$.

R_e: If $X \in \mathcal{P}$, $Y \in \mathcal{Q}$ and $\Phi_e(X) = \Psi_e(Y)$ is total, then it is computable.

(Here as usual $\langle \Phi_e, \Psi_e \rangle$ is a list of all pairs of functionals.)

Discussion

There are two basic ways for constructing minimal pairs: by forcing, and by Lachlan's priority construction.

The forcing argument produces Cohen generic sets. The argument is as follows. When tackling the eth requirement, we look for an *e-split*: a pair (π, υ) of strings such that $\Phi_e(\pi) \perp \Psi_e(\upsilon)$. If we have already declared that σ and τ are initial segments of the sequences A and B that we are building, then we look for an e-split (π, υ) with $\pi \succcurlyeq \sigma$ and $\upsilon \succcurlyeq \tau$. If such a split is found then we declare that $\pi \prec A$ and $\upsilon \prec B$; the requirement is met. If no such split exists and still $\Phi_e(A) = \Psi_e(B)$, then we argue that the common value is computable, by searching all possible extensions.

Lachlan's construction produces c.e. sets. The main idea is freezing one side of a computation below the current length of agreement. That is, as time goes by, we monitor $\Phi_e(A, x)$ and $\Psi_e(B, x)$, and wait for common values to show up. We then allow positive requirements (making A and B noncomputable) to enumerate numbers into one of the two sets but not into both. This one-sided restraint is maintained until we see new agreement. Because one side of the computation was preserved, the new common value is identical to the old one.

The main idea of the Jockusch-Soare construction is to mix these two construction techniques.

To meet one requirement R_e on its own we can try to follow the forcing construction. We look for an e-split (π, υ). If one is found then we declare that $\mathcal{P} \subseteq [\pi]$ and $\mathcal{Q} \subseteq [\upsilon]$, where for a string τ, $[\tau]$ denotes the clopen subset of Cantor space 2^ω consisting of the sequences extending τ. To ensure that $\mathcal{P} \subseteq [\pi]$, for example, we enumerate all n such that $\pi(n) = 0$ into A_0, and all n such that $\pi(n) = 1$ into A_1.

However, we are performing a computable construction, so we cannot ask $\varnothing'$ whether an e-split exists or not. All we can do is wait for one to show up, and then try to take it. The main difficulty in the construction is when we consider how R_e would deal with the action of weaker requirements. Since they cannot wait for R_e to find an e-split, they will grab their own splits when they can. Thus, as time goes by, numbers are enumerated by weaker requirements into the four sets we are enumerating, making $\mathcal{P}$ and $\mathcal{Q}$ shrink in the process. There may be many e-splits, but we may discover them too late: whenever an e-split (π, υ) is discovered at stage s, we already have $[\pi] \cap \mathcal{P}_s = \varnothing$ or $[\upsilon] \cap \mathcal{Q}_s = \varnothing$. We would then like to argue that if $\Phi_e(X) = \Psi_e(Y)$ for $X \in \mathcal{P}$ and $Y \in \mathcal{Q}$ then this common value is computable. However, the forcing argument is useless for this, since e-splits do exist.

In this case we employ Lachlan's technique. Suppose that at no stage s do we find a viable e-split. This means that if we look for strings ζ which are initial segments of $\Phi_{e,s}(X)$ for some $X \in \mathcal{P}_s$ and $\Psi_{e,s}(Y)$ for some $Y \in \mathcal{Q}_s$ then we will not find incomparable such strings. We will then act toward ensuring that ζ is in fact correct, in that $\zeta \prec \Phi_e(X)$ and $\zeta \prec \Psi_e(Y)$ for all $X \in \mathcal{P}$ and $Y \in \mathcal{Q}$ such that $\Phi_e(X) = \Psi_e(Y)$. Some action is required here: if we do nothing, then it is possible that all oracles $X \in \mathcal{P}_s$ and $Y \in \mathcal{Q}_s$ which compute ζ at stage s fall off these classes, and only later, at some stage $t > s$, we find new elements $\tilde{X} \in \mathcal{P}_t$ and $\tilde{Y} \in \mathcal{Q}_t$ which both compute some $\tilde{\zeta}$ incomparable with ζ.

The natural thing to do, at stage s, would be to take some $\pi \prec X$ and $\upsilon \prec Y$ such that $\zeta \preccurlyeq \Phi_e(\pi), \Psi_e(\upsilon)$, and immediately declare that $\mathcal{P} \subseteq [\pi]$ and $\mathcal{Q} \subseteq [\upsilon]$. This option is immediately rejected because we would need to do this for longer and longer such strings ζ—remember that our mission is to compute $\Phi_e(X) = \Psi_e(Y)$. Alternatively, we could just ensure that $[\pi] \cap \mathcal{P} \neq \varnothing$ and $[\upsilon] \cap \mathcal{Q} \neq \varnothing$, or maybe even just one of these; this can be done by *imposing restraint* on weaker requirements, to enumerate into sets only numbers greater than $|\pi|$ or $|\upsilon|$. However again we will want to do this for longer and longer such strings, and we don't want the restraint, even on one side, to go to infinity. The solution, namely Lachlan's, is to impose restraint on one side, wait for recovery, and then maybe impose restraint on the other side. We ensure that $[\pi] \cap \mathcal{P}_r \neq \varnothing$ for $r \geqslant s$; it is possible that υ falls off $\mathcal{Q}$. We wait for a stage $t > s$ at which we get more convergence on some new $\tilde{Y} \in \mathcal{Q}_t$. If this is incomparable with ζ, we found a split and we can win quickly. Otherwise, we see ζ (and more) on both sides, and can injure one of them, while keeping the correctness of $\zeta \prec \Phi_e(X) = \Psi_e(Y)$.

What the restraint means is that when we do see a split (π, υ), we cannot immediately ensure that $\mathcal{P} \subseteq [\pi]$ and $\mathcal{Q} \subseteq [\upsilon]$: this would entail enumerating numbers into both $A_0 \cup A_1$ and $B_0 \cup B_1$, which we promised not to do. We first ensure that $\mathcal{P} \subseteq [\pi]$. We wait for the next stage t at which the node doing the work is accessible. By imposing restraint, we can ensure that if the node was not itself initialised, then we still have $[\upsilon] \cap \mathcal{Q}_t \neq \varnothing$, and so can ensure that $\mathcal{Q} \subseteq [\upsilon]$.

This discussion ignored the fact that we need to make $\mathcal{P}$ and $\mathcal{Q}$ uncountable, that is, make both $A_0 \cup A_1$ and $B_0 \cup B_1$ co-infinite. To ensure that, we will have to impose further restraint on requirements. We will forbid a requirement R_e to enumerate the smallest e-many elements from the complement of $A_{0,s} \cup A_{1,s}$ into $A_0 \cup A_1$, and the same on the B-side.

This means that each requirement R_e will have to be broken into finitely many subrequirements. If R_e is prohibited from enumerating numbers below some number r into sets, then each subrequirement will need to guess what $X\restriction_r$ and $Y\restriction_r$ are. Each such guess determines two clopen sets $\mathcal{C}$ and $\mathcal{D}$ of Cantor space, and there are 2^{2r}-many such pairs. Each subrequirement will be associated with a particular pair $(\mathcal{C}, \mathcal{D})$, and its job will be to find an e-split (π, υ) with $[\pi] \subseteq \mathcal{C}$ and $[\upsilon] \subseteq \mathcal{D}$. When such a split is found, it will want to enumerate numbers from $|\pi| \setminus r$ into $A_0 \cup A_1$ to ensure that if $X \in \mathcal{C} \cap \mathcal{P}$, then $\pi \prec X$; and similarly on the other side.

The last point that needs discussion is the structure of the tree of strategies, in particular, how to deal with strategies and substrategies. Suppose that τ is a node on the tree of strategies working for some requirement R_e; and let r be the restraint imposed on τ, that is, τ is not allowed to enumerate numbers smaller than r into sets. For each pair (ξ, θ) of strings of length r, a subrequirement $R_{\tau,i}$ restricts itself to work within the pair of clopen sets $[\xi]$ and $[\theta]$. Each such subrequirement can have either a Π_2 outcome (agreement goes to infinity) or a Σ_2 outcome, which needs to be guessed by weaker nodes. The correct outcomes are independent between the subrequirements, and so we will add to the tree below τ levels of nodes working for subrequirements.

On the other hand, when such a subrequirement finds a split and wants to act on it, it needs to impose large restraint on every node to its right. If this was done by the individual subrequirement node, this means that restraint would be increased for other nodes σ which work for subrequirements of τ; but that means that subrequirements now have to consider even more clopen sets, and this process would never end. Thus when a subrequirement of τ wishes to act positively, this is actually done by τ and not by its subrequirement.

We note that since the restraint on a node influences the association of nodes to subrequirements below the node, we build the restraint into the tree, rather than dynamically change the tree during the construction. Thus nodes will not be initialised in this construction; rather, many versions of a strategy will each guess the restraint imposed on them.

The tree of strategies

Nodes on the tree of strategies will be finite sequences of numbers and the symbol ∞. With every node ρ we will associate a restraint $r(\rho)$ (imposed on ρ). There will be two kinds of nodes: *primary nodes* τ which work for some requirement R_e; and *auxiliary nodes* σ whose job is to help calculate the restraint imposed by subrequirements. For brevity, for a primary node τ we let $m(\tau) = 2^{2r(\tau)}$. The tree and the restraint are defined together recursively.

We start with the empty string $\langle\rangle$ which is a primary node, working for R_0. We let $r(\langle\rangle) = 0$. Suppose that τ is a primary node, working for a requirement R_e.

- The outcomes of τ are all the numbers $k < \omega$ (ordered naturally). We let $r(\tau\hat{\ }k) = \max\{r(\tau), 2k\}$.
- For each $i = 1, 2, \ldots, m(\tau)$, all extensions of τ of length $|\tau| + i$ are auxiliary nodes associated with a subrequirement $R_{\tau,i}$ (which will be the restriction of R_e to a pair of clopen sets). If σ is such a node then the outcomes of σ are ∞ and all natural numbers, ordered $\infty < 0 < 1 < \cdots$. We let $r(\sigma\hat{\ }\infty) = r(\sigma)$ and $r(\sigma\hat{\ }k) = \max\{r(\sigma), 2k\}$.
- All extensions of τ of length $|\tau| + m(\tau) + 1$ are primary nodes, each working for R_{e+1}.

Notation

To be specific, for a pair of disjoint sets E and F we let

$$\mathsf{Sep}(E, F) = \{X \in 2^\omega : (n \in E \to X(n) = 0) \;\;\&\;\; (n \in F \to X(n) = 1)\}.$$

Recall that for a string $\xi \in 2^{<\omega}$,

$$[\xi] = \{X \in 2^\omega : \xi \prec X\}$$

is the clopen subset of Cantor space 2^ω determined by ξ.

For every $r < \omega$ fix a listing

$$\{(\xi_i(r), \theta_i(r)) : i = 1, 2, \ldots, 2^{2r}\}$$

of all pairs of strings (ξ, θ) of length r. Then, for every primary node τ on the tree of strategies, for $i = 1, 2, \ldots, m(\tau)$, we let $\mathcal{C}_{\tau,i} = [\xi_i(r(\tau))]$ and $\mathcal{D}_{\tau,i} = [\theta_i(r(\tau))]$. The subrequirement $R_{\tau,i}$ is the restriction of R_e to the clopen sets $\mathcal{C}_{\tau,i}$ and $\mathcal{D}_{\tau,i}$.

Construction

We enumerate four sets A_0, A_1, B_0 and B_1, and make sure to keep A_0 and A_1 disjoint, and B_0 and B_1 disjoint. At stage s we let $\mathcal{P}_s = \mathsf{Sep}(A_{0,s}, A_{1,s})$ and $\mathcal{Q}_s = \mathsf{Sep}(B_{0,s}, B_{1,s})$.

At stage 0 nothing happens. At stage $s \geqslant 1$ we describe the path of accessible nodes. The root is always accessible. Let τ be a primary node which is accessible at stage s. If $|\tau| \geqslant s$ then τ does nothing and we end the stage. Suppose that $|\tau| < s$.

Suppose that τ works for R_e. Recall that an *e-split* is a pair (π, υ) of binary strings such that $\Phi_e(\pi) \perp \Psi_e(\upsilon)$. Let $i \leqslant m(\tau)$.

- We say that the subrequirement $R_{\tau,i}$ is *seen to be met* at stage s if there is an e-split (π, υ) (observed by this stage) such that $\mathcal{P}_s \cap \mathcal{C}_{\tau,i} \subseteq [\pi]$ and $\mathcal{Q}_s \cap \mathcal{D}_{\tau,i} \subseteq [\upsilon]$.
- We say that $R_{\tau,i}$ *admits a split* at stage s if there is an e-split (π, υ), observed by this stage, such that $[\pi] \subseteq \mathcal{C}_{\tau,i}$, $[\upsilon] \subseteq \mathcal{D}_{\tau,i}$, $[\pi] \cap \mathcal{P}_s \neq \varnothing$, and $[\upsilon] \cap \mathcal{Q}_s \neq \varnothing$.
- Suppose that $R_{\tau,i}$ does not admit a split at stage s. We then define $\zeta_s(\tau, i)$, the *(τ, i)-agreement* at stage s, to be the longest binary string ζ such that $\zeta \preccurlyeq \Phi_{e,s}(X)$ and $\zeta \preccurlyeq \Psi_{e,s}(Y)$ for some $X \in \mathcal{P}_s \cap \mathcal{C}_{\tau,i}$ and some $Y \in \mathcal{Q}_s \cap \mathcal{D}_{\tau,i}$. Note that since $R_{\tau,i}$ does not admit a split at stage s, no two incomparable strings satisfy the definition of ζ.

At stage s, if there is some subrequirement $R_{\tau,i}$ which admits a split but is not seen to be met at this stage, then we choose the least such i, and we let (π, υ) be the least split admitted by the subrequirement. We then act as follows:

- If $\mathcal{P}_s \cap \mathcal{C}_{\tau,i} \nsubseteq [\pi]$ then we enumerate numbers into $A_{0,s+1}$ and $A_{1,s+1}$ so that $\mathcal{P}_{s+1} \cap \mathcal{C}_{\tau,i} \subseteq [\pi]$. Namely for all $x < |\pi|$, $x \geqslant r(\tau)$, $x \notin A_{0,s} \cup A_{1,s}$, we enumerate x into $A_{0,s+1}$ if $\pi(x) = 0$ and enumerate x into $A_{1,s+1}$ if $\pi(x) = 1$. We declare that τ *acted* at stage s and end the stage.
- If $\mathcal{P}_s \cap \mathcal{C}_{\tau,i} \subseteq [\pi]$ then we act similarly, to ensure that $\mathcal{Q}_{s+1} \cap \mathcal{D}_{\tau,i} \subseteq [\upsilon]$, declare that τ acted and end the stage.

If τ does not act at stage s we extend the path of accessible nodes up to the next primary node.

- We first determine the immediate extension of τ by determining τ's outcome at stage s. The outcome is the greatest stage $t < s$ at which τ acted; if there is no such stage, let t be the least $t \leqslant s$ at which τ was accessible.
- Now let $i \leqslant m(\tau)$ and suppose that a node σ of length $|\tau| + i$ (and so associated with $R_{\tau,i}$) is accessible at stage s.
  - If this is the first stage at which σ is accessible, let the outcome of σ at this stage be ∞.
  - If $R_{\tau,i}$ is seen to be met at stage s, then we let $\sigma\hat{}\infty$ be next accessible.
  - Otherwise, let t be the greatest stage prior to stage s at which $\sigma\hat{}\infty$ was accessible.
    * If $|\zeta_s(\tau, i)| > t$ then we let $\sigma\hat{}\infty$ be next accessible.
    * Otherwise we let $\sigma\hat{}t$ be next accessible.

Verification

Letting nodes guess their restraint implies that no two incomparable nodes can be accessible infinitely many times. Thus the true path consists of those nodes which are accessible infinitely often. To show that the true path is infinite, we will need to show that every primary node on the true path acts only finitely many times.

Note that for all nodes ρ and μ, if $\rho \preccurlyeq \mu$ then $r(\rho) \leqslant r(\mu)$.

Lemma 8.23. *Let μ be a node on the tree of strategies. Suppose that μ is accessible at some stage t; suppose that a node ρ, which lies to the right of μ, is accessible at some stage $s > t$. Then $r(\rho) \geqslant 2t$.*

Proof. Let ν be the longest common initial segment of μ and ρ; let p be the outcome of ν such that $\nu\hat{}p \preccurlyeq \rho$; let q be the outcome of ν such that $\nu\hat{}q \preccurlyeq \mu$. Then $p \neq \infty$, and if $q \neq \infty$ then $q < p$.

We show that $p \geqslant t$; this is sufficient, as $r(\rho) \geqslant r(\nu\hat{}p) \geqslant 2p$.

If ν is an auxiliary node, let E be the set of stages at which $\nu\hat{}\infty$ is accessible. If ν is a primary node, let E be the set of stages at which ν acts. For any finite outcome o of ν, if $\nu\hat{}o$ is accessible at a stage v, then o is the greatest stage in E prior to stage v. If $q = \infty$ then $t \in E$ so $p \geqslant t$. Otherwise, the fact that $q < p$ implies that $E \cap (t, s)$ is nonempty, in which case $p > t$. □

The following is also clear:

Lemma 8.24. *Suppose that a primary node τ acts at stage t. If $\rho \succ \tau$ is accessible at a stage $s > t$ then $r(\rho) \geqslant 2t$.*

For a node ρ on the true path, let $s^*(\rho)$ be the least stage $s > r(\rho)$ at which ρ is accessible. Lemma 8.23 implies:

Corollary 8.25. *If ρ lies on the true path then no node to the left of ρ is accessible after stage $s^*(\rho)$.*

Suppose that τ, a primary node working for R_e, lies on the true path.

Lemma 8.26. *No node stronger than τ ever acts at or after stage $s^*(\tau)$.*

Proof. Let μ be a node stronger than τ and let $t \geqslant s^*(\tau)$. If μ lies to the left of τ then by Corollary 8.25, μ is not accessible at stage t. Suppose that $\mu \prec \tau$. As $r(\tau) < s^*(\tau) \leqslant t$, by Lemma 8.24, if μ acts at stage t then τ cannot be accessible after stage t, contradicting the assumption that τ lies on the true path. □

Lemma 8.27. *Let $i \leqslant m(\tau)$. If the requirement $R_{\tau,i}$ is seen to be met at some stage s, then it is seen to be met at every stage $t > s$.*

Proof. Follows from the definition of "seen to be met," because $\mathcal{P}_t \subseteq \mathcal{P}_s$ and $\mathcal{Q}_t \subseteq \mathcal{Q}_s$. □

Lemma 8.28. *The node τ acts only finitely many times.*

Proof. By induction on $i = 1, 2, \dots, m(\tau)$, we show that τ acts on behalf of the subrequirement $R_{\tau,i}$ only finitely many times. Fix such i and suppose that after stage $s_i \geqslant s^*(\tau)$, τ does not act on behalf of $R_{\tau,j}$ for any $j < i$.

If at some stage $t > s_i$, τ acts on behalf of $R_{\tau,i}$ by enumerating numbers into $B_0 \cup B_1$, then this action (and the fact that we are enumerating numbers into $B_0 \cup B_1$ and not $A_0 \cup A_1$) means that $R_{\tau,i}$ is seen to be met at stage $s+1$; by Lemma 8.27, τ will not act for $R_{\tau,i}$ after stage s.

Suppose that at some stage $u > s_i$, τ acts on behalf of $R_{\tau,i}$ by enumerating numbers into $A_0 \cup A_1$. Let (π, υ) be the e-split prompting this action. The enumeration ensures that $\mathcal{P}_{s+1} \cap \mathcal{C}_{\tau,i} \subseteq [\pi]$. Since $\mathcal{Q}_{s+1} = \mathcal{Q}_s$, we have $\mathcal{Q}_{s+1} \cap [\upsilon] \neq \varnothing$.

Since the pair (π, υ) is observed by stage u, we have $|\upsilon| < u$. Let $t > u$ be the next stage at which τ is accessible. By Lemmas 8.26 and 8.23, no numbers below $2u$, and so below $|\upsilon|$, enter $B_0 \cup B_1$ between stages u and t. As $\mathcal{Q}$ is a separating class, this implies that $\mathcal{Q}_t \cap [\upsilon] \neq \varnothing$. Since $t > s_i$, it follows that at stage t, τ acts on behalf of $R_{\tau,i}$ again, and ensures that $R_{\tau,i}$ is seen to be met from the next stage onwards. □

Lemma 8.28 implies that the true path is infinite.

Lemma 8.29. *The classes $\mathcal{P}$ and $\mathcal{Q}$ are uncountable.*

Proof. Let τ be a primary node on the true path; let s^{**} be the greater between the last stage at which τ acts, and $s^*(\tau)$.

By the convention we already used above, every string examined at a stage s has length $< s$; it follows that $A_{0,s}, A_{1,s}, B_{0,s}, B_{1,s} \subseteq s$.

By Lemmas 8.26, 8.23, and 8.24, no number below $2s^{**}$ is enumerated into any set after stage s^{**}; it follows that no number in the interval $[s^{**}, 2s^{**})$ is ever enumerated into any set. Thus both $A_0 \cup A_1$ and $B_0 \cup B_1$ are disjoint from infinitely many nonempty intervals, and so are co-infinite. □

Lemma 8.30. *Every requirement R_e is met.*

Proof. Let $e < \omega$; let τ be the primary node on the true path which works for R_e.

Let $X \in \mathcal{P}$ and $Y \in \mathcal{Q}$, and suppose that $\Phi_e(X) = \Psi_e(Y)$. There is a unique $i \leqslant m(\tau)$ such that $X \in \mathcal{C}_{\tau,i}$ and $Y \in \mathcal{D}_{\tau,i}$. The subrequirement $R_{\tau,i}$ is never seen to be met, and in fact, by the proof of Lemma 8.28, from some stage onwards, at no stage t at which τ is accessible does $R_{\tau,i}$ admit a split.

Let σ be the auxiliary node on the true path which is associated with $R_{\tau,i}$. The reals X and Y show that $\sigma\hat{}\infty$ lies on the true path. We show that if $s > t$ are stages at which $\sigma\hat{}\infty$ is accessible then $\zeta_t(\tau, i) \prec \zeta_s(\tau, i)$; the fact that no splits are ever observed will imply that $\zeta_t(\tau, i) \prec \Phi_e(X)$ for all such t. Note that the node τ does not act after stage t (or σ would not be on the true path).

As discussed above, the argument is really the Lachlan minimal pair argument. At stage t, at most one node extending σ acts. That node enumerates numbers into $A_0 \cup A_1$, or into $B_0 \cup B_1$, but not both. Without loss of generality, say it is the former. The arguments above show that any node ρ that acts between stages t and s has restraint $r(\rho) \geqslant t$. This implies that if $[v] \subseteq \mathcal{Q}_t \cap \mathcal{D}_{\tau,i}$ has length t and $\Psi_e(v) \succcurlyeq \zeta_t(\tau, i)$ then $[v] \cap \mathcal{Q}_s \neq \varnothing$ as well. □

8.2.2 Adding prompt permissions

To prove Theorem 8.22 we observe that the proof of Lemma 8.28 shows that in fact we can computably bound the number of times a primary node will need to act: at most twice for each $R_{\tau,i}$, once all action for $R_{\tau,j}$ for $j < i$ has ceased. The total is $\sum_{i \leqslant m(\tau)} 2^i = 2^{m(\tau)+1} = 2^{1+2^{2r(\tau)}}$. So we let $h(r) = 2^{1+2^{2r}}$. We need the permissions to be prompt: otherwise the Lachlan mechanism of keeping one side of the computation alive cannot work. Let $\mathbf{d}$ be a c.e. degree which is promptly array noncomputable; by Lemma 8.21 there is some function $g \leqslant_{\mathrm{T}} \mathbf{d}$ which is promptly not h-c.a.

The idea is to use g to permit the action of a node τ. Each time τ wants to act we will seek a change in $g(r(\tau))$. If we do not get it we will of course notice that immediately; we will then essentially want to increase $r(\tau)$ by 1 and try all over again. Of course this means that we need to break the requirement up into more subrequirements. Rather than increase $r(\tau)$ we will incorporate into the tree the guess as to where permission will be given.

Some details

The tree of strategies now consists of three different nodes:

- "super-primary" nodes ς, whose outcomes are guesses as to where g gives permission;
- primary nodes and auxiliary nodes which have the same role as in the previous construction.

Again we define the tree of strategies by recursion, along with the restraint function r. We start with the root, which is a super-primary node working for R_0, again with $r(\langle\rangle) = 0$.

Suppose that ς is a super-primary node, working for R_e. The possible outcomes of ς are the numbers $k \geqslant r(\varsigma)$, ordered naturally. For a super-primary node ς and $k \geqslant r(\varsigma)$, we let $r(\varsigma\hat{\ }k) = k$, and we declare that $\varsigma\hat{\ }k$ is a primary node working for R_e.

Then the definition is as before: the outcomes of a primary node τ are all natural numbers, ordered naturally, with $r(\tau\hat{\ }k) = \max\{r(\tau), 2k\}$. For $i = 1, 2, \ldots, m(\tau)$ (where again $m(\tau) = 2^{2r(\tau)}$), all nodes extending τ of length $|\tau| + i$ are auxiliary nodes working for R_{τ_i}. As above, the outcomes of an auxiliary node σ are $\infty < 0 < 1 < 2 < \cdots$, with $r(\sigma\hat{\ }\infty) = r(\sigma)$ and $r(\sigma\hat{\ }k) = \max\{r(\sigma), 2k\}$. All nodes of length $|\tau| + m(\tau) + 1$ extending τ are super-primary nodes, all working for R_{e+1}.

For each super-primary node ς we will build a (total) counting $\langle o_s^\varsigma\rangle$ down h. By the recursion theorem (and the slow-down lemma) we can find a speed-up function p such that for all ς, the function g promptly p-escapes each $\langle o_s^\varsigma\rangle$, each on infinitely many inputs.

For each super-primary node ς we will have a counter $n_s(\varsigma)$. This is the current input on which we guess that g will give prompt permission.

For every primary node τ we define $\mathcal{C}_{\tau,i}$ and $\mathcal{D}_{\tau,i}$ as above.

Construction

Again, we enumerate four sets A_0, A_1, B_0 and B_1, and make sure to keep A_0 and A_1 disjoint, and B_0 and B_1 disjoint. At stage s we let $\mathcal{P}_s = \mathsf{Sep}(A_{0,s}, A_{1,s})$ and $\mathcal{Q}_s = \mathsf{Sep}(B_{0,s}, B_{1,s})$.

At stage s we start by defining, for every super-primary node ς,

- $n_0(\varsigma) = r(\varsigma)$;
- $o_0^\varsigma(n) = h(n)$ for all $n < \omega$.

No node is accessible at stage 0.

At stage $s > 0$ we start with the root, which is accessible.

Suppose that ς is a super-primary node, accessible at stage s. If $|\varsigma| \geqslant s$ we end the stage; for every super-primary node ϑ we define $n_s(\vartheta) = n_{s-1}(\vartheta)$ and $o_s^\vartheta = o_{s-1}^\vartheta$.

Otherwise, we let $\tau = \varsigma\hat{}(n_{s-1}(\varsigma))$ be next accessible. The definitions of:

- $R_{\tau,i}$ is seen to be met at stage s;
- $R_{\tau,i}$ admits a split at stage s; and
- $\zeta_s(\tau, i)$

are exactly as above.

At stage s, if there is some subrequirement $R_{\tau,i}$ which admits a split but is not seen to be met at this stage, then we choose the least such i, and we let (π, υ) be the least split admitted by the subrequirement. We then act as follows:

- If $\mathcal{P}_s \cap \mathcal{C}_{\tau,i} \nsubseteq [\pi]$ then letting $n = n_{s-1}(\varsigma)$, we:
  - define $o_s^\varsigma(n) = o_{s-1}^\varsigma(n) - 1$, and check to see if $g_{p(s)}(n) \neq g_s(n)$.
  - If so, we enumerate numbers $\geqslant r(\tau)$ into $A_{0,s+1}$ and $A_{1,s+1}$ so that $\mathcal{P}_{s+1} \cap \mathcal{C}_{\tau,i} \subseteq [\pi]$.
    We let $n_s(\varsigma) = n$, $o_s^\varsigma(n) = 0$ for all $m < n$, and $o_s^\varsigma(n) = h(m)$ for all $m > n$. We declare that τ acted at stage s and end the stage. For all super-primary nodes $\vartheta \neq \varsigma$ we let $n_s(\vartheta) = n_{s-1}(\vartheta)$ and $o_s^\vartheta = o_{s-1}^\vartheta$.
  - If not, then declare that $n_s(\varsigma) = n + 1$. For all $m \leqslant n$ we define $o_s^\varsigma(m) = 0$; for all $m > n$ we let $o_s^\varsigma(m) = h(m)$. We treat other super-primary nodes in the same way; we end the stage.
- If $\mathcal{P}_s \cap \mathcal{C}_{\tau,i} \subseteq [\pi]$ then we act exactly as in the first case, trying to ensure that $\mathcal{Q}_{s+1} \cap \mathcal{D}_{\tau,i} \subseteq [\upsilon]$.

If τ does not attempt to act at stage s then we continue to choose accessible nodes until we get to the next super-primary node. The choice of outcomes is precisely as above.

Verification

We follow the verification of the previous construction, noting the differences. Note that again for all nodes ρ and ς, if $\rho \preccurlyeq \varsigma$ then $r(\rho) \leqslant r(\varsigma)$.

First, we need to show that if a primary node τ, the child of a super-primary node ς, is attempting to act at a stage s, then $o_{s-1}^\varsigma(n) > 0$ (where again $n = n_{s-1}(\varsigma) = r(\tau)$). This was discussed above, and was the motivation for the definition of h: the total number of times τ acts is smaller than $h(r(\tau)) = h(n)$; at stage s, $o_s^\varsigma(n)$ is $h(n)$–the number of times τ acted by stage s.

Lemma 8.23 does not hold as stated and needs refinement. For nodes ρ and μ on the tree of strategies, let $\rho \wedge \mu$ be the longest common initial segment of ρ and μ. We write $\mu <^* \rho$ if μ lies to the left of ρ, and $\rho \wedge \mu$ is *not* a super-primary node. The proof of Lemma 8.23 gives:

Lemma 8.31. *Suppose that $\mu <^* \rho$, μ is accessible at some stage t, and that ρ is accessible at some stage $s > t$. Then $r(\rho) \geqslant 2t$.*

The use of this comes from:

Lemma 8.32. *Suppose that μ lies to the left of ρ but that $\mu \not<^* \rho$. If ρ is accessible at some stage t, then μ is not accessible at any stage $s \geqslant t$.*

Proof. Let $\nu = \mu \wedge \rho$. Let $k < m$ be the outcomes of ν such that $\nu\hat{}k \preccurlyeq \mu$ and $\nu\hat{}m \preccurlyeq \rho$. At stage t, we have $n_{t-1}(\nu) = m$, whence for all stages $s \geqslant t$ we will have $n_s(\nu) \geqslant m > k$, so μ will not be accessible after stage s. □

Together, these lemmas give us Corollary 8.25. Lemma 8.24 holds as is, and so we also get Lemma 8.26, with the same proof. Finally, we also conclude another weakened version of Lemma 8.23:

Lemma 8.33. *Suppose that μ lies on the true path, is accessible at some stage t, and some node ρ which lies to the right of μ is accessible at stage $s > t$. Then $r(\rho) \geqslant 2t$.*

Proof. We show that $\mu <^* \rho$, and then appeal to Lemma 8.31. If $\mu \not<^* \rho$ then by Lemma 8.32, μ will not be accessible after stage s, contradicting the assumption that μ lies on the true path. □

Lemma 8.27 holds with the same proof; putting everything together, the proof above gives Lemma 8.28 as well. However, this is not sufficient for showing that the true path is infinite; we need to consider a super-primary node ς which lies on the true path. That is, we need to show that the sequence $\langle n_s(\varsigma)\rangle$ is eventually constant. Since this sequence is non-decreasing, we need to show that it is bounded. This follows from the fact that g does promptly p-escape $\langle o_s^\varsigma\rangle$ on infinitely many inputs. Let k be the least number $n \geqslant r(\varsigma)$ such that g promptly p-escapes $\langle o_s^\varsigma\rangle$ on input n. If $\langle n_s(\varsigma)\rangle$ is unbounded then there is some stage s such that $n_s(\varsigma) = k$. Then for all $t > s$ we also have $n_t(\varsigma) = k$. This is because whenever $\tau = \varsigma\hat{}k$ attempts to act, it does receive permission to act.

The rest of the verification (the proofs of Lemmas 8.29 and 8.30) follows as above, noticing that if τ is a primary node on the true path, then τ will receive permission to act whenever it attempts to act.

8.3 PROMPT PERMISSION AND OTHER CONSTRUCTIONS

Prompt versions of permitting can also be adapted to other constructions we have been discussing. The main example is the embedding of the 1-3-1 lattice. The main idea is that if non-total $<\omega^\omega$-permission is given promptly then when balls enter the permitting bin, instead of appointing a trace for the bottom set B, we ask for prompt permission. If this is not given then the follower is cancelled. This yields:

Theorem 8.34. *If* **d** *is promptly not totally* $<\omega^\omega$*-c.a. then there is an embedding of the 1-3-1 lattice in the c.e. degrees below* **d** *which maps the bottom element to* **0**.

As mentioned in the introduction, a full reversal is impossible, since every high degree bounds such an embedding as well, and some high degrees are not even promptly simple.

Another construction to which such prompt permission can be adapted is the one mentioned in the construction of a noncomputable left-c.e. real ϱ all of whose presentations are computable. This was briefly discussed in Chapter 5, where we mentioned that the construction of such a real is more complicated than the one proving Theorem 1.4(1). In the simplified construction the c.e. set B not only aids in coding permissions, but also in "wiping the deck" concerning earlier promises that interfere with our requirement. In the original construction such a clearing cannot be done, and as a result a more complicated process of proliferating small quanta and their gradual peeling back is employed. The dynamics are similar to the constructions discussed above, and so the following can be established by similar methods.

Theorem 8.35. *If* **d** *is promptly not totally* $<\omega^\omega$*-c.a. then then there is a noncomputable left-c.e. real* $\varrho \leqslant_{\mathrm{T}} \mathbf{d}$ *all of whose presentations are computable.*

Bibliography

[1] Klaus Ambos-Spies and Peter A. Fejer. Embeddings of N_5 and the contiguous degrees. *Ann. Pure Appl. Logic*, 112(2–3):151–188, 2001.

[2] Klaus Ambos-Spies, Carl G. Jockusch, Jr., Richard A. Shore, and Robert I. Soare. An algebraic decomposition of the recursively enumerable degrees and the coincidence of several degree classes with the promptly simple degrees. *Trans. Amer. Math. Soc.*, 281(1):109–128, 1984.

[3] Bernard Anderson and Barbara Csima. A bounded jump for the bounded Turing degrees. *Notre Dame J. Form. Log.*, 55(2):245–264, 2014.

[4] Marat M. Arslanov. The Ershov hierarchy. In *Computability in context*, pages 49–100. Imp. Coll. Press, London, 2011.

[5] Katherine Arthur, Rod G. Downey, and Noam Greenberg. Maximality and collapse in the hierarchy of totally α-c.a. degrees. Submitted.

[6] Chris J. Ash and Julia F. Knight. *Computable structures and the hyperarithmetical hierarchy*, volume 144 of *Studies in Logic and the Foundations of Mathematics*. North-Holland Publishing Co., Amsterdam, 2000.

[7] George Barmpalias, Rod G. Downey, and Noam Greenberg. Working with strong reducibilities above totally ω-c.e. and array computable degrees. *Trans. Amer. Math. Soc.*, 362(2):777–813, 2010.

[8] Émile M. Borel. Les probabilités dénombrables et leurs applications arithmétiques. *Rendiconti del Circolo Matematico di Palermo (1884–1940)*, 27(1):247–271, 1909.

[9] Paul Brodhead, Rod G. Downey, and Keng Meng Ng. Bounded randomness. In *Computation, physics and beyond*, volume 7160 of *Lecture Notes in Comput. Sci.*, pages 59–70. Springer, Heidelberg, 2012.

[10] Cristian S. Calude, Peter H. Hertling, Bakhadyr Khoussainov, and Yongge Wang. Recursively enumerable reals and Chaitin Ω numbers. *Theoret. Comput. Sci.*, 255(1-2):125–149, 2001.

[11] Gregory J. Chaitin. A theory of program size formally identical to information theory. *J. Assoc. Comput. Mach.*, 22:329–340, 1975.

[12] Peter Cholak, Richard Coles, Rod G. Downey, and Eberhard Herrmann. Automorphisms of the lattice of Π^0_1 classes: perfect thin classes and anc degrees. *Trans. Amer. Math. Soc.*, 353(12):4899–4924, 2001.

[13] Peter Cholak, Rod G. Downey, and Richard Shore. Intervals without critical triples. In *Logic Colloquium '95 (Haifa)*, pages 17–43. Springer, Berlin, 1998.

[14] Peter Cholak, Rod G. Downey, and Stephen Walk. Maximal contiguous degrees. *J. Symbolic Logic*, 67(1):409–437, 2002.

[15] Richard J. Coles, Rod G. Downey, and Geoff L. LaForte. Notes on wtt-jump and ordinal notations. Manuscript, 1998.

[16] S. Barry Cooper. *Computability theory.* Chapman & Hall/CRC, Boca Raton, FL, 2004.

[17] Adam R. Day. Indifferent sets for genericity. *J. Symbolic Logic*, 78(1):113–138, 2013.

[18] Richard Dedekind. *Was sind und was sollen die Zahlen?* 8te unveränderte Aufl. Friedr. Vieweg & Sohn, Braunschweig, 1960.

[19] Max W. Dehn. Über unendliche diskontinuierliche Gruppen. *Math. Ann.*, 71(1):116–144, 1911.

[20] David Diamondstone, Noam Greenberg, and Daniel Turetsky. Natural large degree spectra. *Computability*, 2(1):1–8, 2013.

[21] Rod G. Downey. Lattice nonembeddings and initial segments of the recursively enumerable degrees. *Ann. Pure Appl. Logic*, 49(2):97–119, 1990.

[22] Rod G. Downey. Computability theory, algorithmic randomness and Turing's anticipation. In *Turing's legacy: developments from Turing's ideas in logic*, volume 42 of *Lect. Notes Log.*, pages 90–123. Assoc. Symbol. Logic, La Jolla, CA, 2014.

[23] Rod G. Downey and Noam Greenberg. Totally $<\omega^\omega$ computably enumerable and m-topped degrees. In *Theory and applications of models of computation*, volume 3959 of *Lecture Notes in Comput. Sci.*, pages 46–60. Springer, Berlin, 2006.

[24] Rod G. Downey and Noam Greenberg. Turing degrees of reals of positive effective packing dimension. *Inform. Process. Lett.*, 108(5):298–303, 2008.

[25] Rod G. Downey, Noam Greenberg, and Rebecca Weber. Totally ω-computably enumerable degrees and bounding critical triples. *J. Math. Log.*, 7(2):145–171, 2007.

[26] Rod G. Downey and Denis R. Hirschfeldt. Algorithmic randomness. *Communications of the Association for Computing Machinery*, 62(5):70–80, 2019.

[27] Rod G. Downey and Denis R. Hirschfeldt. *Algorithmic randomness and complexity*. Theory and Applications of Computability. Springer, New York, 2010.

[28] Rod G. Downey, Denis R. Hirschfeldt, André Nies, and Frank Stephan. Trivial reals. In *Proceedings of the 7th and 8th Asian Logic Conferences*, pages 103–131. Singapore Univ. Press, Singapore, 2003.

[29] Rod G. Downey and Carl G. Jockusch, Jr. T-degrees, jump classes, and strong reducibilities. *Trans. Amer. Math. Soc.*, 301(1):103–136, 1987.

[30] Rod G. Downey, Carl G. Jockusch, Jr., and Michael Stob. Array nonrecursive sets and multiple permitting arguments. In *Recursion theory week (Oberwolfach, 1989)*, volume 1432 of *Lecture Notes in Math.*, pages 141–173. Springer, Berlin, 1990.

[31] Rod G. Downey, Carl G. Jockusch, Jr., and Michael Stob. Array nonrecursive degrees and genericity. In *Computability, enumerability, unsolvability*, volume 224 of *London Math. Soc. Lecture Note Ser.*, pages 93–104. Cambridge Univ. Press, Cambridge, 1996.

[32] Rod G. Downey and Geoffrey L. LaForte. Presentations of computably enumerable reals. *Theoret. Comput. Sci.*, 284(2):539–555, 2002. Computability and complexity in analysis (Castle Dagstuhl, 1999).

[33] Rod G. Downey and Steffen Lempp. Contiguity and distributivity in the enumerable Turing degrees. *J. Symb. Logic*, 62(4):1215–1240, 1997.

[34] Rod G. Downey and Richard A. Shore. Degree-theoretic definitions of the low_2 recursively enumerable sets. *J. Symbolic Logic*, 60(3):727–756, 1995.

[35] Rod G. Downey and Richard A. Shore. Lattice embeddings below a nonlow_2 recursively enumerable degree. *Israel J. Math.*, 94:221–246, 1996.

[36] Rod G. Downey and Sebastiaan A. Terwijn. Computably enumerable reals and uniformly presentable ideals. *MLQ Math. Log. Q.*, 48(suppl. 1):29–40, 2002. Dagstuhl Seminar on Computability and Complexity in Analysis, 2001.

[37] Richard L. Epstein, Richard Haas, and Richard L. Kramer. Hierarchies of sets and degrees below $\mathbf{0}'$. In *Logic Year 1979–80 (Proc. Seminars and Conf. Math. Logic, Univ. Connecticut, Storrs, Conn., 1979/80)*, volume 859 of *Lecture Notes in Math.*, pages 32–48. Springer, Berlin, 1981.

[38] Yu. L. Ershov, S. S. Goncharov, A. Nerode, J. B. Remmel, and V. W. Marek, editors. *Handbook of recursive mathematics. Vol. 2*, volume 139 of *Studies in Logic and the Foundations of Mathematics*. North-Holland, Amsterdam, 1998. Recursive algebra, analysis and combinatorics.

[39] Yuri L. Ershov. A certain hierarchy of sets. I. *Algebra i Logika*, 7(1):47–74, 1968.

[40] Yuri L. Ershov. A certain hierarchy of sets. II. *Algebra i Logika*, 7(4): 15–47, 1968.

[41] Yuri L. Ershov. A certain hierarchy of sets. III. *Algebra i Logika*, 9:34–51, 1970.

[42] Santiago Figueira, Denis Hirschfeldt, Joseph S. Miller, Keng Meng Ng, and André Nies. Counting the changes of random Δ_2^0 sets. In *Programs, proofs, processes*, volume 6158 of *Lecture Notes in Comput. Sci.*, pages 162–171. Springer, Berlin, 2010.

[43] Johanna N. Y. Franklin, Noam Greenberg, Frank Stephan, and Guohua Wu. Anti-complex sets and reducibilities with tiny use. *J. Symbolic Logic*, 78(4):1307–1327, 2013.

[44] Johanna N. Y. Franklin and Keng Meng Ng. ω-change randomness and weak Demuth randomness. *J. Symb. Log.*, 79(3):776–791, 2014.

[45] E. Mark Gold. Language identification in the limit. *Inform. and Control*, 10(5):447–474, 1967.

[46] Noam Greenberg. The role of true finiteness in the admissible recursively enumerable degrees. *Mem. Amer. Math. Soc.*, 181(854):vi+99, 2006.

[47] Noam Greenberg, Denis R. Hirschfeldt, and André Nies. Characterizing the strongly jump-traceable sets via randomness. *Adv. Math.*, 231 (3–4):2252–2293, 2012.

[48] Noam Greenberg and André Nies. Benign cost functions and lowness properties. *J. Symbolic Logic*, 76(1):289–312, 2011.

[49] Leo A. Harrington and Saharon Shelah. The undecidability of the recursively enumerable degrees. *Bull. Amer. Math. Soc. (N.S.)*, 6(1):79–80, 1982.

[50] Grete Hermann. Die Frage der endlich vielen Schritte in der Theorie der Polynomideale. *Math. Ann.*, 95(1):736–788, 1926.

[51] Shamil Ishmukhametov. Weak recursive degrees and a problem of Spector. In *Recursion theory and complexity (Kazan, 1997)*, volume 2 of *de Gruyter Ser. Log. Appl.*, pages 81–87. de Gruyter, Berlin, 1999.

[52] Carl G. Jockusch, Jr., and Robert I. Soare. A minimal pair of Π_1^0 classes. *J. Symbolic Logic*, 36:66–78, 1971.

[53] Bjorn Kjos-Hanssen, Joseph S. Miller, and Reed Solomon. Lowness notions, measure and domination. *J. Lond. Math. Soc. (2)*, 85(3):869–888, 2012.

[54] Stephen C. Kleene. On notations for ordinal numbers. *J. Symbolic Logic*, 3(4):150–155, 1938.

[55] Stephen C. Kleene and Emil L. Post. The upper semi-lattice of degrees of recursive unsolvability. *Ann. of Math. (2)*, 59:379–407, 1954.

[56] Andrei N. Kolmogorov. Three approaches to the quantitative definition of information. *Internat. J. Comput. Math.*, 2:157–168, 1968.

[57] Martin Kummer. Kolmogorov complexity and instance complexity of recursively enumerable sets. *SIAM J. Comput.*, 25(6):1123–1143, 1996.

[58] Antonín Kučera and Theodore A. Slaman. Randomness and recursive enumerability. *SIAM J. Comput.*, 31(1):199–211, 2001.

[59] Alistair H. Lachlan. Lower bounds for pairs of recursively enumerable degrees. *Proc. London Math. Soc. (3)*, 16:537–569, 1966.

[60] Alistair H. Lachlan. On some games which are relevant to the theory of recursively enumerable sets. *Ann. of Math. (2)*, 91:291–310, 1970.

[61] Alistair H. Lachlan. Embedding nondistributive lattices in the recursively enumerable degrees. In *Conference in Mathematical Logic—London '70 (Proc. Conf., Bedford Coll., London, 1970)*, pages 149–177. Lecture Notes in Math., Vol. 255. Springer, Berlin, 1972.

[62] Alistair H. Lachlan and Robert I. Soare. Not every finite lattice is embeddable in the recursively enumerable degrees. *Adv. in Math.*, 37(1):74–82, 1980.

[63] Steffen Lempp and Manuel Lerman. A finite lattice without critical triple that cannot be embedded into the enumerable Turing degrees. *Ann. Pure Appl. Logic*, 87(2):167–185, 1997. Logic Colloquium '95 Haifa.

[64] Steffen Lempp, Manuel Lerman, and D. Reed Solomon. Embedding finite lattices into the computably enumerable degrees—a status survey. In *Logic Colloquium '02*, volume 27 of *Lect. Notes Log.*, pages 206–229. Assoc. Symbol. Logic, La Jolla, CA, 2006.

[65] Manuel Lerman. Admissible ordinals and priority arguments. In *Cambridge Summer School in Mathematical Logic (Cambridge, 1971)*, pages 311–344. Lecture Notes in Math., Vol. 337. Springer, Berlin, 1973.

[66] Manuel Lerman. *Degrees of unsolvability*. Perspectives in Mathematical Logic. Springer-Verlag, Berlin, 1983. Local and global theory.

[67] Manuel Lerman. The embedding problem for the recursively enumerable degrees. In *Recursion theory (Ithaca, N.Y., 1982)*, volume 42 of *Proc. Sympos. Pure Math.*, pages 13–20. Amer. Math. Soc., Providence, RI, 1985.

[68] Leonid A. Levin. Some theorems on the algorithmic approach to probability theory and information theory (1971 dissertation directed by A. N. Kolmogorov). *Ann. Pure Appl. Logic*, 162(3):224–235, 2010. Translated from the Russian original.

[69] Ming Li and Paul Vitányi. *An introduction to Kolmogorov complexity and its applications*. Texts and Monographs in Computer Science. Springer-Verlag, New York, 1993.

[70] Donald A. Martin. Classes of recursively enumerable sets and degrees of unsolvability. *Z. Math. Logik Grundlagen Math.*, 12:295–310, 1966.

[71] Michael McInerney. *Topics in algorithmic randomness and computability theory*. PhD thesis, Victoria University of Wellington, 2016.

[72] André Nies. Lowness properties and randomness. *Adv. Math.*, 197(1):274–305, 2005.

[73] André Nies. Reals which compute little. In *Logic Colloquium '02*, volume 27 of *Lect. Notes Log.*, pages 261–275. Assoc. Symbol. Logic, La Jolla, CA, 2006.

[74] André Nies. *Computability and randomness*, volume 51 of *Oxford Logic Guides*. Oxford University Press, Oxford, 2009.

[75] André Nies, Richard A. Shore, and Theodore A. Slaman. Interpretability and definability in the recursively enumerable degrees. *Proc. London Math. Soc. (3)*, 77(2):241–291, 1998.

[76] Piergiorgio Odifreddi. Strong reducibilities. *Bull. Amer. Math. Soc. (N.S.)*, 4(1):37–86, 1981.

[77] Piergiorgio Odifreddi. *Classical recursion theory. Vol. II*, volume 143 of *Studies in Logic and the Foundations of Mathematics*. North-Holland Publishing Co., Amsterdam, 1999.

[78] Emil L. Post. Recursively enumerable sets of positive integers and their decision problems. *Bull. Amer. Math. Soc.*, 50:284–316, 1944.

[79] Hartley Rogers, Jr. *Theory of recursive functions and effective computability*. McGraw-Hill Book Co., New York-Toronto, Ont.-London, 1967.

[80] Gerald E. Sacks. *Degrees of unsolvability.* Princeton University Press, Princeton, N.J., 1963.

[81] Gerald E. Sacks. On the degrees less than $0'$. *Ann. of Math. (2)*, 77:211–231, 1963.

[82] Gerald E. Sacks. The recursively enumerable degrees are dense. *Ann. of Math. (2)*, 80:300–312, 1964.

[83] Gerald E. Sacks. *Higher recursion theory.* Perspectives in Mathematical Logic. Springer-Verlag, Berlin, 1990.

[84] Joseph R. Shoenfield. On degrees of unsolvability. *Ann. of Math. (2)*, 69:644–653, 1959.

[85] Joseph R. Shoenfield. Undecidable and creative theories. *Fund. Math.*, 49:171–179, 1960/1961.

[86] Joseph R. Shoenfield. Applications of model theory to degrees of unsolvability. In J. W. Addison, L. Henkin, and A. Tarski, editors, *The Theory of Models, Proceedings of the 1963 International Symposium at Berkeley*, Studies in Logic and the Foundations of Mathematics, pages 359–363, Amsterdam, 1965.

[87] Richard A. Shore. The recursively enumerable α-degrees are dense. *Ann. Math. Logic*, 9(1–2):123–155, 1976.

[88] Richard A. Shore. Natural definability in degree structures. In *Computability theory and its applications (Boulder, CO, 1999)*, volume 257 of *Contemp. Math.*, pages 255–271. Amer. Math. Soc., Providence, RI, 2000.

[89] Stephen G. Simpson. *Subsystems of second order arithmetic.* Perspectives in Mathematical Logic. Springer-Verlag, Berlin, 1999.

[90] Robert I. Soare. Recursive theory and Dedekind cuts. *Trans. Amer. Math. Soc.*, 140:271–294, 1969.

[91] Robert I. Soare. *Recursively enumerable sets and degrees.* Perspectives in Mathematical Logic. Springer-Verlag, Berlin, 1987. A study of computable functions and computably generated sets.

[92] Robert M. Solovay. Draft of paper (or series of papers) related to Chaitin's work. IBM Thomas J. Watson Research Center, Yorktown Heights, NY, 215 pages, 1975.

[93] Clifford Spector. Recursive well-orderings. *J. Symb. Logic*, 20:151–163, 1955.

[94] Frank Stephan and Guohua Wu. Presentations of K-trivial reals and Kolmogorov complexity. In *Computability in Europe 2005: New*

Computational Paradigms, volume 3526 of *Lecture Notes in Comput. Sci.*, pages 461–469. Springer, 2005.

[95] Sebastiaan A. Terwijn and Domenico Zambella. Computational randomness and lowness. *J. Symbolic Logic*, 66(3):1199–1205, 2001.

[96] Steven K. Thomason. Sublattices of the recursively enumerable degrees. *Z. Math. Logik Grundlagen Math.*, 17:273–280, 1971.

[97] Alan M. Turing. On computable numbers with an application to the Entscheidungsproblem. *Proc. Lond. Math. Soc. (2)*, 42:230–265, 1936. A correction, 43:544–546.

[98] Alan M. Turing. Systems of Logic Based on Ordinals. *Proc. London Math. Soc. (2)*, 45(3):161–228, 1939.

[99] Alan M. Turing. A note on normal numbers. In J. L. Britton, editor, *Collected Works of A. M. Turing: Pure Mathematics*, pages 117–119, with notes by the editor in 263–265. North Holland, 1992.

[100] Richard von Mises. Grundlagen der wahrscheinlichkeitsrechnung. *Mathematische Zeitschrift*, 5(1):52–99, 1919.

[101] Stephen M. Walk. *Toward the definability of the array noncomputable degrees*. PhD thesis, University of Notre Dame, 1999.

[102] Klaus Weihrauch. *Computable analysis*. Texts in Theoretical Computer Science. An EATCS Series. Springer-Verlag, Berlin, 2000. An introduction.

[103] Barry Weinstein. *On embedding of the lattice 1-3-1 into the recursively enumerable degrees*. PhD thesis, University of California, Berkeley, 1988.

[104] Guohua Wu. Prefix-free languages and initial segments of computably enumerable degrees. In *Computing and combinatorics (Guilin, 2001)*, volume 2108 of *Lecture Notes in Comput. Sci.*, pages 576–585. Springer, Berlin, 2001.

[105] C. E. Mike Yates. A minimal pair of recursively enumerable degrees. *J. Symbolic Logic*, 31:159–168, 1966.

Lightning Source UK Ltd.
Milton Keynes UK
UKHW021835011020
370870UK00004B/228

9 780691 199665